VOLUME ONE HUNDRED AND SIXTY ONE

INTERNATIONAL REVIEW OF NEUROBIOLOGY

Effects of Peri-Adolescent Licit and Illicit Drug Use on the Developing CNS Part II

INTERNATIONAL REVIEW OF NEUROBIOLOGY

VOLUME 161

VOLUME ONE HUNDRED AND SIXTY ONE

INTERNATIONAL REVIEW OF NEUROBIOLOGY

Effects of Peri-Adolescent Licit and Illicit Drug Use on the Developing CNS Part II

Edited by

RICHARD L. BELL

Department of Psychiatry;
Paul and Carole Stark Neurosciences Research Institute,
Indiana University School of Medicine,
Indianapolis, IN, United States

SHAFIQUR RAHMAN

Department of Pharmaceutical Sciences,
South Dakota State University,
Brookings, SD, United States

ELSEVIER

ACADEMIC PRESS
An imprint of Elsevier

Academic Press is an imprint of Elsevier
50 Hampshire Street, 5th Floor, Cambridge, MA 02139, United States
525 B Street, Suite 1650, San Diego, CA 92101, United States
The Boulevard, Langford Lane, Kidlington, Oxford OX5 1GB, United Kingdom
125 London Wall, London, EC2Y 5AS, United Kingdom

First edition 2021

ISBN: 978-0-323-99260-2
ISSN: 0074-7742

For information on all Academic Press publications
visit our website at https://www.elsevier.com/books-and-journals

Publisher: Zoe Kruze
Acquisitions Editor: Sam Mahfoudh
Developmental Editor: Federico Paulo Mendoza
Production Project Manager: Abdulla Sait
Cover Designer: Matthew Limbert

Typeset by STRAIVE, India

VOLUME ONE HUNDRED AND SIXTY ONE

INTERNATIONAL REVIEW OF NEUROBIOLOGY

Effects of Peri-Adolescent Licit and Illicit Drug Use on the Developing CNS Part II

Edited by

RICHARD L. BELL

Department of Psychiatry;
Paul and Carole Stark Neurosciences Research Institute,
Indiana University School of Medicine,
Indianapolis, IN, United States

SHAFIQUR RAHMAN

Department of Pharmaceutical Sciences,
South Dakota State University,
Brookings, SD, United States

ELSEVIER

ACADEMIC PRESS
An imprint of Elsevier

Academic Press is an imprint of Elsevier
50 Hampshire Street, 5th Floor, Cambridge, MA 02139, United States
525 B Street, Suite 1650, San Diego, CA 92101, United States
The Boulevard, Langford Lane, Kidlington, Oxford OX5 1GB, United Kingdom
125 London Wall, London, EC2Y 5AS, United Kingdom

First edition 2021

ISBN: 978-0-323-99260-2
ISSN: 0074-7742

For information on all Academic Press publications
visit our website at https://www.elsevier.com/books-and-journals

Publisher: Zoe Kruze
Acquisitions Editor: Sam Mahfoudh
Developmental Editor: Federico Paulo Mendoza
Production Project Manager: Abdulla Sait
Cover Designer: Matthew Limbert

Typeset by STRAIVE, India

Contents

Contributors *ix*
Preface *xi*

1. Peri-adolescent exposure to (meth)amphetamine in animal models **1**

T.J. Phillips and S.J. Aldrich

1. Introduction 2
2. Effects of (meth)amphetamines in the peri-adolescent period 31
3. Conclusions and future directions 44

Acknowledgments 45
Conflict of interest 46
References 46

2. The impact of adolescent nicotine exposure on alcohol use during adulthood: The role of neuropeptides **53**

G. Chen, M. Ghazal, S. Rahman, and K. Lutfy

1. Introduction 54
2. Adolescence: A vulnerable developmental period 56
3. Adolescent nicotine/alcohol exposure and alterations in neuropeptides 64
4. Conclusion 72

Acknowledgments 73
References 73

3. Cannabis exposure during adolescence: A uniquely sensitive period for neurobiological effects **95**

K.Z. Peters, N.E. Zlebnik, and J.F. Cheer

1. Introduction 96
2. Cannabinoids and the eCB system 98
3. The endogenous cannabinoid system across the life span: Adolescence as a uniquely vulnerable period 100
4. Interactions between eCBs and dopamine: Effects on late maturing systems 103
5. Adolescent THC exposure, effects on subcortical dopamine via alterations in prefrontal GABA circuits 107
6. Conclusions 110

References 110

4. The stoned age: Sex differences in the effects of adolescent cannabinoid exposure on prefrontal cortex structure and function in animal models **121**

D.E. Ginder, H.R. Wright, and R.J. McLaughlin

1. Introduction 122
2. The adolescent brain is exquisitely sensitive to exogenous cannabinoids 124
3. Effects of adolescent cannabinoid exposure on prefrontal cortex (PFC)-dependent behavior 126
4. Effects of adolescent cannabinoid exposure on prefrontal cortex (PFC) structure and function 129
5. Conclusions 136

References 138

5. Adolescent opioid abuse: Role of glial and neuroimmune mechanisms **147**

S. Rahman, Z.I. Rahman, P.J. Ronan, K. Lutfy, and R.L. Bell

1. Introduction 148
2. Glial cells and brain development 150
3. Neuroimmune response and adolescent brain development 151
4. Opioid exposure and glial activation 152
5. Opioid exposure and glial neuroimmune signaling 154
6. Glial targets, opioid abuse, and therapeutic significance 157
7. Conclusions 158

References 158

6. Adolescent neuroimmune function and its interaction with alcohol **167**

T.L. Doremus-Fitzwater and T. Deak

1. Introduction 168
2. Neuroimmune development and functioning during adolescence 172
3. Early adolescent sensitivity to ethanol-induced alterations in neuroimmune processes 178
4. Impact of stress and immune stimuli on adolescent neuroimmune functioning 187
5. Conclusions 191

Acknowledgments 192

Conflicts of interest 192

References 192

7. Alcohol, inflammation, and blood-brain barrier function in health and disease across development 209

A.S. Vore and T. Deak

1. Introduction 211
2. Form and function of the BBB: Gatekeeper of neurovascular unit function 214
3. Alcohol effects on BBB permeability 227
4. Alterations in BBB permeability exacerbate brain pathology 232

Funding 238

References 238

8. Age modifies the effect of ethanol on behavior: Investigations in adolescent, adult and aged rats 251

D.B. Matthews and B.M. Imhoff

1. Effect of acute ethanol administration on behavior 253
2. Effects of chronic ethanol administration and corresponding withdrawal state on behavior 260
3. Conclusion 267

Funding 267

References 267

9. Early life stress and susceptibility to addiction in adolescence 277

K.E. Tschetter, L.B. Callahan, S.A. Flynn, S. Rahman, T.P. Beresford, and P.J. Ronan

1. Introduction 278
2. Adolescence 278
3. Early life stress 279
4. Investigation of molecular mechanisms underlying the ELS-induced addiction susceptibility phenotype in adolescence 280
5. Conclusions 295

Acknowledgments 295

Conflict of interest 295

References 295

10. Future directions for part II: Substance use disorders in adolescence—A vision for a better future 303

S. Rahman and R.L. Bell

1. Future directions and strategies 303
2. Additional stakeholders for a lasting change 305

References 306

Contributors

S.J. Aldrich
Department of Behavioral Neuroscience, Oregon Health & Science University, Portland, OR, United States

R.L. Bell
Department of Psychiatry, Indiana University School of Medicine, Indianapolis, IN, United States

T.P. Beresford
Laboratory for Clinical and Translational Research in Psychiatry, Rocky Mountain Regional, VA Medical Center; Department of Psychiatry, University of Colorado School of Medicine, Aurora, CO, United States

L.B. Callahan
Research Service, Sioux Falls VA Healthcare System; Department of Psychiatry; Division of Basic Biomedical Sciences, University of South Dakota Sanford School of Medicine, Sioux Falls, SD, United States

J.F. Cheer
Department of Anatomy and Neurobiology; Department of Psychiatry; Program in Neuroscience, University of Maryland School of Medicine, Baltimore, MD, United States

G. Chen
College of Osteopathic Medicine of the Pacific; Department of Pharmaceutical Sciences, College of Pharmacy, Western University of Health Sciences, Pomona, CA, United States

T. Deak
Developmental Exposure Alcohol Research Center (DEARC); Binghamton University-SUNY; Behavioral Neuroscience Program, Department of Psychology, Binghamton, NY, United States

T.L. Doremus-Fitzwater
Department of Psychology, Ithaca College, Ithaca; Developmental Exposure Alcohol Research Center (DEARC), Binghamton, NY, United States

S.A. Flynn
Department of Psychiatry, University of South Dakota Sanford School of Medicine, Sioux Falls, SD, United States

M. Ghazal
Department of Pharmaceutical Sciences, College of Pharmacy, Western University of Health Sciences, Pomona, CA, United States

D.E. Ginder
Department of Psychology, Washington State University, Pullman, WA, United States

B.M. Imhoff
Department of Psychology, University of Wisconsin – Eau Claire, Eau Claire, WI, United States

K. Lutfy
Department of Pharmaceutical Sciences, College of Pharmacy, Western University of Health Sciences, Pomona, CA, United States

D.B. Matthews
Department of Psychology, University of Wisconsin – Eau Claire, Eau Claire, WI, United States

R.J. McLaughlin
Department of Psychology; Department of Integrative Physiology and Neuroscience, Washington State University, Pullman, WA, United States

K.Z. Peters
Department of Anatomy and Neurobiology, University of Maryland School of Medicine, Baltimore, MD, United States; Sussex Neuroscience, School of Psychology, University of Sussex, Falmer, United Kingdom

T.J. Phillips
Department of Behavioral Neuroscience, Oregon Health & Science University; Veterans Affairs Portland Health Care System, Portland, OR, United States

S. Rahman
Department of Pharmaceutical Sciences, South Dakota State University, Brookings, SD, United States

Z.I. Rahman
University of Minnesota Medical School, Minneapolis, MN, United States

P.J. Ronan
Research Service, Sioux Falls VA Healthcare System; Department of Psychiatry and Basic Biomedical Sciences; Division of Basic Biomedical Sciences, University of South Dakota Sanford School of Medicine, Sioux Falls, SD; Laboratory for Clinical and Translational Research in Psychiatry, Rocky Mountain Regional, VA Medical Center, Aurora, CO, United States

K.E. Tschetter
Research Service, Sioux Falls VA Healthcare System; Department of Psychiatry; Division of Basic Biomedical Sciences, University of South Dakota Sanford School of Medicine, Sioux Falls, SD, United States

A.S. Vore
Behavioral Neuroscience Program, Department of Psychology, Developmental Exposure Alcohol Research Center, Binghamton, NY, United States

H.R. Wright
Department of Integrative Physiology and Neuroscience, Washington State University, Pullman, WA, United States

N.E. Zlebnik
Department of Anatomy and Neurobiology, University of Maryland School of Medicine, Baltimore, MD, United States

Preface

Volume 161 of the *International Review of Neurobiology* is the second of a two-part Special Issue on the *Effects of Peri-Adolescent Licit and Illicit Drug Use on the Developing CNS*. In the first part of this Special Issue (Volume 160), Chapter 1, by Coleman and colleagues, examined the findings on the long-term consequences of adolescent exposure to ethanol on the neuroimmune system and its control of cellular events within the mesocorticolimbic reward neurocircuitry; Chapter 2, by Brocato and Wolstenholme, evaluated the literature on the long-term neuroepigenetic effects of adolescent ethanol exposure both directly and indirectly via ethanol metabolic products; Chapter 3, by Lannoy and Sullivan, discussed the role of familial genetics and the trajectory of peri-adolescent alcohol use in the context of neuroimaging; Chapter 4, by Dannenhoffer et al., extended the clinical focus to cross-translational studies on peri-adolescent ethanol-associated higher-order cognitive deficits, cognition subserved by the prefrontal cortex; Chapter 5, by Hauser and colleagues, reviewed the common co-abuse of ethanol and nicotine/tobacco during adolescent development and its role in long-term deficits in cholinergic activity and function; Chapter 6, by Holmgren and Wills, described the role of hyperglutamatergic activity after chronic ethanol exposure during the peri-adolescent/adult windows of development, with a particular focus on the extended amygdala and its role in anhedonia during withdrawal; Chapter 7, by Walker and colleagues, discussed the consistent observation that chronic ethanol exposure, during peri-adolescence and adulthood, disrupts hippocampal physiology and function; Chapter 8, by Wooden and collaborators, expanded the discussion on hippocampal deficits, induced by chronic alcohol, to include marked alterations in adult neurogenesis; Chapter 9, by Robinson et al., emphasized the limited number of studies that have used both sexes when evaluating the long-term effects of peri-adolescent ethanol exposure, noting these deficits are found in the basic research literature using both animal and human subjects.

While Part I focused primarily, but not exclusively, on the long-term effects of peri-adolescent ethanol exposure, this volume expands the topics to include the effects of multiple substances of abuse. Chapter 1, by Phillips and Aldrich, examines the literature on the chronic effects of peri-adolescent exposure to amphetamine-like compounds; Chapter 2, by Chen et al., discusses the role of neuropeptide (dys)function in the long-term effects of adolescent nicotine exposure; Chapter 3, by Peters and colleagues, highlights the profound effects of cannabis exposure during adolescence, noting the development of the endogenous cannabinoid system; Chapter 4, by Ginder and associates, extends the discussion on peri-adolescent cannabinoid exposure with a focus on prefrontal cortical function and sex differences; Chapter 5, by Rahman et al., reviews the literature on glial and neuroimmune function and how peri-adolescent opioid use/exposure affects the same; Chapter 6, by Doremus-Fitzwater and Deak, extends this discussion on neuroimmune function to include the deleterious effects of peri-adolescent ethanol exposure; Chapter 7, by Vore and Deak, surveys the existing literature on how central and peripheral immune activity alter blood–brain barrier function, and the role of ethanol in exacerbating these effects; Chapter 8, by Matthews and Imhoff, discusses research from their own and colleagues' laboratories on the behavioral effects of ethanol across the life span; Chapter 9, by Tschetter et al., reviews research on how early-life stressors often predispose a subject to develop addictive behaviors during adolescence and emerging adulthood; and Chapter 10, by Rahman and Bell, provide a summary and future directions to fill gaps in the existing literature on addiction, developmental stage-of-life, sex-of-the-subject, and environmental factors.

We would like to thank, once again, all of the authors for their contributions. These contributions highlight the importance of continued research on the neurobiological underpinnings of adolescent addiction. We would also like to thank the IRN Serial Editors: Prof. Patricia Janak and Prof. Jenner for their invitation and backing of this Special Issue. We also appreciate the tremendous assistance provided by Acquisition Editor: Sam Mahfoudh; Developmental Editor: Federico Paulo Mendoza; Project Manager: Abdulla Sait; and others at Elsevier.

Richard L. Bell
Department of Psychiatry
Indiana University School of Medicine
Indianapolis, IN, United States

Shafiqur Rahman
Department of Pharmaceutical Sciences
South Dakota State University
Brookings, SD, United States

CHAPTER ONE

Peri-adolescent exposure to (meth)amphetamine in animal models

T.J. Phillips[a,b,*] and S.J. Aldrich[a]

[a]Department of Behavioral Neuroscience, Oregon Health & Science University, Portland, OR, United States
[b]Veterans Affairs Portland Health Care System, Portland, OR, United States
*Corresponding author: e-mail address: phillipt@ohsu.edu

Contents

1. Introduction 2
2. Effects of (meth)amphetamines in the peri-adolescent period 31
   2.1 Neurotoxicity 31
   2.2 Vulnerability to addiction 33
   2.3 Effects on comorbid disorders 36
   2.4 Effects on motivational circuitry 38
   2.5 Behavioral reward and motivation 39
   2.6 Cognitive impacts 40
   2.7 Social behavior 43
   2.8 Effects on motivation for other drugs 43
3. Conclusions and future directions 44
Acknowledgments 45
Conflict of interest 46
References 46

Abstract

Experimentation with psychoactive drugs is often initiated in the peri-adolescent period, but knowledge of differences in the outcomes of peri-adolescent- vs adult-initiated exposure is incomplete. We consider the existing animal research in this area for (meth)amphetamines. Established for a number of phenotypes, is lower sensitivity of peri-adolescents than adults to acute effects of (meth)amphetamines, including neurotoxic effects of binge-level exposure. More variable are data for long-term consequences of peri-adolescent exposure on motivational and cognitive traits. Moreover, investigations often exclude an adult-initiated exposure group critical for answering questions about outcomes unique to peri-adolescent initiation. Despite this, it is clear from the animal research that (meth)amphetamine exposure during the peri-adolescent period, whether self- or other-administered, impacts brain motivational circuitry and cognitive function, and alters adult sensitivity to other drugs and natural

International Review of Neurobiology, Volume 161
ISSN 0074-7742
https://doi.org/10.1016/bs.irn.2021.06.011

rewards. Such consequences occurring in humans have the potential to predispose toward unfortunate and potentially disastrous family, social and livelihood outcomes.

Abbreviations

DA dopamine
DAT DA transporter
DOPAC 3,4-dihydroxyphenylacetic acid
MDMA 3,4-methylenedioxymethamphetamine
PND postnatal day
TH tyrosine hydroxylase
VMAT-2 vesicular monoamine transporter-2

1. Introduction

In humans, a general definition of peri-adolescence, often lumped into the juvenile period, is the time of transition from childhood to adolescence, though precise age demarcations are elusive. In rodents, peri-adolescence has more systematically been based on the time between the appearance of gonadotropin cycling and reproductive maturity, approximately postnatal days (PND) 28–42, which is the period including a few days preceding and after the onset of puberty (~40 days in rodents) (Spear & Brake, 1983). One reason for an interest in the impact of psychostimulant drug use during this period is that these drugs have neurotoxic effects that could be age-dependent. Another is that, even in the absence of apparent neurotoxicity, use during this period can have immediate and long-lasting developmental, psychological and social consequences (Embry, Hankins, Biglan, & Boles, 2009). Although other periods of development are relevant, the peri-adolescent period is a time when drug experimentation is likely to occur (Chambers, Taylor, & Potenza, 2003; Johnston, O'Malley, & Backman, 1998; Park & Haning, 2016; Richmond-Rakerd, Slutske, & Wood, 2017; Saw et al., 2017), and for some drugs, there is evidence that age at first use partially accounts for differences in substance use trajectory, with poorer outcomes associated with earlier use (Richmond-Rakerd et al., 2017; Westbrook, Carrica, Banks, & Gulley, 2020), including higher rates of relapse, compared to adult-onset users (Poudel & Gautam, 2017). Thus, use during the peri-adolescent period increases risk for addiction and for poor recovery, including potential long-term neurocognitive deficits, since neurocognitive development is not complete by this period (Casey, 2015; Murty, Calabro, & Luna, 2016; Park & Haning, 2016; Rawson, Gonzales, McCann, & Ling, 2007; Sanders, 2013; Stock & Beste, 2017). The mean age of first methamphetamine use may be higher by a year or 2, when compared to first use of alcohol, nicotine and inhalants, although some data suggest age of use is dependent upon drug use sequence. Thus, Yimsaard, Maes, Verachai, and Kalayasiri (2018) found that age of onset was around 16 years when methamphetamine, rather than another drug, was the first drug used, whereas age of onset for methamphetamine use was about 20 years when other drugs were first used. In this paper, we examine the rodent research that has focused on severity and duration of neurotoxicity and on potential long-term consequences on cognition and other traits, when exposure occurs during the peri-adolescent period.

Another reason it is important to examine drug use during the peri-adolescent stage of development is to determine whether drug sensitivity diverges from that found in adulthood for effects that impact the trajectory of drug use. Logically, initial drug sensitivity is likely to impact continued use and the probability of addiction. Thus, higher sensitivity to rewarding effects would be expected to increase use, whereas higher sensitivity to aversive effects would be expected to decrease use. If the peri-adolescent period is a time when the balance between these effects is skewed toward reward, risk for addiction could be increased by initiation of use during this developmental period. We examine the literature for any evidence of this.

Some caveats at this point are pertinent. First, there is considerable inconsistency in both human and animal studies with regard to classification of the individuals examined to particular developmental stages. We provide tables with details of drug, age and population of exposure for the studies that we describe for behavioral (Table 1) and neurobiological (Table 2) findings. Next, in some cases, the outcomes cannot be specifically attributed to the peri-adolescent period, as comparisons are not made to other age groups; rather they describe the outcomes of control vs drug treatment during the peri-adolescent period on a phenotype subsequently measured. We comment on this throughout. Finally, we do not attempt to summarize the entire literature in this mini-review, and refer the reader to other reviews that have discussed the development of motivation neurocircuitry and short- and long-term effects of (meth)amphetamine exposure during the (peri) adolescent period (Buck & Siegel, 2015; Chambers et al., 2003; Hammond, Mayes, & Potenza, 2014; Luikinga, Kim, & Perry, 2018; Spear, 2016; Teixeira-Gomes et al., 2015).

Table 1 Study details, and behavioral and physiological outcomes for cited rodent papers of (meth)amphetamine exposure to peri-adolescent rodents.

Drug, exposure details, age	Sex, strain, species	Behavior	Key findings	Citations
SAL or D-AMPH (2 or 10 mg/kg, IP) administered once immediately before testing (PND 33–43 or PND >60)	Male CD-1 mice	Locomotor activity and stereotypy	Peri-adolescent mice had reduced sensitivity to the locomotor stimulant (2 mg/kg) and stereotypic (10 mg/kg) effects of D-AMPH	Adriani and Laviola (2000)
MA (0.05 mg/kg, IV); 2-h daily sessions during acquisition, then 2-h or 6-h sessions × 15 d (PND 26–28 through PND 40–43 or PND 93–103 through PND 107–117)	Male Wistar rats	IV self-administration	Peri-adolescent and adult rats had similar MA infusions and active lever presses, when given shorter access (2-h). Peri-adolescent rats earned significantly more MA infusions, resulting in greater intake, and had significantly more active lever presses than adult rats, when given long access (6-h)	Anker, Baron, Zlebnik, and Carroll (2012)
SAL or MA (7.5 mg/kg, IP) administered 4 × at a 2-h interval × 2 d (PND 30–31 or PND 96–97)	Male C57BL/6J mice	Locomotor activity, Porsolt forced swim test, and social interaction in sequence; PND 47–49 or PND 113–115	MA treatment had no affect on percent distance moved in the center of the arena or time immobile in the swim test for peri-adolescents. MA-treated adults had a significantly higher percent distance in the center and greater time immobile, compared to SAL-treated adults	Baker, Magnuson, Dahly, and Siegel (2018)

SAL or D-AMPH (0.5, 1.5 and 5.0 mg/kg, IP); administered once daily × 2 d; PND 35 or PND 80	Male Long Evans rats	Locomotor activity and stereotypy	Peri-adolescent mice had reduced sensitivity to the locomotor stimulant and stereotypic effects of D-AMPH	Bolanos, Glatt, and Jackson (1998)
SAL or D-AMPH (1 mg/kg, IP) administered once daily × 15 d (PND 28–42)	Male C57BL/6 mice	Spatial-discrimination-reversal task and extradimensional shift task for visual discrimination; starting on PND 90	D-AMPH-treated mice required more trials to reach criterion for correct responding	Boomhower and Newland (2017)
SAL or MA (7.5 mg/kg, IP) administered 4 × daily at a 2-h interval × 2 d (PND 30–31); or SAL or MA (2 mg/kg, IP) × 4 d each for CPP (PND 41–48)	Male C57BL/6J mice	Locomotor activity, novel object recognition, Porsolt forced swim test, Morris water test in sequence; PND 41–50; separate group CPP; PND 41–49	Mice that received MA alone spent a greater percentage of time in the center of the open field compared to controls. There were no treatment effects for the remaining traits	Buck, Morris, Weber, Raber, and Siegel (2017)
SAL or MA (10 mg/kg, SC) administered 4 × at a 2-h interval (PND 20, PND 40 or PND 60) at an ambient temperature of 22 °C (all ages) or 30 °C (PND 20 and PND 40 only)	Male Sprague-Dawley rats	Body temperature every 30 min after each treatment	Adult (PND 60) MA-treated rats experienced hyperthermia at both ambient temperatures, but infant (PND 20) and peri-adolescent (PND 40) MA-treated rats, exerienced hyperthermia only at 30 °C	Cappon, Morford, and Vorhees (1997)

Continued

Table 1 Study details, and behavioral and physiological outcomes for cited rodent papers of (meth)amphetamine exposure to peri-adolescent rodents.—cont'd

Drug, exposure details, age	Sex, strain, species	Behavior	Key findings	Citations
SAL or D,L-AMPH (1 mg/kg, IP) administered 72 h after no stress or repeated stress (PND 28–34 or PND 60–66)	Male Wistar rats	Locomotor activity; PND 37 or PND 69	Nonstressed peri-adolescent rats exhibited reduced sensitivity to D, L-AMPH-induced stimulation compared to adults. Stress sensitized both age groups to D, L-AMPH-induced stimulation, but peri-adolescent mice exhibited a lower stimulant response	Cruz, Marin, Leão, and Planeta (2012)
SAL or MA (1 or 3 mg/kg; IP) administered daily × 10 d (PND 35–44 or PND 55–64). Morphine (1.8, 3.2, 5.6, 10 and 18 mg/kg, SC; quarter log cumulative dosing) on PND 60 or PND 80	Male Sprague-Dawley rats	Morphine-induced antinociception; PND 60 or PND 80	Morphine-induced antinociception was reduced by MA pre-exposure in adult, but not peri-adolescent rats	Cyr, Ingram, Aicher, and Morgan (2016)
Maternal separation stress or no stress (PND 2–14) and then SAL or MA (1 mg/kg, IP) administered 4 × in a CPP procedure (PND 33–36)	Male Sprague-Dawley rats	CPP; PND 37 and 51	Peri-adolescent rats displayed MA-induced CPP that was attenuated by maternal separation	Dimatelis, Russell, Stein, and Daniels (2012)

SAL or D-AMPH (10 mg/kg, IP) administered daily × 10 d (PND 28–37 or PND 84–93)	Male Sprague-Dawley rats	Elevated plus maze, Morris water maze, Y-maze, and prepulse inhibition in sequence; PND 38–78 or PND 94–134	D-AMPH treated peri-adolescent, but not adult rats exhibited deficient retention test peformance in the Y-maze compared to their controls. There were no significant treatment effects for the other tasks	Featherby, van den Buuse, Lubman, and Lawrence (2008)
SAL or MDMA (20 mg/kg, IP) administered 4 × at a 2-h interval on a single d (PND 28 or PND 84) in mice housed 1, 5 or 10 per cage or tested at an ambient temperature of 21 or 27 °C	Male C57BL/6 mice	Body temperature before treatment and 1 h after each treatment	There were more marked effects of crowding and of warmer test environment on MDMA-induced hyperthermia in adult, compared to peri-adolescent mice	Frau, Simola, Porceddu, and Morelli (2016)
SAL or MA (1 or 5 mg/kg, IP); 1 mg/kg administered once daily × 2 d (PND 40–41 or PND 80–81); 5 mg/kg administered 4 × daily at a 2-h interval × 2 d (PND 40–41 or PND 80–81); then SAL or MA (1 mg/kg, IP) before testing (PND 80 or PND 120)	Male C57BL/6J, DBA/2J, and 129S6SvEv/Tac mice	Locomotor activity; PND 80 or PND 120	High dose MA pre-treatment in peri-adolescent, but not adult, mice resulted in a sensitized response to MA only in the C57BL/6J strain	Good and Radcliffe (2011)

Continued

Table 1 Study details, and behavioral and physiological outcomes for cited rodent papers of (meth)amphetamine exposure to peri-adolescent rodents.—cont'd

Drug, exposure details, age	Sex, strain, species	Behavior	Key findings	Citations
SAL or AMPH (3 mg/kg, IP) administered once daily × 10 d (PND 28–37)	PTN KO and WT mice	Passive avoidance, Y-maze, elevated plus maze, forced swim in squence; PND 38–71	AMPH pretreatment induced significant memory deficits in PTN KO, but not WT mice in the passive avoidance and Y-maze tests; there were no significant effects of AMPH on elevated plus maze or forced swim behaviors	Gramage, Del Olmo, Fole, Martín, and Herradón (2011)
AMPH (2.5 mg/kg, IP) administered once daily × 10 d (PND 40–49)	Male Wistar rats	Barnes maze and locomotor activity; PND 66–87	AMPH-pretreated rats exhibited deficits in spatial memory and reversal learning. There were no effects on locomotor activity	Grochecki et al. (2021)
SAL or MA (3 mg/kg, SC) administered once (PND 40) or on 3 d (PND 38–40), immediately after consuming saccharin	Male Sprague-Dawley rats	CTA; PND 38–40. Two-bottle choice water vs. saccharin; PND 41 or PND 90	1 and 3 MA treatment trials incuded CTA to saccharin that was sustained at PND 41 and 90, as indicated by reduced saccharin preference in the MA-treated mice, compared to SAL-treated controls	Harrod, Lacy, and Ballina (2010)
SAL or D,L-AMPH (1, 4 and 8 mg/kg, IP) administered once immediately after consuming sucrose (PND 18, PND 35 or PND 52)	Male and female Sprague-Dawley rats	CTA for sucrose assessed in a two-bottle choice water vs sucrose test; PND 24–27, PND 41–44 or PND 58–61	Peri-adolescent rats displayed resistance to AMPH-induced CTA, compared to infant and adult rats	Infurna and Spear (1979)

SAL or MA (7.5 mg/kg, IP) administered 4 × daily at a 2-h interval (PND 30–31)	Male and female C57BL/6J mice	Locomotor activity, novel object recognition, social interaction, Porsolt forced swim, Morris water maze in sequence; PND 41–52 or PND 62–73	Mice exposed to MA spent more time immobile in the Porsolt swim test, regardless of age of testing. There were no treatment effects for the other traits	Joca, Zuloaga, Raber, and Siegel (2014)
SAL or D-AMPH (0.5 mg/kg, IP) administered once daily (M-F; PND 28–55), then IV cocaine self-administration	Male WKY/Cr, WIS/Dr, and SHR/Cr rats	IV cocaine self-administration; beginning on PND 77	SHR rats acquired cocaine self-administration faster than the other rat strains. Prior D-AMPH treatment in SHR and WIS was associated with reduced cocaine self-administration. Prior D-AMPH treatment in WKY rats was associated with faster acquisition of cocaine self-administration, but not greater intake	Jordan, Lemay, Dwoskin, and Kantak (2016)
SAL or D-AMPH (0.5 mg/kg, IP) administered once daily (M-F; PND 28–55), 30 min prior to daily set-shifting testing, then IV cocaine self-administration	Male WKY/Cr, WIS/Dr, and SHR/Cr rats	Set shifting; PND 28–55 and IV cocaine self-administration; beginning on PND 77	Peri-adolescent SHR rats had poorer baseline performance and D-AMPH given prior to testing improved performance only in SHR rats. SHR self-administered more cocaine. There were no effects of prior D-AMPH exposure on cocaine self-administration, but prior D-AMPH exposure reduced cocaine seeking during reinstatement in SHR rats	Jordan, Taylor, Dwoskin, and Kantak (2016)

Continued

Table 1 Study details, and behavioral and physiological outcomes for cited rodent papers of (meth)amphetamine exposure to peri-adolescent rodents.—cont'd

Drug, exposure details, age	Sex, strain, species	Behavior	Key findings	Citations
SAL or D-AMPH (3 mg/kg, IP) administered once daily every other d for 10 total treatments, (beginning PND 27–45 or PND 37–55)	Male and female Sprague-Dawley rats	Sucrose preference, locomotor activity, elevated plus maze in sequence; PND 46–65 or PND 56–75	D-AMPH pre-treatment reduced sucrose preference in both age groups; this effect was sustained for longer in the PND 37 treatment group. There were no statistically treatment effects on locomotor activity. The PND 27 drug-treated group exhibited increased open-arm exploration in the elevated plus maze, compared to their SAL control	Kang, Wu, Galvez, and Gulley (2016)
SAL or MA (10 mg/kg, SC) administered 4 × at a 2-h interval (PND 40 or PND 90) or SAL or MA (15 mg/kg, SC) administered a single time (PND 40 or PND 90)	Male Sprague-Dawley rats	Body temperature 30 min after each treatment, every 2 h for 8 h	There were no significant age differences in the hyperthermic response to MA	Kokoshka, Fleckenstein, Wilkins, and Hanson (2000)

SAL or MA (4 mg/kg, IP) administered immediately before activity test and then alternate treatment given in home cage 30 min post-test; there was also a SAL-SAL control. MA (2 mg/kg, IP) administered to all groups 24 h later, immediately before testing; PND 16–17, PND 20–21, PND 24–25, or PND 34–35	Male and female Sprague-Dawley rats	Locomotor activity	Locomotor sensitization occurred in the youngest two groups, but not pre-adolescent (PND 24–25) and peri-adolescent (PND 34–35) rats	Kozanian, Gutierrez, Mohd-Yusof, and McDougall (2012)
SAL or D-AMPH (0.7 mg/kg, IP) administered once daily × 7 d (PND 35–41), then IV cocaine self-administration	Male and female Long-Evans rats	Locomotor activity; PND 31–41 and IV cocaine self-administration; beginning on PND 79	D-AMPH rats exhibited significant stimulation, compared to SAL-treated rats and exhibited locomotor sensitization. Rats exposed to D-AMPH more rapidly acquired cocaine self-administration	Lacy, Schorsch, and Austin (2018)

Continued

Table 1 Study details, and behavioral and physiological outcomes for cited rodent papers of (meth)amphetamine exposure to peri-adolescent rodents.—cont'd

Drug, exposure details, age	Sex, strain, species	Behavior	Key findings	Citations
SAL or D-AMPH (2 or 10 mg/kg, IP) administered once daily × 3 d (PND 33–43 through PND 35–45 or PND 61–71 through PND 63–73). AMPH (2 mg/kg, IP) administered to all groups 5 d later (PND 40–50 or PND 68–78), immediately before testing	Male Sprague-Dawley rats	Locomotor activity and stereotypy	D-AMPH pretreated adult rats exhibited locomotor sensitization to the challenge dose, compared to the SAL group challenged with D-AMPH. D-AMPH pretreated peri-adolescent mice exhibited reduced locomotor activity after the challenge dose, compared to the SAL group challenged with D-AMPH. D-AMPH pretreated peri-adolescent rats exhibited stereotypy on challenge d, not observed in SAL controls or adults	Laviola, Pascucci, and Pieretti (2001)
SAL or MA (4 mg/kg, IP) administered once daily × 10 d (PND 28–37)	Male and female Sprague-Dawley rats	Sucrose food pellet reward; PND 150–170	Male peri-adolescent rats exposed to MA, but not female rats, had reduced lever pressing for sucrose pellets, when tested in adulthood	Lee and Kim (2019)
SAL or MA (2 mg/kg, oral) administered once daily × 5 d (PND 36–41)	Male Long Evans rats	Locomotor activity; elevated plus maze and radial arm maze in sequence; PND 56–68	MA decreased time in the open arms and decreased time in the center of the open field. More reference memory errors were committed by MA treated mice	Loxton and Canales (2017)

SAL or MA (escalating doses from 0.5 to 3 mg/kg, IP) administered 2–4 × daily (2–6 h interval) × 9 weekdays (PND 35–45 ± 2 d or PND 70–80 ± 2 d); or MA (0.03 mg/kg per infusion, IV) available for 2 h daily × 11 d (PND 35–49 or PND 70–84)	Male Sprague-Dawley rats	Fear learning and memory; PND 50 or PND 85	MA treatments impaired extinction retrieval in adult, but not peri-adolescent rats. MA self-administration disrupted acquisition of conditioned freezing in both age groups when measured at PND 85, but not in peri-adolescent mice when measured at PND 50	Luikinga, Perry, Madsen, Lawrence, and Kim (2019)
SAL or MA (7.5 mg/kg, SC) administered biweekly on consecutive d × 6 weeks (beginning on PND 40). SAL or MA (7.5 mg/kg, SC) administered 4 × at a 2-h interval (PND 90)	Male Sprague-Dawley rats	Body temperature; PND 90	MA treatment beginning in peri-adolescence reduced the hyperthermia induced by binge MA exposure	McFadden et al. (2011)
SAL or AMPH (2 or 10 mg/kg, IP) administered once daily × 10 d (PND 33–42). then SAL or AMPH challenge immediately before testing (PND 61–70)	Male Sprague-Dawley rats	Locomotor activity	Locomotor sensitization occurred across repeated injections and persisted when testing occurred after a 4-week drug-free period	McPherson and Lawrence (2006)

Continued

Table 1 Study details, and behavioral and physiological outcomes for cited rodent papers of (meth)amphetamine exposure to peri-adolescent rodents.—cont'd

Drug, exposure details, age	Sex, strain, species	Behavior	Key findings	Citations
SAL or MA (2 mg/kg, IP) administered once immediately before testing (PND 31–42)	Male Wistar rats	Locomotor activity and social preference in sequence	MA had a locomotor stimulant effect, but no significant effect on social preference	Motbey, Hunt, Bowen, Artiss, and McGregor (2012)
SAL or MA (0.2–6 mg/kg, SC) administered twice daily at a 6-h interval in an escalating dose regimen (PND 19–34)	Flinders Sensitive and Resistant rats	Novel object recognition, social interaction, locomotor activity, and forced swim test in sequence; PND 35 and PND 60	SAL controls, but not MA exposed rats, exhibited greater percent time exploring the novel object, regardless of line and test age. Prior MA exposure decreased social interaction, regardless of line and test age. Prior MA exposure decreased locomotor activity at PND 35, but not PND 60, regardless of line. MA exposure decreased immobility at PND 35 and increased immobility at PND 60, regardless of line	Mouton, Harvey, Cockeran, and Brink (2016)
SAL or MA (5 mg/kg, IP) administered once daily × 10 d (PND 30–39 or PND 50–59). D-AMPH (0.25, 0.5 and 1 mg/kg, IP) administered twice per week as challenge doses during five-choice serial reaction time task (5-CSRTT) testing (PND 188–197)	Male Lister Hooded rats	Locomotor activity after first and final pretreatments and then 5-CSRTT testing in sequence	Peri-adolescent rats exposed to MA exhibited a high level of stimulation on PND 30, but not PND 39 after repeated exposure. The older age group exhibited mild stimulation at PND 50, but no stimulation at PND 59, after repeated expsoure. Attentional deficits in the 5-CSRTT were present only in the older age group	Nazari, Perez-Fernandez, Flores, Moreno, and Sánchez-Santed (2020)

SAL or MA (2.5, 5 and 10 mg/kg, IP) administered once immediately before testing (PND 24, PND 33 or PND 60). SAL or MA (5 mg/kg, IP) daily × 7 d (PND 24, PND 33 or PND 60)	Male CD-1 mice	Locomotor activity and stereotypy immediately after acute treatment or on d 1 and 7 during repeated treatment	There were no age-related differences in acute stimulant response to MA or in response during repeated treatments	Niculescu, Ehrlich, and Unterwald (2005)
SAL or MA (24 mg/kg, IP) administered once daily × 14 d (beginning on PND 34–48)	Male C57BL/6J mice	Novel object recognition and spatial recognition memory, before treatment, 7 and 14 days into treatment, and 7, 14 and 21 d after treatment	Memory was not altered by active MA exposure. Novel spatial memory, but not novel object memory, was impaired 7, 14 and 21 d after MA treatment, compared to the SAL-treated group	North et al. (2013)
SAL or MA (0.125 mg/kg, IP) administered once daily, with or without a social partner × 2 d (PND 34–35)	Male Sprague-Dawley rats	CPP; PND 36 preference test	The presence of a social partner enhanced MA-induced CPP	Reyna, Madden, Thiel, and Pentkowski (2021)

Continued

Table 1 Study details, and behavioral and physiological outcomes for cited rodent papers of (meth)amphetamine exposure to peri-adolescent rodents.—cont'd

Drug, exposure details, age	Sex, strain, species	Behavior	Key findings	Citations
SAL or MA (15 mg/kg; SC) administered twice per week for 6 weeks (PND 40–82), then SAL or MA (10 mg/kg, SC) 4 × at a 2-h interval (PND 46 or PND 90). Separate group SAL or MA (10 mg/kg, SC) administered 4 × at a 2-h interval (PND 40)	Male Sprague-Dawley rats	Body temperature; PND 46 or PND 90	Biweekly MA pretreatment attenuated MA-induced hyperthermia	Riddle, Kokoshka, Wilkins, Hanson, and Fleckenstein (2002)
SAL or AMPH (4 mg/kg, IP) once daily × 5 d (PND 27–31). AMPH (2 mg/kg, IP) challenge (PND 33)	Male and female Wistar rats	Locomotor activity; PND 27–31 and 33 and ethanol drinking; PND 35, 37 and 39	AMPH stimulated locomotor activity, with tolerance across days. AMPH pre-exposure increased voluntary ethanol intake in males, but not females	Ruiz, Calliari, Genovese, Scorza, and Pautassi (2018)
SAL or D-AMPH (1 and 10 mg/kg, IP) once daily × 10 d (PND 22–31). SAL or MA (0.5 mg/kg, IP) challenge (PND 90)	Male and female C57BL/6J mice	Locomotor activity; PND 90	Prior exposure to D-AMPH, regardless of dose, resulted in sensitization to MA	Shanks et al. (2015)

SAL or D-AMPH (3 mg/kg, IP) administered once daily every other d for 10 total treatments, (beginning (PND 27–45). SAL and D-AMPH challenge (3 mg/kg, IP; PND 70–0)	Male Sprague-Dawley rats	Locomotor activity and stereotypy; PND 70–80	Peri-adolescent D-AMPH exposure reduced the stimulant response, but increased stereotypy to D-AMPH challenge in adulthood	Sherrill and Gulley (2018)
SAL or D-AMPH (3 mg/kg, IP) administered once daily every other d for 10 total treatments, (beginning (PND 37–55 or PND 98–116). SAL or D-AMPH (3 mg/kg, IP) challenge	Male Sprague-Dawley rats	Locomotor activity and stereotypy; day 1 and 10 of treatment and on a challenge d after cognitive testing. Operant delayed matching-to-position cognitive task; beginning on PND 85 or PND 120	Locomotor stimulant responses to D-AMPH were comparable in rats of the two age groups on day 1 and 10 and locomotor reduction was comparable on the challenge day. Stereotypy was comparable on day 1 and on the challenge day, but greater in adult than peri-adolescent rats on day 10. Cognitive deficits were greater, when tested in adulthood in rats that had been treated with D-AMPH during peri-adolescent, compared to adult period	Sherrill, Stanis, and Gulley (2013)
SAL or MA (4 mg/kg, IP) administered once immediately before testing (PND 42–43 or PND 108–109)	Male C57BL/6J mice	Locomotor activity, stereotypy and rearing	MA-induced locomotor stimulation and reductions in percent time spent in the center of the open field that were not age-dependent	Struntz and Siegel (2018)

Continued

Table 1 Study details, and behavioral and physiological outcomes for cited rodent papers of (meth)amphetamine exposure to peri-adolescent rodents.—cont'd

Drug, exposure details, age	Sex, strain, species	Behavior	Key findings	Citations
SAL or MA (5, 10 and 15 mg/kg, SC) administered 4 × at a 2-h interval (PND 40 or PND 90)	Male Sprague-Dawley rats	Body temperature; PND 40 or PND 90	Repeated doses of MA caused equal hyperthermic responses in peri-adolescent and adult rats	Truong et al. (2005)
SAL or AMPH (1, 2 and 5 mg/kg, IP) administered once immediately before testing (PND 28, PND 42 or PND 65)	Male Sprague-Dawley rats	Locomotor activity and stereotypy	Locomotor stimulant and stereotypic effects of MA were equivalent across age groups	Walker et al. (2010)
MA (0.1 mg/kg per infusion, IV) self-administration × 7 d in 2-h sessions (beginning on PND 40 or PND 90), followed by 14 daily 6-h sessions, then 4 daily 2-h extinction sessions, then MA (1 mg/kg, IP) priming injection; IV self-administration	Male and female Sprague-Dawley rats	IV self-administration	Peri-adolescent rats earned more MA than adult rats during long-, but not short-access sessions, and had greater cumulative MA intake. Extinction and reinstatement behaviors were equivalent	Westbrook and Gulley (2020)

Saccharin or MA (0.1 mg/kg per infusion, IV) self-administration × 7 d in 2-h sessions (beginning on PND 40 or PND 90), followed by 14 daily 6-h sessions; then abstinence intermixed with cognitive testing; IV self-administration beginning on PND 40 or PND 90	Male and female Sprague-Dawley rats	IV self-administration, novel object recognition and object-in-place recognition	Peri-adolescent rats earned more saccharin and MA than adult rats during long-, but not short-access sessions, and had greater cumulative saccharin and MA intake. Peri-adolescent-onset MA rats failed to demonstrate significant object-in-place recognition, a deficit not found in adult-onset MA rats. There was no impact on novel object recognition	Westbrook, Dwycr, Cortes, and Gulley (2020)
SAL or MA (escalating doses from 0.3 to 3 mg/kg, SC) administered once daily × 10 d (PND 41–50)	Male Long Evans rats	Visual discrimination, reversal learning; beginning on PND 55. Two-bottle choice MA drinking, and two-bottle choice quinine drinking; beginning PND 119–173	MA pre-exposed rats had deficient visual discrimination and reversal learning, compared to the SAL control. MA pre-exposed rats consumed more MA, compared to the SAL control and consumed less quinine. Learning deficits were correlated with amount of MA consumed in adulthood	Ye, Pozos, Phillips, and Izquierdo (2014)

Continued

Table 1 Study details, and behavioral and physiological outcomes for cited rodent papers of (meth)amphetamine exposure to peri-adolescent rodents.—cont'd

Drug, exposure details, age	Sex, strain, species	Behavior	Key findings	Citations
SAL or MA (0.125 and 0.5 mg/kg, IP) once daily × 5 d for locomotor activity testing; SAL or MA (0.06, 0.125, 0.5 and 1 mg/kg, IP) once daily × 3 or 5 d for CPP (both PND 34–41 or PND 66–73)	Male Sprague-Dawley rats	Locomotor activity and CPP	Adult MA-exposed rats were more sensitive to the locomotor stimulant and rewarding effects of MA than peri-adolescents	Zakharova, Leoni, Kichko, and Izenwasser (2009)
SAL or MA (1, 2 and 4 mg/kg, IP) administered once immediately before testing (PND 30–35 or PND 69–74)	Male C57BL/6J mice	Locomotor activity and brain MA concentration	Adults were more sensitive to the stimulant effects of MA than peri-adolescents. The behavioral difference was not a result of differences in brain MA concentrations	Zombeck, Gupta, and Rhodes (2009)
SAL or MA (2 and 4 mg/kg, IP) administered once immediately before testing (PND 30–35 or PND 69–74)	Male C57BL/6J mice	Locomotor activity and Fos	Adults were more sensitive to the stimulant effects of MA than peri-adolescents. Fos induction in peri-adolescents was significantly greater with respect to amount of locomotor activity in several brain regions (see publication for details)	Zombeck, Lewicki, Patel, Gupta, and Rhodes (2010)

CPP, conditioned place preference; d, day or days; D-AMPH, d-amphetamine; D,L-AMPH, d,l-amphetamine; h, hour; IP, intraperitoneally; IV, intravenous; KO, knockout; MA, methamphetamine; MDMA, 3,4-methylenedioxymethamphetamine; PND, postnatal day; PTN, pleiotrophin; SAL, saline; SC, subcutaneously; WT, wild-type.

Table 2 Study details and neurobiological outcomes for cited rodent papers of (meth)amphetamine exposure to peri-adolescent rodents

Drug, exposure details and age	Sex, strain, species	Neurobiological measure	Key findings	Citations
SAL or D-AMPH (2 or 10 mg/kg, IP) administered once immediately prior to testing (PND 33–43 or PND >60)	Male CD-1 mice	Plasma CORT levels	Peri-adolescent mice had elevated basal CORT. Adults had larger increases in CORT after activity testing and D-AMPH exposure	Adriani and Laviola (2000)
SAL or MA (7.5 mg/kg, IP) administered 4 × at a 2-h interval × 2 d (PND 30–31 or PND 96–97)	Male C57BL/6J mice	Plasma CORT levels; PND 47–49 or PND 113–115	No significant effects on CORT levels	Baker et al. (2018)
SAL or MA (7.5 mg/kg, IP) administered 4 × daily at a 2-h interval × 2 d (PND 30–31)	Male C57BL/6J mice	Plasma CORT levels; PND 50 and MAP-2 protein levels in the nucleus accumbens; PND 51 or 52	No effect of MA treatment on CORT or MAP-2 levels	Buck et al. (2017)
SAL or D-AMPH (2.5 mg/kg, IP) administered once daily × 7 d (PND 35–42). SAL or D-AMPH (2.5 mg/kg, IP) administered 30 min before euthanasia	Male C57BL/6 mice	*Bdnf*, *Arc* and *cFos* expression; PND 72	Adult levels of *Bdnf*, *Arc*, and *cFos* expression were impacted by peri-adolescent D-AMPH exposure (see paper for details)	Calabrese et al. (2013)

Continued

Table 2 Study details and neurobiological outcomes for cited rodent papers of (meth)amphetamine exposure to peri-adolescent rodents—cont'd

Drug, exposure details and age	Sex, strain, species	Neurobiological measure	Key findings	Citations
SAL or MA (10 mg/kg, SC) administered 4 × at a 2-h interval (PND 20, PND 40 or PND 60) at an ambient temperature of 22 °C (all ages) or 30 °C (PND 20 and PND 40 only)	Male Sprague-Dawley rats	Neostriatal DA, 5-HT, 5-HIAA, GFAP	At 22 °C, MA treatment reduced DA and increased GFAP only in adult rats. At 30 °C, MA treatment reduced DA and increased GFAP in peri-adolescent (PND 40), but not infant (PND 20) rats	Cappon et al. (1997)
SAL or D,L-AMPH (1 mg/kg, IP) administered 72 h after no stress or repeated stress (PND 28–34 or PND 60–66)	Male Wistar rats	DA, DOPAC, HVA levels; PND 37 or PND 69	In adults, there were no stress or treatment effects on DOPAC or HVA levels; D,L-AMPH increased DA levels in the nucleus accumbens and ventral tegmental area, but there were no effects of stress group. In peri-adolescents, there were no stress or treatment effects on DOPAC or HVA levels; D,L-AMPH increased DA levels in the nucleus accumbens and in the amygdala only in the non-stressed group (see paper for details)	Cruz et al. (2012)

SAL or AMPH (5 mg/kg, IP) administered once daily × 7 d (PND 24–30, PND 33–39 or PND 60–66)	Male CD-1 mice	detlaFosB, DA, DAT, DARPP-32; PND 31, PND 40 or PND 67	Baseline levels of deltaFosB were lower in peri-adolescent (PND 24 and PND 36), compared to adult mice. Chronic administration of AMPH significantly upregulated deltaFosB in the nucleus accumbens and caudate putamen only in peri-adolescent PND 33–39 treated mice. DAT and DARPP-32 were not significantly impacted by AMPH treatment	Ehrlich, Sommer, Canas, and Unterwald (2002)
SAL or D-AMPH (10 mg/kg, IP) administered daily × 10 d (PND 28–37 or PND 84–93). SAL or D-AMPH challenge administered 36 d after behavioral testing, 1 h before euthanasia (PND 141 or PND 197)	Male Sprague-Dawley rats	CREB and TrkB by IHC; PND 141 or PND 197	D-AMPH induced downregulation of hippocampal CREB mRNA in peri-adolescent, but not adult rats. No differences were found for TrkB mRNA	Featherby et al. (2008)

Continued

Table 2 Study details and neurobiological outcomes for cited rodent papers of (meth)amphetamine exposure to peri-adolescent rodents—cont'd

Drug, exposure details and age	Sex, strain, species	Neurobiological measure	Key findings	Citations
SAL or MDMA (20 mg/kg, IP) administered 4 × at a 2-h interval on a single d (PND 28 or PND 84) in mice housed 1, 5 or 10 per cage or tested at an ambient temperature of 21 or 27 °C	Male C57BL/6 mice	Caudate putamen or substantia nigra nissl staining and IHC for CD11b, GFAP and TH; PND 30 or PND 86	Crowding increased MDMA-induced glia activation and DA neurotoxicity, similarly in the two age groups. Adult mice were generally more sensitive than peri-adolescent mice to MDMA-induced astroglial activation. Increased ambient temperature increased MDMA-induced microglial activation with no age dependence; but astroglia activation occurred only in adult mice. TH depletion by MDMA was not impacted by ambient temperature	Frau et al. (2016)
SAL or MA (5 mg/kg, IP) administered 4 × daily at a 2-h interval × 2 d (PND 40–41 or PND 80–81). SAL or MA (1 mg/kg, IP) challenge (PND 80 or PND 120) 1 h before euthanasia	Male C57BL/6J, DBA/2J and 129S6SvEv/Tac mice	Striatal and prefrontal cortex TH, DA, DOPAC	MA treatment produced significant losses in TH, DA and DOPAC that were strain-dependent. In general, MA effects were greater in adult than peri-adolescent mice	Good, Liang, Patel, and Radcliffe (2011)

SAL or AMPH (3 mg/kg, IP) administered once daily × 10 d (PND 28–37)	PTN KO and WT mice	LTP after a series of behavioral tests; PND 64	Hippocampal CA1 LTP was greater in PTN KO than WT mice. Peri-adolescent AMPH exposure increased LTP in WT, but not PTN KO mice	Gramage et al. (2011)
AMPH (2.5 mg/kg, IP) administered once daily × 10 d (PND 40–49)	Male Wistar rats	GluN1, GluN2A, GluN2B, PSD-95, and MMP-9; PND 49, 50, 63 and 87 for MMP-9; PND 87 for all	AMPH during peri-adolescence reduced hippocampal GluN2A expression on PND 87, with no other significant effects	Grochecki et al. (2021)
SAL or MA (7.5 mg/kg, IP) administered 4 × daily at a 2-h interval (PND 30–31)	Male and female C57BL/6J mice	Hypothalamic vasporessin; PND 53–54 or PND 74–75	Hypothalamic vasopressin was lower after MA compared to saline exposure in tissues obtained on PND 53–54; this effect was specific to males. No difference was found between the SAL and MA groups for tissues obtained on PND 74–75	Joca et al. (2014)
SAL or D-AMPH (3 mg/kg, IP) administered once daily every other d for 10 total treatments (PND 27–45 or PND 37–55)	Male and female Sprague-Dawley rats	Puberty onset in female and D1 receptor expression; PND 66 or PND 76	Puberty onset was accelerated by D-AMPH exposure. D1 receptor expression was reduced by D-AMPH in the prefrontal cortex, but not nucleus accumbens, specifically for male rats exposed PND 27–45	Kang et al. (2016)

Continued

Table 2 Study details and neurobiological outcomes for cited rodent papers of (meth)amphetamine exposure to peri-adolescent rodents—cont'd

Drug, exposure details and age	Sex, strain, species	Neurobiological measure	Key findings	Citations
SAL or MA (10 mg/kg, SC) administered 4 × at a 2-h interval or SAL or MA (15 mg/kg, SC) administered a single time (PND 40 or PND 90)	Male Sprague-Dawley rats	DA, DAT, TH, TPH, and striatal and plasma MA concentrations; PND 40 or 47 or PND 90 or 97	Striatal DAT, TH and TPH were reduced in adult, but not peri-adolescent, MA-exposed rats 7 d after treatment. DAT and TPH activity were reduced 1 h after both binge and single MA treatments, compared to saline in rats of both ages. Adult rats had more than twice the striatal and plasma levels of MA 1 h after binge administration	Kokoshka et al. (2000)
SAL or D-AMPH (2 or 10 mg/kg, IP) administered once daily × 3 d (PND 33–43 through PND 35–45 or PND 61–71 through PND 63–73). AMPH (2 mg/kg, IP) administered to all groups 5 d later (PND 40–50 or PND 68–78), immediately before testing	Male Sprague-Dawley rats	In vivo microdialysis of striatal DA; PND 40–50 or PND 68–78	D-AMPH pre-treated peri-adolescent, but not adult rats had higher DA levels after D-AMPH challenge	Laviola et al. (2001)

SAL or MA (7.5 mg/kg, SC) administered biweekly on consecutive d × 6 weeks (beginning on PND 40). SAL or MA (7.5 mg/kg, SC) administered 4 × at a 2-h interval, 1 h or 7 d before euthanasia (PND 90)	Male Sprague-Dawley rats	Vesicular DA uptake; PND 90 (1 h post binge treatment) or PND 97 (7 d post binge treatment) in 23 °C and 25 °C environments	Peri-adolescent MA exposure reduced deficits in DA reuptake caused by binge MA treatment, both 1 h and 7 d after binge MA. When hyperthermia was sustained by testing at 25 °C, the protective effect of peri-adolescent MA exposure was absent	McFadden et al. (2011)
SAL or MA (24 mg/kg, IP) administered once daily × 14 d (beginning on PND 34–48)	Male C57BL/6J mice	Cortical, hippocampal and striatal DAT and TH; 1 d after treatment. Hippocampal synaptic plasticity by ex vivo electrophysiology; 1 and 21 d after treatment	TH was significant reduced by MA in all brain regions. MA increased the high molecular weight DAT in the cortex and striatum and decreased the low molecular weight DAT in the hippocampus and striatum	North et al. (2013)
SAL or D-AMPH (3 mg/kg, IP) administered once daily every other d for 10 total treatments (PND 27–45 or PND 85–103)	Male Sprague-Dawley rats	Medial prefrontal cortex pyramidal cell slice electrophysiology inhibitory transmission	When slices from ~PND 133 for both age groups were examined, there was no treatment effect on baseline sIPSC frequency and D-AMPH application-induced increased inhibitory tone in slices from control rats that was absent in slices from D-AMPH pre-treated rats, regardless of age group. When slices were obtained at a common 3–4 weeks after pre-treatment, D-AMPH application suppressed sIPSC activity (rather than increased) only in slices from peri-adolescent exposed rats	Paul, Kang, Cox, and Gulley (2016)

Continued

Table 2 Study details and neurobiological outcomes for cited rodent papers of (meth)amphetamine exposure to peri-adolescent rodents—cont'd

Drug, exposure details and age	Sex, strain, species	Neurobiological measure	Key findings	Citations
SAL or MA (10 and 20 mg/kg, IP) administered 4 × daily at a 2-h interval (PND 20, 40, 60 and 80)	Male Sprague-Dawley rats	Caudate-putamen TH and GFAP by IHC; PND 23, 43, 63 and 83	TH depletion and GFAP increase in PND 60 and 80 MA-treated rats. No TH depletion, but GFAP increase by MA in PND 40 rats. No TH depletion or GFAP increase by MA in PND 20 rats	Pu and Vorhees (1993)
SAL or MA (5, 10 and 15 mg/kg, IP) administered 4 × at a 2-h interval, 1 h or 7 d before euthanasia (PND 40 or PND 90)	Male Sprague-Dawley rats	Vesicular DA uptake, VMAT-2 expression, and striatal MA levels; PND 40, 47, 90 or 97	VMAT-2 and striatal MA levels were similar in SAL controls, regardless of age. DA reuptake was greater in SAL-treated adult than peri-adolescent rats and was decreased in both age groups 1 h after binge MA administration. At 7 h after binge MA, DA uptake was decreased only in adult rats	Rau, Truong, Wilkins, Fleckenstein, and Hanson (2006)
SAL or MA (15 mg/kg; SC) administered twice per week for 6 weeks (PND 40–82), then SAL or MA (10 mg/kg, SC) 4 × at a 2-h interval (PND 46 or PND 90). Separate group SAL or MA (10 mg/kg, SC) administered 4 × at a 2-h interval (PND 40)	Male Sprague-Dawley rats	Vesicular DA uptake and DAT binding; PND 47 or PND 97, 1 week after binge MA treatment	Rats treated with binge MA in adulthood, but not peri-adolescence, had significantly decreased striatal DA reuptake and DAT binding. The effect in adult rats was attenuated by peri-adolescent MA pre-treatment	Riddle et al. (2002)

SAL or AMPH (4 mg/kg, IP) once daily × 5 d (PND 27–31). AMPH (2 mg/kg, IP) challenge (PND 33)	Male and female Wistar rats	Insular cortex DA, 5-HT and NE levels; PND 36	AMPH reduced insular levels of DA in both sexes and of NE in females only. There was no significant effect on 5-HT	Ruiz et al. (2018)
SAL or D-AMPH (3 mg/kg, IP) administered once daily every other d for 10 total treatments, (beginning (PND 27–45). SAL and D-AMPH challenge (3 mg/kg, IP; PND 70–80)	Male Sprague-Dawley rats	In vivo prelimbic cortex putative pyramidal neuron responses to D-AMPH challenge; PND 70–80	D-AMPH pre-exposure resulted in a greater probability of prelimbic neuron reductions in responding after D-AMPH challenge in adulthood	Sherrill and Gulley (2018)
SAL or MA (4 mg/kg, IP) administered once immediately before testing (PND 42–43 or PND 108–109)	Male C57BL/6J mice	TH levels and phosphorylation; plasma CORT levels; PND 42–43 or PND 108–109	There were no significant effects of MA treatment	Struntz and Siegel (2018)
SAL or MA (5, 10 and 15 mg/kg, SC) administered 4 × at a 2-h interval (PND 40 or PND 90)	Male Sprague-Dawley rats	VMAT-2 immunoreactivity and vesicular DA uptake; PND 47 or PND 97	VMAT-2 immunoreactivity and only in the vesicular subcellular fraction and DA uptake were greater in SAL adult than SAL peri-adolescent rats. Adult rats exhibited greater DA deficits, compared to peri-adolescent	Truong et al. (2005)
MA (aqueous, 40 μm) added to an in vitro suspension of striatal tissue (PND 38–42 or PND 88–92)	Male Sprague-Dawley rats	DA transport and efflux velocity; VMAT-2 immunoreactivity	Tissue from peri-adolescent rats had faster initial DA transport velocity, MA-induced DA efflux, and DAT immunoreactivity. There was no age effect for VMAT-2 immunoreactivity	Volz, Farnsworth, Rowley, Hanson, and Fleckenstein (2009)

Continued

Table 2 Study details and neurobiological outcomes for cited rodent papers of (meth)amphetamine exposure to peri-adolescent rodents—cont'd

Drug, exposure details and age	Sex, strain, species	Neurobiological measure	Key findings	Citations
SAL or AMPH (1, 2 and 5 mg/kg, IP) administered once immediately before testing (PND 28, PND 42 or PND 65)	Male Sprague-Dawley rats	Caudate DA overflow; PND 28 or PND 65	There were no effects of age on AMPH-induced caudate DA overflow	Walker et al. (2010)
SAL or MA (2 and 4 mg/kg, IP) administered once immediately before testing (PND 30–35 or PND 69–74)	Male C57BL/6J mice	Fos levels	Fos levels were similar for the two age groups for most regions and doses. Level of locomotor activity was correlated with Fos in several brain areas within both age groups. Fos induction in peri-adolescents was significantly greater for a given amount of locomotor activity in some brain regions (see publication for details)	Zombeck et al. (2010)

AMPH, amphetamine; *Arc*, acitivty-regulated cytoskeleton associated protein gene; *Bdnf*, brain derived neurotrophic factor gene; CA-1, hippocampal field; CD11b, complement type 3 receptor; *cFos*, an inducible early gene; CORT, corticosterone; CREB, cAMP response element-binding protein; d, day; D1, dopamine 1 receptor; D-AMPH, d-amphetamine; D,L-AMPH, d,l-amphetamine; DA, dopamine; DARPP-32, dopamine and cAMP-regulated phosphoprotein Mr 32000; DAT, dopamine transporter; DOPAC, 3,4-dihydroxyphenylacetic acid; *Fos*, an inducible early gene; GFAP, glial fibrillary acidic protein; GluN1, NMDA receptor subunit 1; GluN2A, NMDA receptor subunit 2A; GluN2B, NMDA receptor subunit 2B; h, hour; 5-HIAA, 5-hydroxyindoleacetic acid; 5-HT, serotonin; HVA, homovanillic acid; IHC, immunohistochemistry; IP, intraperitoneally; KO, knockout; LTP, long-term potentiation; MA, methamphetamine; MAP-2, microtubule-associated protein 2; MDMA, methylenedioxymethamphetamine; MMP-9, matrix metalloproteinase-9; mRNA, messenger ribonucleic acid; NE, norepinephrine; PND, postnatal day; PSD-95, postsynaptic density protein 95; PTN, pleiotrophin; SAL, saline; SC, subcutaneously; TH, tyrosine hydroxylase; TPH, tryptophan hydroxylase; TrkB, tropomyosin receptor kinase B; VMAT-2, vesicular monoamine transporter 2; WT, wild-type.

2. Effects of (meth)amphetamines in the peri-adolescent period

2.1 Neurotoxicity

(Meth)amphetamines are neurotoxic at certain doses or durations of exposure. In many such studies, "binge-like" exposure has been used, typically 2–4 injections administered at a 2 h interval on 1–2 days (see Table 2), and then impacts are examined at various time points post-treatment. Most often, toxic effects on dopaminergic systems have been measured, and in several direct comparisons of adult and peri-adolescent rats, methamphetamine-induced markers of dopamine (DA) system toxicity were found in adult but not young rats, when examined 3–7 days after pre-exposure (Cappon et al., 1997; Kokoshka et al., 2000; Pu & Vorhees, 1993; Rau et al., 2006; Riddle et al., 2002; Truong et al., 2005). In one of these studies, serotonergic deficits were examined and found to be comparable for rats of the two age groups at a 7 days post-exposure time point; 1 h after treatment, toxic effects on both dopaminergic and serotonergic systems were comparable in the two age groups (Kokoshka et al., 2000). After binge-like treatment, adult rats in the Kokoshka et al. (2000) study had striatal and plasma methamphetamine levels that were at least twice those measured in peri-adolescent rats. Thus, it is possible that pharmacokinetic differences contributed to the greater DA toxicity in adult rats found 1 week after treatment. On the other hand, Rau et al. (2006) found comparable levels of striatal methamphetamine after binge treatment, yet found greater DA toxicity in adult, compared to peri-adolescent, rats.

It should be acknowledged that, in some cases, greater methamphetamine-induced decreases in markers of toxicity in adult animals correspond with greater basal levels of the marker under study (e.g., Rau et al., 2006; Truong et al., 2005). Further, differences in sensitivity between adult and peri-adolescent rodents to binge-like methamphetamine exposure may be impacted by genetic differences. For example, one study measured DA, the DA metabolite, 3,4-dihydroxyphenylacetic acid (DOPAC), and the rate-limiting enzyme for catecholamine biosynthesis, tyrosine hydroxylase (TH), in three inbred mouse strains 40 days after peri-adolescent or adult methamphetamine treatment. Although age-related outcomes were strain-dependent, in general, methamphetamine-induced striatal DA, DOPAC and TH depletion were greater in adult than peri-adolescent mice (Good et al., 2011).

One factor that is well known to impact the neurotoxic effects of amphetamines is ambient temperature and hyperthermia. In some cases, differences in level of hyperthermia contributed to differences in neurotoxicity induced by amphetamines in peri-adolescent vs adult animals (Cappon et al., 1997), whereas in other studies this did not appear to be the case (Kokoshka et al., 2000; Truong et al., 2005). For example, Cappon et al. (1997) found that methamphetamine administered at 22 °C induced hyperthermia, reduced neostriatal DA and increased glial fibrillary acidic protein content in adult, but not peri-adolescent, rats. However, when the ambient temperature was increased to 30 °C, robust hyperthermia and neurotoxicity occurred in the peri-adolescent rats as well. It should be noted that a PND 20 group experienced hyperthermia at 30 °C, but neurotoxic effects of MA were not found. The drug 3,4-methylenedioxymethamphetamine (MDMA), which in mice is a neurotoxin for DA, but not other catecholamines (Mueller, Maldonado-Adrian, Yuan, McCann, & Ricaurte, 2013), was examined for its effects on microglial and glial activation and DA neurotoxicity under different conditions of crowding or ambient temperature. Greater cage density was associated with higher levels of MDMA-induced microglial and glial activation, as well as greater reductions in TH, regardless of age of treatment. However, overall, adult mice were more sensitive than peri-adolescent mice to MDMA-induced astroglial activation (Frau et al., 2016). An increase in ambient temperature from 21 to 27 °C comparably increased MDMA-induced microglial activation in mice of both age groups, but increased astroglial activation only in the adult mice. There was no significant impact of increased ambient temperature on MDMA-induced TH depletion, but relationships of some markers of toxicity with degree of measured hyperthermia were found.

Some research has examined the impact of peri-adolescent drug exposure on neurotoxic effects in adulthood. For example, Riddle et al. (2002) found that binge-like methamphetamine exposure did not induce DA system neurotoxicity measured 7 days later in adult rats if they had received biweekly treatments with methamphetamine, beginning on PND 40. Acute methamphetamine-induced hyperthermia was also attenuated in the adult rats by the biweekly methamphetamine pre-exposure in peri-adolescence, an outcome consistent with another study (McFadden et al., 2011). McFadden et al. (2011) also found that peri-adolescent methamphetamine exposure provided protection against a DA reuptake deficit after binge methamphetamine exposure, unless hyperthermia was sustained by testing at an increased room temperature. Neither of these studies included a group in which drug treatment was initiated in adulthood; thus, it is possible that the observed effects are not specific to methamphetamine treatment during the peri-adolescent period.

Overall, the existing literature indicates that peri-adolescence, when compared to adulthood, is a period of reduced sensitivity to the neurotoxic effects of methamphetamine. Because the vesicular monoamine transporter-2 (VMAT-2) and DA transporter (DAT) are the primary regulators of cytoplasmic DA, their potential involvement in this age-related difference is relevant. Truong et al. (2005) found greater decreases in VMAT-2 in adult compared to peri-adolescent rats; however, the older rats also had higher basal VMAT-2 levels, raising the possibility of a floor effect. Through studies of VMAT-2 that cofractionate with synaptosomal membranes in an osmotic lysis preparation, and of plasmalemmal DAT, Volz et al. (2009) obtained data suggesting that younger rats may have a higher sequestration capacity for cytoplasmic DA that could contribute to their reduced sensitivity to the neurotoxic effects of methamphetamine.

2.2 Vulnerability to addiction

2.2.1 Initial drug effects

Initial sensitivity to positive and negative drug effects are likely to play a role in subsequent use. Thus, low sensitivity to stimulant/euphoric effects could reduce the probability of further use or lead to higher use in an effort to experience such effects. High sensitivity to aversive effects is likely to inhibit use. Locomotor stimulation in rodents by acute drug treatment is used as a putative measure of euphoria or reward pathway sensitivity. There is evidence that peri-adolescent animals exhibit stimulant responses to amphetamines (Motbey et al., 2012; Walker et al., 2010), but the more significant question is whether sensitivity differs at this stage, compared to in adulthood. A number of studies have compared animals of these ages and have generally found reduced sensitivity of peri-adolescent mice and rats to the stimulant effects of lower (meth)amphetamine doses and to the stereotypic effects of higher doses (Adriani & Laviola, 2000; Bolanos et al., 1998; Cruz et al., 2012; Zakharova et al., 2009; Zombeck et al., 2009; Zombeck et al., 2010); although there are studies that found equivalent sensitivity (Niculescu et al., 2005; Struntz & Siegel, 2018; Walker et al., 2010). For example, Walker et al. (2010) found neither an age-dependent behavioral nor amphetamine-induced caudate DA overflow difference. Zombeck et al. (2009) concluded the age-related difference in methamphetamine-induced stimulation could not be attributed to level of brain methamphetamine;

but in a subsequent paper, Zombeck et al. (2010) obtained Fos-activation data suggesting a stronger Fos induction for a given amount of locomotor activity in peri-adolescent mice in some brain areas. To the best of our knowledge, these findings have not been followed up. Adriani and Laviola (2000) found higher d-amphetamine-induced stimulation in adult mice associated with larger increases in corticosterone levels. Overall, the general consensus is that there is reduced sensitivity to acute (meth)amphetamine effects on locomotor and stereotypic behaviors during the peri-adolescent period. This suggests that a larger dose would be required at this developmental stage to produce a stimulant/euphoric effect.

Few studies have examined age-dependent differences in sensitivity to aversive effects of (meth)amphetamine, or the impact of peri-adolescent drug exposure on aversion measured at a later time. In an early study, peri-adolescent rats exhibited reduced sensitivity to amphetamine-induced conditioned taste aversion compared to both infant and adult rats conditioned with a single exposure to amphetamine, given immediately after consuming sucrose (Infurna & Spear, 1979). Harrod et al. (2010) conditioned peri-adolescent rats using 1 or 3 pairings of methamphetamine with saccharin and measured saccharin preference 1 day later or in adulthood. Robust, sustained saccharin aversion was found at both test times, regardless of number of conditioning trials. When the acute effect of methamphetamine on percent time spent in the center of an open field was examined, peri-adolescent and adult mice exhibited equivalent decreases, suggesting similar sensitivity to methamphetamine-induced anxiety (Struntz & Siegel, 2018). However, another interpretation is that the 4 mg/kg dose used in this study, which was highly stimulating, equivalently increased thigmotaxis. Additional research is needed in this area to determine whether there is age-dependent differential sensitivity to potentially aversive behavioral and physiological (e.g., body temperature, heart rate, blood pressure) effects of acute (meth)amphetamine that could impact further use.

2.2.2 Repeated drug effects

Behavioral sensitization associated with repeated drug exposure has been used as an indirect reflection of neuroadaptation in circuitry relevant to drug reward. Lasting sensitization may impact relapse to drug use (Robinson, Fischer, Ahuja, Lesser, & Maniates, 2016; Steketee & Kalivas, 2011). The impact of peri-adolescent (meth)amphetamine exposure on sensitization of the locomotor stimulant response to repeated drug or stress exposures has been examined. Zakharova et al. (2009) first examined acute sensitivity to the stimulant effects of methamphetamine in peri-adolescent and adult rats and then changes across four additional daily treatments. Methamphetamine had an initial stimulant effect in both age groups that was greater in adult rats, but neither age group exhibited sensitization to the low doses of methamphetamine used (see Table 1). When mice were treated repeatedly across multiple days with d-amphetamine during the peri-adolescent period, compared to a saline-treated control, they exhibited a sensitized locomotor response to an acute methamphetamine treatment in adulthood (Shanks et al., 2015). However, when "one-trial" locomotor sensitization to methamphetamine was examined by treating rats at multiple stages of development with methamphetamine on just 2 consecutive days (spanning PND 12–35), only the two youngest age groups (treated on PND 12–13 and PND 16–17) exhibited sensitization (Kozanian et al., 2012), suggesting a change in neuroadaptive response to methamphetamine between the preweanling to peri-adolescent period.

Laviola et al. (2001) treated peri-adolescent and adult rats with saline or d-amphetamine for 3 days and then challenged all rats with d-amphetamine the next day. These investigators only reported data for the challenge day, when they observed locomotor sensitization in adult, but not peri-adolescent, rats. Stereotypic head-bobbing was also scored, for which peri-adolescent, but not adult, rats exhibited sensitization. The d-amphetamine pre-exposed peri-adolescent mice had sensitized DA responses to drug challenge that may explain their higher levels of stereotypy, which likely competed with the expression of stimulation. Drugs of abuse and exposure to acute and chronic stressors are known to impact both brain reward and brain stress systems (Burke & Miczek, 2014; Torres-Berrio, Cuesta, Lopez-Guzman, & Nava-Mesa, 2018). When the potential cross-sensitizing effect of repeated stress was considered, sensitization of the locomotor response to amphetamine was found in both peri-adolescent and adult rats; however, increased reward circuitry DA levels to amphetamine challenge were dependent upon prior stressor exposure (Cruz et al., 2012).

A measure of sensitization that models abstinence and relapse is acute drug response measured at a time point after withdrawal from chronic exposure. For example, rats treated with amphetamine for 10 consecutive days during peri-adolescence exhibited locomotor sensitization, both at the end of the consecutive treatments and when re-tested after drug challenge 4 weeks later (McPherson & Lawrence, 2006). However, this study did not include a group in which treatment was initiated in adulthood. Good and Radcliffe (2011) examined strain-dependent responses to methamphetamine

40 days after mice had been exposed to methamphetamine during peri-adolescence or as adults and found a sensitized response only in one strain and only if they were treated during peri-adolescence. This particular study suggests that lasting behavioral neuroadaptation depends on age of first exposure; however, there is a rich research literature in adult rodents demonstrating lasting sensitization to psychostimulants (e.g., Bingor, Azriel, Amiad, & Yaka, 2021; Lominac et al., 2014; Vanderschuren & Kalivas, 2000; Wu et al., 2016). Although disparate results could be explained by strain and species differences, there have not been many studies measuring magnitude of response following long periods of withdrawal after repeated or chronic (meth)amphetamine treatment, directly comparing animals initially treated during peri-adolescence vs adulthood.

Unlike studies of acute sensitivity, many of which have focused on differences between peri-adolescent-treated and adult-treated animals, most of the studies examining effects of repeated (meth)amphetamine exposure have concentrated on the later impact of peri-adolescent-initiated exposure. Results appear inconsistent; however, some of the variation is likely attributable to the large assortment of exposure doses, numbers of treatments, treatment intervals, and duration of abstinence before later testing, rather than lack of replication. More research that includes adult comparison groups will be necessary to draw definitive conclusions about the specific impact of peri-adolescent exposure.

2.3 Effects on comorbid disorders

Psychiatric disorders, including, but not limited to, anxiety and mood disorders, are often found to be comorbid with (meth)amphetamine use, abuse, and addiction (Akindipe, Wilson, & Stein, 2014). The multiple hypotheses proposed to explain the relationship between drug use and psychiatric disorders have been discussed in several reviews, and they span drug use as a coping mechanism for pre-existing mental illnesses to drug use causing these psychiatric disorders (e.g., Baker & Dawe, 2005; Brook, Richter, & Rubenstone, 2000; Volkow, 2004). Peri-adolescence is a period of vulnerability for drug-related psychiatric disorders that may increase addiction vulnerability. Some longitudinal data imply that drug use during (peri)adolescence increases risk for the development of a psychiatric disorder (Brook, Cohen, & Brook, 1998), and younger age of methamphetamine use has been found to be a predictor of comorbid psychiatric disorders (Akindipe et al., 2014).

Because peri-adolescence is a period of high vulnerability to drug experimentation, this early exposure has the potential for long-term effects that impact comorbid psychiatric disorders. This has been examined using animal models of anxiety-like and depression-like behaviors. For example, peri-adolescent mice treated with a binge regimen of methamphetamine spent a greater percentage of their time in the center of an open field, compared to saline controls, when tested about 10 days later, suggesting a reduction in anxiety-like behavior (Buck et al., 2017). These results conflict with those for an elevated plus maze study in peri-adolescent rats repeatedly treated with saline or methamphetamine by oral gavage; in this case, less time was spent in the open arms in the methamphetamine group, when they were tested 15 days after treatment (Loxton & Canales, 2017). However, when mice were tested in the Porsolt forced swim test after the treatment regimen used by Buck et al. (2017), the methamphetamine-treated mice exhibited increased percent time immobile, whether tested 2 or 5 weeks after exposure (Joca et al., 2014), an outcome thought to reflect greater depression-like behavior. A reduction in hypothalamic vasopressin associated with methamphetamine exposure was also found in this study, but only in male mice. Likewise, twice daily methamphetamine treatment during peri-adolescence resulted in increased immobility in adult rats, although there was no difference in magnitude of this effect between the Flinders Sensitive and Resistant lines, bred to model depression and as a control, respectively (Overstreet & Wegener, 2013). None of these studies included a group that was treated during adulthood and then tested later. However, when Baker et al. (2018) compared peri-adolescent and adult mice, exposed to saline or a binge regimen of methamphetamine, for percent distance moved in the center of the arena about 2 weeks after treatment, there was no impact of prior methamphetamine exposure in peri-adolescent mice, whereas adult methamphetamine pre-exposed mice had a higher percent distance, compared to saline controls. These mice were also tested for total distance traveled, percent time in the center of the open field (in addition to distance moved), social interaction and plasma corticosterone levels, and no significant treatment effects were found. However, for the Porsolt forced swim test, methamphetamine-exposed adult, but not peri-adolescent, mice had a greater percent time immobile, compared to their respective saline-treated controls. Thus, in this study comparing animals of the two ages, peri-adolescent mice were resistant to later effects of methamphetamine on behaviors thought to reflect anxiety and depression. More data are needed in this area.

2.4 Effects on motivational circuitry

Primary reward and motivation circuitry continues to develop during the (peri)adolescent period (Casey, 2015; Laviola, Adriani, Terranova, & Gerra, 1999; Murty et al., 2016; Sanders, 2013). Although some evidence suggests peri-adolescence to be a period of reduced sensitivity to the toxic effects of amphetamines, changes in this circuitry may promote novelty-seeking and incentive motivational processes that augment drug use and increase vulnerability to addiction (Casey, 2015; Chambers et al., 2003; Hammond et al., 2014). For example, with repeated exposure, peri-adolescent rats exhibit greater d-amphetamine-induced striatal DA sensitization (Laviola et al., 2001) and greater accumbens and caudate putamen levels of the transcription factor, deltaFosB (Ehrlich et al., 2002), compared to adult rats. When mice were repeatedly administered saline or d-amphetamine across the peri-adolescent period and brain gene expression changes examined 1 month later, long-lasting effects of stimulant exposure were found that suggested impacts on brain processes facilitating sensitization (Calabrese et al., 2013). In rats repeatedly treated with saline or d-amphetamine during peri-adolescence or as adults and then euthanized at a common age (~80 days later for peri-adolescent and ~40 days later for adult), there was no effect of drug pre-exposure on the frequency of baseline inhibitory postsynaptic currents recorded in prefrontal cortex pyramidal cells, *in vitro*. However, d-amphetamine application to the slices increased the frequency of these inhibitory currents in saline controls, but not in d-amphetamine pre-exposed rats. When the time after pre-exposure was held constant for the two age groups (about 3 weeks after treatment for both groups), d-amphetamine uniquely suppressed (rather than increased) inhibitory currents in slices from peri-adolescent rats. This suggests that a dampening of inhibitory influences occurs in the presence of drug exposure in this brain region at an earlier withdrawal time point, when d-amphetamine pre-exposure occurs during the peri-adolescent period (Paul et al., 2016). This could be a mechanism relevant to sensitization of reward and motivational processes. Other research has documented changes in the mesocorticolimbic DA system, leading to heightened sensitivity to d-amphetamine challenge in adulthood (~PND 70) in rats repeatedly treated with d-amphetamine across the peri-adolescent period (Sherrill & Gulley, 2018).

Overall, there appear to be some unique effects of (meth)amphetamine in reward circuitry brain regions when exposure occurs during the peri-adolescent period. Studies have mainly examined molecules known to regulate psychostimulant actions. The differential effects in peri-adolescent vs adult animals suggest that these neuroadaptive changes may have a role in enhanced addictive tendencies during the (peri)adolescent period.

2.5 Behavioral reward and motivation

The peri-adolescent period may be a time of enhanced sensitivity to rewarding effects of (meth)amphetamine. Longer access to drug self-administration has been found to increase the amount of drug self-administered and lead to higher levels of drug seeking in reinstatement sessions following extinction (Knackstedt & Kalivas, 2007; Mandt, Copenhagen, Zahniser, & Allen, 2015; Rogers, De Santis, & See, 2008). Moreover, rats given prolonged access to cocaine are resistant to the suppressive effects of aversive stimuli on drug-seeking behavior (e.g., Pelloux, Everitt, & Dickinson, 2007). These outcomes have been characterized as compulsive drug use. When peri-adolescent and adult rats were allowed to self-administer methamphetamine under short- and long-access conditions, peri-adolescent rats earned more reinforcers than adults during long-, but not short-access, and had greater cumulative intake (Anker et al., 2012; Westbrook & Gulley, 2020). These data may indicate a decreased ability of peri-adolescent-onset rats to control their drug-taking. This outcome was replicated in another study by this research group that examined short- and long-access saccharin self-administration. Peri-adolescent rats also self-administered more saccharin than adult rats, suggesting greater avidity for this naturally rewarding substance (Westbrook, Dwyer, et al., 2020).

Behavioral sensitization associated with repeated drug administration was discussed above, but the impact of repeated drug exposure on behavioral reward is important in considerations of addiction risk. As mentioned in Section 2.2.2., in a study by Zakharova et al. (2009), low doses of methamphetamine did not induce locomotor sensitization. However, these investigators also examined conditioned place preference and found that a low dose of 0.5 mg/kg conditioned a place preference after fewer conditioning trials in rats exposed to methamphetamine during peri-adolescence, compared to rats exposed as adults. Although neither peri-adolescent nor adult rats exhibited measurable behavioral sensitization to this methamphetamine dose, neuroadaptations in reward circuitry may have occurred that were detectable by measuring conditioned reward, but not by measuring changes in locomotor behavior. Interestingly, when the impact of early life stress on sensitivity to methamphetamine-induced conditioned place preference was examined in PND 37 and 51 (repeated measure) male Sprague-Dawley rats, place preference was attenuated by prior stress (Dimatelis et al., 2012). Whether this might translate to increased or decreased risk for later drug use is unknown.

In another approach to investigating changes in reward circuitry, the effect of peri-adolescent d-amphetamine exposure on later (1 day, 1 week, and 3 week) preference for the natural reward, sucrose, was examined. Sucrose preference was reduced by prior d-amphetamine exposure, when examined 1 day after the treatments ended. This effect of prior drug exposure was attenuated, but still apparent 3 weeks later (Kang et al., 2016). The authors suggested that this reflects lasting anhedonia. Similarly, peri-adolescent Sprague-Dawley rats treated with methamphetamine daily for 10 days during peri-adolescence exhibited reduced responding for a sucrose food pellet reward in adulthood; however, this effect was found only in males and an adult treatment group was not included (Lee & Kim, 2019). This outcome is consistent with results described in Section 2.3., suggesting that prior peri-adolescent methamphetamine exposure produced a sustained increase in depression-like behavior (Joca et al., 2014). Thus, there may be persistent peri-adolescent exposure effects that impact motivational behavior directed toward natural and drug rewards, and tie in with depression-like behaviors that impact drug use.

2.6 Cognitive impacts

In adult humans and animals, excessive MA use/exposure can have adverse neurocognitive impacts, lasting long after abstinence (McCann et al., 2008; Nordahl, Salo, & Leamon, 2003). Likewise, persistent changes in brain function have been documented after psychostimulant exposure during the (peri)adolescent period (Salmanzadeh et al., 2020; Sherrill et al., 2013). Animal models allow for investigation of bidirectional causality (Melugin, Nolan, & Siciliano, 2021); thus, whether cognitive issues predate and lead to drug use or are a consequence of drug use. However, the literature for the peri-adolescent period has been largely focused on the cognitive consequences of (meth)amphetamine exposure during that period. For example, peri-adolescent, but not adult, rats treated daily with d-amphetamine had deficient retention test performance in a Y-maze test (Featherby et al., 2008); downregulation of hippocampal CREB mRNA was also unique to this group. A subsequent study by another lab, using the same peri-adolescent treatment schedule (though 3, rather than 10, mg/kg amphetamine), investigated the potential role of pleiotrophinin, a cytokine with a role in synaptic plasticity, in memory. Prior amphetamine exposure induced transient memory deficits in pleiotrophinin knockout, but not wild-type, mice that was accompanied by the absence of an increase in hippocampal CA1 long-term potentiation only in wild-type mice (Gramage et al., 2011). The absence of a memory deficit in wild-type mice in the Gramage et al. (2011) study appears not to replicate findings in rats (Featherby et al., 2008); however, the dose used in mice was more than 3 × lower than that used in rats. The data do suggest that pleiotrophinin increases vulnerability to cognitive deficits induced by peri-adolescent amphetamine exposure.

In a study using an escalating regimen of methamphetamine treatments to study effects on fear learning and memory, extinction retrieval was impaired in adult-treated, but not peri-adolescent-treated, rats when measures began the day after the final treatment. These investigators also examined the impact of voluntary methamphetamine self-administration during the peri-adolescent and adult periods on later performance in the fear and memory task. When measures began the day after the final treatment, an impairment in the acquisition of conditioned freezing occurred only in the adult group. However, when the peri-adolescent treatment group was instead tested in adulthood, a deficit similar to the adult treatment group was observed (Luikinga et al., 2019). Thus, in this study, methamphetamine treatment impaired fear conditioning when measurements occurred in adulthood, but not late adolescence. Deficits in adulthood after peri-adolescent drug exposure were also found for studies of attentional learning (Nazari et al., 2020), spatial memory and reversal learning (Grochecki et al., 2021) in rats. Grochecki et al. (2021) measured levels of several proteins in the prefrontal cortex and hippocampus and found that peri-adolescent amphetamine reduced expression of the hippocampal glutamate NMDA receptor subunit GluN2A, as well as postsynaptic protein PSD-95, but not GluN1, GluN2B, or the matrix metalloproteinase, MMP-9. Because the GluN2A NMDA receptor subunit appears to preferentially affect short-term memory and the rapid acquisition of spatial information (Franchini, Carrano, Di Luca, & Gardoni, 2020), Grochecki et al. (2021) suggested that the decline in GluN2A in amphetamine-exposed rats may infer decreased efficacy of synaptic transmission during synaptic plasticity associated with learning and memory processing. Because NMDA receptors possessing the GluN2A subunit are vital in memory consolidation processes

(Cercato et al., 2017; Ge et al., 2010), reduced efficacy of glutamate activity would be expected to hinder memory consolidation that might be more likely detected with elapsed time, providing a potential mechanism of action for the reduced cognitive abilities seen only later in adulthood.

Several studies have been performed in the C57BL/6J mouse strain. In mice of this strain repeatedly treated with saline or a high dose of methamphetamine across the peri-adolescent period, novel spatial memory, but not novel object memory, was persistently disrupted by prior methamphetamine exposure (North et al., 2013). However, when methamphetamine was administered in a binge regimen on 2 consecutive days to peri-adolescent mice, no significant impact on spatial ability was found 2 weeks later (Buck et al., 2017). Some studies have found cognitive deficits for extremely low doses. Thus, when C57BL/6 mice were repeatedly treated once daily with saline or 1 mg/kg d-amphetamine during peri-adolescence, and then tested for spatial and visual discrimination beginning in adulthood, the amphetamine-treated mice required more trials to reach criterion for correct responding (Boomhower & Newland, 2017). Also, the Grochecki et al. (2021) paper that was already discussed, used a lower 2.5 mg/kg dose of amphetamine in peri-adolescent rats treated once daily and found deficits in spatial memory and reversal learning in adulthood (Grochecki et al., 2021).

There is considerable variation in dose, pattern of drug administration, species (strain), cognitive test, and age of cognitive testing after prior drug exposure across studies. Demonstrated peri-adolescent (meth)amphetamine-induced deficits do not appear to be attributable to particular aspects of learning and memory, although there are some reports suggesting that memory deficits at this developmental stage may be impacted by task novelty (e.g., Featherby et al., 2008).

A final unique example is one in which rats were administered daily saline or escalating doses of methamphetamine, beginning in the peri-adolescent period, and were then tested for visual discrimination learning and methamphetamine intake in a two-bottle choice test in adulthood. The low dose escalating methamphetamine exposure led to a generalized learning impairment, and the amount of voluntary methamphetamine intake in adulthood, when the animals were offered water vs methamphetamine in water, was positively correlated with the degree of learning impairment (Ye et al., 2014). One potential explanation for the relationship between these traits is that (peri)adolescent methamphetamine exposure impacts mechanisms that contribute to both learning and methamphetamine intake. The data from this study indicate that prior exposure effects on learning have some predictive value for subsequent voluntary methamphetamine intake. Overall, long-lasting deficits in learning and memory performance have been a common outcome in studies examining the effects of peri-adolescent exposure to (meth)amphetamine, as summarized in greater detail by Salmanzadeh et al. (2020).

2.7 Social behavior

Effects of (meth)amphetamine exposure during the peri-adolescent period on social behaviors have not often been investigated. Motbey et al. (2012) examined social preference after a single acute treatment with methamphetamine in peri-adolescent rats and found no significant effect. Baker et al. (2018) exposed mice to a binge regimen of methamphetamine during peri-adolescence or as adults and found no effects of drug vs saline on social interaction. However, when Flinders Sensitive and Resistant rat lines were treated with methamphetamine during a timeframe beginning before and part way through peri-adolescence (PND 19–34), social behavior measured the following day and once the animals reached adulthood was decreased in comparison to a saline control (Mouton et al., 2016). There was no difference in magnitude of the methamphetamine treatment effect between the rat lines. Finally, a recent study investigated the impact of the presence or absence of a social partner on methamphetamine-induced conditioned reward in peri-adolescent rats, and found rats exposed to methamphetamine in the presence of a social partner to exhibit heightened conditioned preference, compared to rats exposed in the absence of a partner (Reyna et al., 2021). Though the literature is extremely limited in this area, the data so far indicate some long-lasting impacts of peri-adolescent methamphetamine exposure on social interaction, as well as a potential role for social interaction in heightening sensitivity to the rewarding effects of methamphetamine.

2.8 Effects on motivation for other drugs

Alterations in sensitivity and self-administration and seeking of other drugs after peri-adolescent exposure to (meth)amphetamines have been examined in a few studies. Thus, when rats were repeatedly treated with saline or methamphetamine across the peri-adolescent period or in adulthood, morphine-induced antinociception was reduced by prior methamphetamine in adult rats only, when all rats were tested 2 weeks later (Cyr et al., 2016). Thus, prior history of methamphetamine exposure could have implications for analgesic responses to opioids. When the impact of peri-adolescent amphetamine exposure on ethanol intake was examined, drug-exposed male, but not female, rats exhibited increased ethanol intake compared to saline-exposed rats (Ruiz et al., 2018). Results for neurotransmitter levels (DA, serotonin and norepinephrine) in the insular cortex did not correspond with the behavioral outcomes. Cocaine self-administration and seeking were examined in a multi-strain study. Rats were administered a low 0.5 mg/kg dose of d-amphetamine, beginning in peri-adolescence, on week days to model doses used for the treatment of attention deficit hyperactivity disorder. After about 3 weeks of abstinence, adult rats began IV cocaine self-administration. The SHR strain acquired cocaine self-administration more rapidly, and the effects of prior d-amphetamine exposure on cocaine self-administration were strain-dependent with reduced self-administration in some strains, including SHR, and faster acquisition, but not greater intake, in the WKY strain (Jordan, Lemay, et al., 2016). In a companion paper by the same research group, using the same general methods except that d-amphetamine peri-adolescent exposure occurred in concert with set-shifting testing, SHR rats exhibited poorer baseline set-shifting performance that was improved by d-amphetamine given 30 min prior to testing. SHR rats self-administered more cocaine, which was not reduced by prior d-amphetamine exposure, although cocaine seeking was reduced (Jordan, Taylor, et al., 2016). Thus, the overall outcome was either a reduction or no impact of prior d-amphetamine treatment on later cocaine intake and seeking. However, in a study in a different rat strain, daily 0.7 mg/kg d-amphetamine treatment during a similar peri-adolescent period, led to faster acquisition of cocaine self-administration, after 5–6 weeks of abstinence, compared to a saline control (Lacy et al., 2018). Overall, the data are mixed, and the literature in this area is small, making it difficult to determine whether cocaine reward/reinforcement is consistently reduced or augmented by (meth)amphetamine pre-exposure.

3. Conclusions and future directions

A review of animal models of peri-adolescent substance use in 2003 indicated that research in this area was sparse at that time (Smith, 2003), although use of addictive drugs in peri-adolescent and adolescent humans, including illicit amphetamines, was common (e.g., Johnston et al., 1998). The research conducted on the effects of (meth)amphetamine exposure during the peri-adolescent period has increased, but the prior conclusion remains: "...the biological characteristics of peri-adolescence which predispose toward consumption to intoxication, and the mechanisms underlying drug progression and persisting drug effects, still remain to be well defined." (Smith, 2003). Although more studies have emerged that directly compare the impacts of prior peri-adolescent to prior adult drug exposure, in our opinion, this practice should be a consistent feature of studies if a goal is to determine effects that are unique to the peri-adolescent period. One goal of the current mini-review was to examine the rodent research literature to determine whether severity and duration of neurotoxicity is greater when exposure occurs during the peri-adolescent vs adult period. This literature is reasonably large, and appears to support the opposite relationship—reduced susceptibility when exposure occurs during peri-adolescence. However, some of the data indicate that even in the absence of apparent neurotoxicity, use during this period can have immediate and long-lasting developmental, psychological, cognitive and social consequences. Finally, with regard to risk for addiction, it is important to consider sensitivity to rewarding and aversive drug effects, as well as the impact of drug exposure on brain motivational circuitry. The peri-adolescent period appears to be a time of reduced sensitivity to acute (meth)amphetamine effects on locomotor and stereotypic behaviors, but increased sensitivity to motivational behavioral and neurochemical effects. Sensitivity to aversive effects may be reduced, though the literature in this area needs broadening. Although some have used drug-induced stimulation as a surrogate for reward circuitry activation, the data are mixed regarding the association between magnitude of stimulation and risk for drug use. In fact, in mice bred for high vs low levels of methamphetamine-induced stimulation, the high stimulation line avoided consuming methamphetamine, suggesting that genetic factors resulting in extreme sensitivity to stimulation may protect against use (Kamens et al., 2005). Reduced initial sensitivity to locomotor effects (including stereotypy) and aversive effects, in combination with heightened sensitivity to rewarding effects, may set the stage for increased (meth)amphetamine use when initiated during peri-adolescence and thus, amplified risk for addiction.

Acknowledgments

Funding was provided by NIDA U01DA041579 (T.J.P.), NIDA R01DA046081 (T.J.P.), Department of Veterans Affairs Merit Review Grant I01BX002106 (T.J.P.), and the VA Research Career Scientist Program (T.J.P.) during the writing of the manuscript. The contents of this article do not represent the views of the U.S. Department of Veterans Affairs or the United States government.

Conflict of interest

The authors declare that they have no conflict of interest.

References

Adriani, W., & Laviola, G. (2000). A unique hormonal and behavioral hyporesponsivity to both forced novelty and d-amphetamine in periadolescent mice. *Neuropharmacology, 39*, 334–346.

Akindipe, T., Wilson, D., & Stein, D. J. (2014). Psychiatric disorders in individuals with methamphetamine dependence: Prevalence and risk factors. *Metabolic Brain Disorders, 29*, 351–357.

Anker, J. J., Baron, T. R., Zlebnik, N. E., & Carroll, M. E. (2012). Escalation of methamphetamine self-administration in adolescent and adult rats. *Drug and Alcohol Dependence, 124*, 149–153.

Baker, A., & Dawe, S. (2005). Amphetamine use and co-occurring psychological problems: Review of the literature and implications for treatment. *Australian Psychologist, 40*, 88–95.

Baker, E. P., Magnuson, E. C., Dahly, A. M., & Siegel, J. A. (2018). The effects of enriched environment on the behavioral and corticosterone response to methamphetamine in adolescent and adult mice. *Developmental Psychobiology, 60*, 664–673.

Bingor, A., Azriel, M., Amiad, L., & Yaka, R. (2021). Potentiated response of ERK/MAPK signaling is associated with prolonged withdrawal from cocaine behavioral sensitization. *Journal of Molecular Neuroscience* (Jan 21, online ahead of print).

Bolanos, C. A., Glatt, S. J., & Jackson, D. (1998). Subsensitivity to dopaminergic drugs in periadolescent rats: A behavioral and neurochemical analysis. *Developmental Brain Research, 111*, 25–33.

Boomhower, S. R., & Newland, M. C. (2017). Effects of adolescent exposure to methylmercury and d-amphetamine on reversal learning and an extradimensional shift in male mice. *Experimental and Clinical Psychopharmacology, 25*, 64–73.

Brook, J. S., Cohen, P., & Brook, D. W. (1998). Longitudinal study of co-occurring psychiatric disorders and substance use. *Journal of the American Academy of Child & Adolescent Psychiatry, 37*, 322–330.

Brook, J. S., Richter, L., & Rubenstone, E. (2000). Consequences of adolescent drug use on psychiatric disorders in early adulthood. *Annals of Medicine, 32*, 401–407.

Buck, J. M., Morris, A. S., Weber, S. J., Raber, J., & Siegel, J. A. (2017). Effects of adolescent methamphetamine and nicotine exposure on behavioral performance and MAP-2 immunoreactivity in the nucleus accumbens of adolescent mice. *Behavioural Brain Research, 323*, 78–85.

Buck, J. M., & Siegel, J. A. (2015). The effects of adolescent methamphetamine exposure. *Frontiers in Neuroscience, 9*, 151.

Burke, A. R., & Miczek, K. A. (2014). Stress in adolescence and drugs of abuse in rodent models: Role of dopamine, CRF, and HPA axis. *Psychopharmacology, 231*, 1557–1580.

Calabrese, F., Richetto, J., Racagni, G., Feldon, J., Meyer, U., & Riva, M. A. (2013). Effects of withdrawal from repeated amphetamine exposure in peri-puberty on neuroplasticity-related genes in mice. *Neuroscience, 250*, 222–231.

Cappon, G. D., Morford, L. L., & Vorhees, C. V. (1997). Ontogeny of methamphetamine-induced neurotoxicity and associated hyperthermic response. *Developmental Brain Research, 103*, 155–162.

Casey, B. J. (2015). Beyond simple models of self-control to circuit-based accounts of adolescent behavior. *Annual Review of Psychology, 66*, 295–319.

Cercato, M. C., Vazquez, C. A., Kornisiuk, E., Aguirre, A. I., Colettis, N., Snitcofsky, M., et al. (2017). GluN1 and GluN2A NMDA receptor subunits increase in the hippocampus during memory consolidation in the rat. *Frontiers in Behavioral Neuroscience, 10*, 1–14.

Chambers, R. A., Taylor, J. R., & Potenza, M. N. (2003). Developmental neurocircuitry of motivation in adolescence: A critical period of addiction vulnerability. *American Journal of Psychiatry*, *160*, 1041–1052.
Cruz, F. C., Marin, M. T., Leão, R. M., & Planeta, C. S. (2012). Stress-induced cross-sensitization to amphetamine is related to changes in the dopaminergic system. *Journal of Neural Transmission (Vienna)*, *119*, 415–424.
Cyr, M. C., Ingram, S. L., Aicher, S. A., & Morgan, M. M. (2016). Chronic psychostimulant exposure to adult, but not periadolescent rats reduces subsequent morphine antinociception. *Pharmacology, Biochemistry and Behavior*, *101*, 538–543.
Dimatelis, J. J., Russell, V. A., Stein, D. J., & Daniels, W. M. (2012). The effects of lobeline and naltrexone on methamphetamine-induced place preference and striatal dopamine and serotonin levels in adolescent rats with a history of maternal separation. *Metabolic Brain Disease*, *27*, 351–361.
Ehrlich, M. E., Sommer, J., Canas, E., & Unterwald, E. M. (2002). Periadolescent mice show enhanced deltaFosB upregulation in response to cocaine and amphetamine. *Journal of Neuroscience*, *22*, 9155–9159.
Embry, D., Hankins, M., Biglan, A., & Boles, S. (2009). Behavioral and social correlates of methamphetamine use in a population-based sample of early and later adolescents. *Addictive Behaviors*, *34*, 343–351.
Featherby, T., van den Buuse, M., Lubman, D. I., & Lawrence, A. J. (2008). Persistent downregulation of hippocampal CREB mRNA parallels a Y-maze deficit in adolescent rats following semi-chronic amphetamine administration. *British Journal of Pharmacology*, *154*, 417–428.
Franchini, L., Carrano, N., Di Luca, M., & Gardoni, F. (2020). Synaptic GluN2A-containing NMDA receptors: From physiology to pathological synaptic plasticity. *International Journal of Molecular Sciences*, *21*, 1538.
Frau, L., Simola, N., Porceddu, P. F., & Morelli, M. (2016). Effect of crowding, temperature and age on glia activation and dopaminergic neurotoxicity induced by MDMA in the mouse brain. *Neurotoxicology*, *56*, 127–138.
Ge, Y., Dong, Z., Bagot, R. C., Howland, J. G., Phillips, A. G., Wong, T. P., et al. (2010). Hippocampal long-term depression is required for the consolidation of spatial memory. *Proceedings of the National Academy of Sciences of the United States of America*, *107*, 16697–16702.
Good, R. L., Liang, L. P., Patel, M., & Radcliffe, R. A. (2011). Mouse strain- and age-dependent effects of binge methamphetamine on dopaminergic signaling. *Neurotoxicology*, *32*, 751–759.
Good, R. L., & Radcliffe, R. A. (2011). Methamphetamine-induced locomotor changes are dependent on age, dose and genotype. *Pharmacology, Biochemistry and Behavior*, *98*, 101–111.
Gramage, E., Del Olmo, N., Fole, A., Martín, Y. B., & Herradón, G. (2011). Periadolescent amphetamine treatment causes transient cognitive disruptions and long-term changes in hippocampal LTP depending on the endogenous expression of pleiotrophin. *Addiction Biology*, *18*, 19–29.
Grochecki, P., Smaga, I., Lopatynska-Mazurek, M., Gibula-Tarlowska, E., Kedzierska, E., Listos, J., et al. (2021). Effects of mephedrone and amphetamine exposure during adolescence on spatial memory in adulthood: Behavioral and neurochemical analysis. *International Journal of Molecular Sciences*, *22*, 589.
Hammond, C. J., Mayes, L. C., & Potenza, M. N. (2014). Neurobiology of adolescent substance use and addictive behaviors: Prevention and treatment implications. *Adolescent Medicine: State of the Art Reviews*, *25*, 15–32.
Harrod, S. B., Lacy, R. T., & Ballina, L. E. (2010). Persistent expression of methamphetamine-induced CTA in periadolescent rats. *Pharmacology, Biochemistry and Behavior*, *96*, 515–520.

Infurna, R. N., & Spear, L. P. (1979). Developmental changes in amphetamine-induced taste aversions. *Pharmacology, Biochemistry and Behavior, 11*, 31–35.
Joca, L., Zuloaga, D. G., Raber, J., & Siegel, J. A. (2014). Long-term effects of early adolescent methamphetamine exposure on depression-like behavior and the hypothalamic vasopressin system in mice. *Developmental Neuroscience, 36*, 108–118.
Johnston, L. D., O'Malley, P. M., & Backman, J. G. (1998). *National Survey Results on drug use from the monitoring of the future study. 1975–1997.* NIDA: NIH publication (no. 98-4345).
Jordan, C. J., Lemay, C., Dwoskin, L. P., & Kantak, K. M. (2016). Adolescent d-amphetamine treatment in a rodent model of attention deficit/hyperactivity disorder: Impact on cocaine abuse vulnerability in adulthood. *Psychopharmacology, 233*, 3891–3903.
Jordan, C. J., Taylor, D. M., Dwoskin, L. P., & Kantak, K. M. (2016). Adolescent d-amphetamine treatment in a rodent model of ADHD: Pro-cognitive effects in adolescence without an impact on cocaine cue reactivity in adulthood. *Behavioural Brain Research, 297*, 165–179.
Kamens, H. M., Burkhart-Kasch, S., McKinnon, C. S., Li, N., Reed, C., & Phillips, T. J. (2005). Sensitivity to psychostimulants in mice bred for high and low stimulation to methamphetamine. *Genes, Brain and Behavior, 4*, 110–125.
Kang, S., Wu, M. M., Galvez, R., & Gulley, J. M. (2016). Timing of amphetamine exposure in relation to puberty onset determines its effects on anhedonia, exploratory behavior, and dopamine D1 receptor expression in young adulthood. *Neuroscience, 339*, 72–84.
Knackstedt, L. A., & Kalivas, P. W. (2007). Extended access to cocaine self-administration enhances drug-primed reinstatement but not behavioral sensitization. *The Journal of Pharmacology and Experimental Therapeutics, 322*, 1103–1109.
Kokoshka, J. M., Fleckenstein, A. E., Wilkins, D. G., & Hanson, G. R. (2000). Age-dependent differential responses of monoaminergic systems to high doses of methamphetamine. *Journal of Neurochemistry, 75*, 2095–2102.
Kozanian, O. O., Gutierrez, A., Mohd-Yusof, A., & McDougall, S. A. (2012). Ontogeny of methamphetamine- and cocaine-induced one-trial behavioral sensitization in preweanling and adolescent rats. *Behavioural Pharmacology, 23*, 367–379.
Lacy, R. T., Schorsch, H. K., & Austin, R. P. (2018). Adolescent d-amphetamine exposure enhances the acquisition of cocaine self-administration in male and female rats. *Experimental and Clinical Psychopharmacology, 26*, 18–28.
Laviola, G., Adriani, W., Terranova, M. L., & Gerra, G. (1999). Psychobiological risk factors for vulnerability to psychostimulants in human adolescents and animal models. *Neuroscience and Biobehavioral Reviews, 23*, 933–1010.
Laviola, G., Pascucci, T., & Pieretti, S. (2001). Striatal dopamine sensitization to D-amphetamine in periadolescent but not in adult rats. *Pharmacology, Biochemistry and Behavior, 68*, 115–124.
Lee, J. H., & Kim, D. G. (2019). Diminished food-related motivation in adult rats treated with methamphetamine during adolescence. *Neuroreport, 30*, 1143–1147.
Lominac, K. D., McKenna, C. L., Schwartz, L. M., Ruiz, P. N., Wroten, M. G., Miller, B. W., et al. (2014). Mesocorticolimbic monoamine correlates of methamphetamine sensitization and motivation. *Frontiers in Systems Neuroscience, 8*, 70.
Loxton, D., & Canales, J. J. (2017). Long-term cognitive, emotional and neurogenic alterations induced by alcohol and methamphetamine exposure in adolescent rats. *Progress in Neuro-Psychopharmacology & Biological Psychiatry, 74*, 1–8.
Luikinga, S. J., Kim, J. H., & Perry, C. J. (2018). Developmental perspectives on methamphetamine abuse: Exploring adolescent vulnerabilities on brain and behavior. *Progress in Neuropsychopharmacology & Biological Psychiatry, 87*, 78–84.
Luikinga, S. J., Perry, C. J., Madsen, H. B., Lawrence, A. J., & Kim, J. H. (2019). Effects of methamphetamine exposure on fear learning and memory in adult and adolescent rats. *Neurochemical Research, 44*, 2081–2091.

Mandt, B. H., Copenhagen, L. I., Zahniser, N. R., & Allen, R. M. (2015). Escalation of cocaine consumption in short and long access self-administration procedures. *Drug and Alcohol Dependence, 149*, 166–172.

McCann, U. D., Kuwabara, H., Kumar, A., Palermo, M., Abbey, R., Brasic, J., et al. (2008). Persistent cognitive and dopamine transporter deficits in abstinent methamphetamine users. *Synapse, 62*, 91–100.

McFadden, L. M., Hoonakker, A. J., Vieira-Brock, P. L., Stout, K. A., Sawada, N. M., Ellis, J. D., et al. (2011). Methamphetamine treatment during development attenuates the dopaminergic deficits caused by subsequent high-dose methamphetamine administration. *Synapse, 65*, 771–777.

McPherson, C. S., & Lawrence, A. J. (2006). Exposure to amphetamine in rats during peri-adolescence establishes behavioural and extrastriatal neural sensitization in adulthood. *International Journal of Neuropsychopharmacology, 9*, 377–392.

Melugin, P. R., Nolan, S. O., & Siciliano, C. A. (2021). Bidirectional causality between addiction and cognitive deficits. *International Review of Neurobiology, 157*, 371–407.

Motbey, C. P., Hunt, G. E., Bowen, M. T., Artiss, S., & McGregor, I. S. (2012). Mephedrone (4-methylmethcathinone,'meow'): Acute behavioural effects and distribution of Fos expression in adolescent rats. *Addiction Biology, 17*, 409–422.

Mouton, M., Harvey, B. H., Cockeran, M., & Brink, C. B. (2016). The long-term effects of methamphetamine exposure during pre-adolescence on depressive-like behavior in a genetic animal model of depression. *Metabolic Brain Disease, 31*, 63–74.

Mueller, M., Maldonado-Adrian, C., Yuan, J., McCann, U. D., & Ricaurte, G. A. (2013). Studies of (±)-3,4-methylenedioxymethamphetamine (MDMA) metabolism and disposition in rats and mice: Relationship to neuroprotection and neurotoxicity profile. *The Journal of Pharmacology and Experimental Therapeutics, 344*, 479–488.

Murty, V., Calabro, F., & Luna, B. (2016). The role of experience in adolescent cognitive development: Integration of executive, memory, and mesolimbic systems. *Neuroscience and Biobehavioral Reviews., 70*, 46–58.

Nazari, A., Perez-Fernandez, C., Flores, P., Moreno, M., & Sánchez-Santed, F. (2020). Age-dependent effects of repeated methamphetamine exposure on locomotor activity and attentional function in rats. *Pharmacology, Biochemistry and Behavior, 191*, 172879.

Niculescu, M., Ehrlich, M. E., & Unterwald, E. M. (2005). Age-specific behavioral responses to psychostimulants in mice. *Pharmacology, Biochemistry and Behavior, 82*, 280–288.

Nordahl, T. E., Salo, R., & Leamon, M. (2003). Neuropsychological effects of chronic methamphetamine use on neurotransmitters and cognition: A review. *Journal of Neuropsychiatry and Clinical Neuroscience, 15*, 317–325.

North, A., Swant, J., Salvatore, M. F., Gamble-George, J., Prins, P., Butler, B., et al. (2013). Chronic methamphetamine exposure produces a delayed, long-lasting memory deficit. *Synapse, 67*, 245–257.

Overstreet, D. H., & Wegener, G. (2013). The flinders sensitive line rat model of depression–25 years and still producing. *Pharmacological Reviews, 65*, 143–155.

Park, T. M., & Haning, W. F. (2016). Stimulant use disorders. *Child and Adolescent Psychiatric Clinics of North America, 25*, 461–471.

Paul, K., Kang, S., Cox, C. L., & Gulley, J. M. (2016). Repeated exposure to amphetamine during adolescence alters inhibitory tone in the medial prefrontal cortex following drug re-exposure in adulthood. *Behavioural Brain Research, 309*, 9–13.

Pelloux, Y., Everitt, B. J., & Dickinson, A. (2007). Compulsive drug seeking by rats under punishment: Effects of drug taking history. *Psychopharmacology, 194*, 127–137.

Poudel, A., & Gautam, S. (2017). Age of onset of substance use and psychosocial problems among individuals with substance use disorders. *BMC Psychiatry, 17*, 10.

Pu, C., & Vorhees, C. V. (1993). Developmental dissociation of methamphetamine-induced depletion of dopaminergic terminals and astrocyte reaction in rat striatum. *Developmental Brain Research, 72*, 325–328.

Rau, K. S., Truong, J. G., Wilkins, D. G., Fleckenstein, A. E., & Hanson, G. R. (2006). Age-dependent effects of methamphetamine on VMAT-2. *Annals of the New York Academy of Sciences, 1074*, 154–159.

Rawson, R. A., Gonzales, R., McCann, M., & Ling, W. (2007). Use of methamphetamine by young people: Is there reason for concern? *Addiction, 102*, 1021–1022.

Reyna, N. C., Madden, J. T., Thiel, K. J., & Pentkowski, N. S. (2021). Methamphetamine and social rewards interact to produce enhanced conditioned place preference in male adolescent rats. *Pharmacology, Biochemistry and Behavior, 201*, 173091.

Richmond-Rakerd, L. S., Slutske, W. S., & Wood, P. K. (2017). Age of initiation and substance use progression: A multivariate latent growth analysis. *Psychology of Addictive Behaviors, 31*, 664–675.

Riddle, E. L., Kokoshka, J. M., Wilkins, D. G., Hanson, G. R., & Fleckenstein, A. E. (2002). Tolerance to the neurotoxic effects of methamphetamine in young rats. *European Journal of Pharmacology, 435*, 181–185.

Robinson, M. J. F., Fischer, A. M., Ahuja, A., Lesser, E. N., & Maniates, H. (2016). Roles of "wanting" and "liking" in motivating behavior: Gambling, food, and drug addictions. *Current Topics in Behavioral Neurosciences, 27*, 105–136.

Rogers, J. L., De Santis, S., & See, R. E.l. (2008). Extended methamphetamine self-administration enhances reinstatement of drug seeking and impairs novel object recognition in rats. *Psychopharmacology, 199*, 615–624.

Ruiz, P., Calliari, A., Genovese, P., Scorza, C., & Pautassi, R. M. (2018). Amphetamine, but not methylphenidate, increases ethanol intake in adolescent male, but not in female, rats. *Brain and Behavior, 8*, e00939.

Salmanzadeh, H., Ahmadi-Soleimani, S. M., Pachenari, N., Azadi, M., Halliwell, R. F., Rubino, T., et al. (2020). Adolescent drug exposure: A review of evidence for the development of persistent changes in brain function. *Brain Research Bulletin, 156*, 105–117.

Sanders, R. A. (2013). Adolescent psychosocial, social, and cognitive development. *Pediatrics in Review, 34*, 354–359.

Saw, Y. M., Saw, T. N., Yasuoka, J., Chan, N., Kham, N. P. E., Khine, W., et al. (2017). Gender difference in early initiation of methamphetamine use among current methamphetamine users in muse, northern Shan State, Myanmar. *Harm Reduction Journal, 14*, 21.

Shanks, R. A., Ross, J. M., Doyle, H. H., Helton, A. K., Picou, B. N., Schulz, J., et al. (2015). Adolescent exposure to cocaine, amphetamine, and methylphenidate cross-sensitizes adults to methamphetamine with drug- and sex-specific effects. *Behavioural Brain Research, 281*, 116–124.

Sherrill, L. K., & Gulley, J. M. (2018). Effects of amphetamine exposure during adolescence on behavior and prelimbic cortex neuron activity in adulthood. *Brain Research, 1694*, 111–120.

Sherrill, L. K., Stanis, J. J., & Gulley, J. M. (2013). Age-dependent effects of repeated amphetamine exposure on working memory in rats. *Behavioural Brain Research, 242*, 84–94.

Smith, R. F. (2003). Animal models of periadolescent substance abuse. *Neurotoxicology and Teratology, 25*, 291–301.

Spear, L. P. (2016). Consequences of adolescent use of alcohol and other drugs: Studies using rodent models. *Neuroscience and Biobehavioral Reviews, 70*, 228–243.

Spear, L. P., & Brake, S. C. (1983). Periadolescence: Age-dependent behavior and psychopharmacological responsivity in rats. *Developmental Psychobiology, 16*, 83–109.

Steketee, J. D., & Kalivas, P. W. (2011). Drug wanting: Behavioral sensitization and relapse to drug-seeking behavior. *Pharmacological Reviews, 63*, 348–365.

Stock, A.-K., & Beste, C. (2017). On the necessity of translational cognitive-neurotoxicological research in methamphetamine abuse and addiction. *Archives of Toxicology, 91*, 2707–2709.

Struntz, K. H., & Siegel, J. A. (2018). Effects of methamphetamine exposure on anxiety-like behavior in the open field test, corticosterone, and hippocampal tyrosine hydroxylase in adolescent and adult mice. *Behavioural Brain Research, 348*, 211–218.

Teixeira-Gomes, A., Costa, V. M., Feio-Azevedo, R., de Lourdes Bastos, L., Carvalho, F., & Capela, J. P. (2015). The neurotoxicity of amphetamines during the adolescent period. *International Journal of Developmental Neuroscience*, *41*, 44–62.

Torres-Berrio, A., Cuesta, S., Lopez-Guzman, S., & Nava-Mesa, M. O. (2018). Interaction between stress and addiction: Contributions from Latin-American neuroscience. *Frontiers in Psychology*, *9*, 2639.

Truong, J. G., Wilkins, D. G., Baudys, J., Crouch, D. J., Johnson-Davis, K. L., Gibb, J. W., et al. (2005). Age-dependent methamphetamine-induced alterations in vesicular monoamine transporter-2 function: Implications for neurotoxicity. *The Journal of Pharmacology and Experimental Therapeutics*, *314*, 1087–1092.

Vanderschuren, L. J. M. J., & Kalivas, P. W. (2000). Alterations in dopaminergic and glutamatergic transmission in the induction and expression of behavioral sensitization: A critical review of preclinical studies. *Psychopharmacology*, *51*, 99–120.

Volkow, N. D. (2004). The reality of comorbidity: Depression and drug abuse. *Biological Psychiatry*, *56*, 714–717.

Volz, T. J., Farnsworth, S. J., Rowley, S. D., Hanson, G. R., & Fleckenstein, A. E. (2009). Age-dependent differences in dopamine transporter and vesicular monoamine transporter-2 function and their implications for methamphetamine neurotoxicity. *Synapse*, *63*, 147–151.

Walker, Q. D., Morris, S. E., Arrant, A. E., Nagel, J. M., Parylak, S., Zhou, G., et al. (2010). Dopamine uptake inhibitors but not dopamine releasers induce greater increases in motor behavior and extracellular dopamine in adolescent rats than in adult male rats. *The Journal of Pharmacology and Experimental Therapeutics*, *335*, 123–132.

Westbrook, S. R., Carrica, L. K., Banks, A., & Gulley, J. M. (2020). AMPed-up adolescents: The role of age in the abuse of amphetamines and its consequences on cognition and prefrontal cortex development. *Pharmacology, Biochemistry and Behavior*, *198*, 173016.

Westbrook, S. R., Dwyer, M. R., Cortes, L. R., & Gulley, J. M. (2020). Extended access self-administration of methamphetamine is associated with age- and sex-dependent differences in drug taking behavior and recognition memory in rats. *Behavioural Brain Research*, *390*, 112659.

Westbrook, S. R., & Gulley, J. M. (2020). Effects of the GluN2B antagonist, Ro 25-6981, on extinction consolidation following adolescent- or adult-onset methamphetamine self-administration in male and female rats. *Behavioural Pharmacology*, *31*, 748–758.

Wu, J., Zhu, D., Zhang, J., Li, G., Liu, Z., & Sun, J. (2016). Melatonin treatment during the incubation of sensitization attenuates methamphetamine-induced locomotor sensitization and MeCP2 expression. *Progress in Neuro-Psychopharmacology & Biological Psychiatry*, *65*, 145–152.

Ye, T., Pozos, H., Phillips, T. J., & Izquierdo, A. (2014). Long-term effects of exposure to methamphetamine in adolescent rats. *Drug and Alcohol Dependence*, *138*, 17–23.

Yimsaard, P., Maes, M. M., Verachai, V., & Kalayasiri, R. (2018). Pattern of methamphetamine use and the time lag to methamphetamine dependence. *Journal of Addiction Medicine*, *12*, 92–98.

Zakharova, E., Leoni, G., Kichko, I., & Izenwasser, S. (2009). Differential effects of methamphetamine and cocaine on conditioned place preference and locomotor activity in adult and adolescent male rats. *Behavioural Brain Research*, *198*, 45–50.

Zombeck, J. A., Gupta, T., & Rhodes, J. S. (2009). Evaluation of a pharmacokinetic hypothesis for reduced locomotor stimulation from methamphetamine and cocaine in adolescent versus adult male C57BL/6J mice. *Psychopharmacology*, *201*, 589–599.

Zombeck, J. A., Lewicki, A. D., Patel, K., Gupta, T., & Rhodes, J. S. (2010). Patterns of neural activity associated with differential acute locomotor stimulation to cocaine and methamphetamine in adolescent versus adult male C57BL/6J mice. *Neuroscience*, *165*, 1087–1099.

CHAPTER TWO

The impact of adolescent nicotine exposure on alcohol use during adulthood: The role of neuropeptides

G. Chen[a,b], M. Ghazal[b], S. Rahman[c], and K. Lutfy[b,*]

[a]College of Osteopathic Medicine of the Pacific, Western University of Health Sciences, Pomona, CA, United States
[b]Department of Pharmaceutical Sciences, College of Pharmacy, Western University of Health Sciences, Pomona, CA, United States
[c]Department of Pharmaceutical Sciences, South Dakota State University, Brookings, SD, United States
*Corresponding author: e-mail address: klutfy@westernu.edu

Contents

1. Introduction 54
2. Adolescence: A vulnerable developmental period 56
   2.1 Hedonic homeostasis and adolescent nicotine use 56
   2.2 Adolescent nicotine use and subsequent drug use 58
   2.3 Adolescent nicotine use and subsequent alcohol use 59
   2.4 Nicotine and alcohol co-abuse and codependence 61
3. Adolescent nicotine/alcohol exposure and alterations in neuropeptides 64
   3.1 Corticotropin releasing hormone (CRH, also known as CRF) 64
   3.2 Neuropeptide Y (NPY) 67
   3.3 Glucagon-like peptide-1 (GLP-1) 68
   3.4 Pituitary adenylyl cyclase activating polypeptide (PACAP) 69
   3.5 Opioid peptides 70
4. Conclusion 72
Acknowledgments 73
References 73

Abstract

Nicotine and alcohol abuse and co-dependence represent major public health crises. Indeed, previous research has shown that the prevalence of alcoholism is higher in smokers than in non-smokers. Adolescence is a susceptible period of life for the initiation of nicotine and alcohol use and the development of nicotine-alcohol codependence. However, there is a limited number of pharmacotherapeutic agents to treat addiction to nicotine or alcohol alone. Notably, there is no effective medication to treat this comorbid disorder. This chapter aims to review the early nicotine use and its impact on subsequent alcohol abuse during adolescence and adulthood as well as the role of

International Review of Neurobiology, Volume 161
ISSN 0074-7742
https://doi.org/10.1016/bs.irn.2021.07.006

neuropeptides in this comorbid disorder. The preclinical and clinical findings discussed in this chapter will advance our understanding of this comorbid disorder's neurobiology and lay a foundation for developing novel pharmacotherapies to treat nicotine and alcohol codependence.

Abbreviations

ACTH adrenocorticotropic hormone
GABA γ-amino butyric acid
BNST bed nucleus of the stria terminalis
CRH corticotropin releasing hormone
CRF corticotropin releasing factor
Ex4 exendin-4
HPA hypothalamic-pituitary-adrenal
GLP-1 glucagon-like peptide 1
mPFC medial prefrontal cortex
mRNA messenger ribonucleic acid
NPY neuropeptide Y
nAChRs nicotinic acetylcholine receptors
NAc nucleus accumbens
PVN paraventricular nucleus of the hypothalamus
PVT paraventricular nucleus of the thalamus
PACAP pituitary adenylyl cyclase activating polypeptide
POMC proopiomelanocortin
VTA ventral tegmental area

1. Introduction

Nicotine addiction is characterized by loss of control in nicotine use and nicotine-seeking behaviors accompanied by negative health and socioeconomic consequences. According to earlier estimates, annual productivity losses due to smoking were about $100 billion worldwide. It is predicted that decreases in tobacco products use may reduce the US economic burden by over $50 billion (McGinnis & Foege, 1999). Nearly fifty million people in the US continue to use one or the other form of nicotine-containing products (Fiore, Schroeder, & Baker, 2014). Smoking tobacco affects not only the user but also others who are second- and even third-hand bystanders. Tobacco smoking leads to respiratory and cardiovascular diseases, such as lung cancer, stroke, hypertension, myocardial infarction, and sudden death (Murray & Lopez, 1997a, 1997b, 1997c; Robbins, Manson, Lee, Satterfield, & Hennekens, 1994). Over a billion smokers exist worldwide, and half of these subjects die due to smoking-related diseases (Saha, Bhalla, Whayne Jr., & Gairola, 2007). A study on adolescent smoking behaviors found that of those surveyed, 85% were regular smokers, with 75% and 61% smoking ten or more cigarettes daily, respectively (Myers & Brown, 1994). At the 48-month follow-up, heavy smokers reported more respiratory problems, indicating that adolescent nicotine use has negative health consequences relatively early in adulthood. Additionally, a substantial portion of these individuals smoked without using other addictive drugs, showing that even adolescents not partaking in other drug use were still at risk for long-term nicotine consumption and health problems (Myers & Brown, 1994; Myers & Kelly, 2006).

Previous research has consistently demonstrated a link between psychiatric disorders, particularly major depressive disorder, nicotine, and other drug use (Brown, Lewinsohn, Seeley, & Wagner, 1996). Levels of nicotine dependence are significantly higher among populations with mood disorders (Grant, Hasin, Chou, Stinson, & Dawson, 2004; Leonard et al., 2001). These individuals may smoke to "self-medicate" and improve their mood due to nicotine's ability to alleviate negative affective states (Breslau, Novak, & Kessler, 2004; Breslau, Peterson, Schultz, Chilcoat, & Andreski, 1998; Dani & Harris, 2005; Markou, Kosten, & Koob, 1998). On the other hand, smokers have a higher risk of developing depression than nonsmokers; thus, tobacco use may cause mood disorders (Breslau et al., 1998; John, Meyer, Rumpf, & Hapke, 2004; Klungsoyr, Nygard, Sorensen, & Sandanger, 2006). About 70% of adolescents who smoke daily have psychiatric disorders and higher rates of comorbid diseases than adults (Choi, Patten, Gillin, Kaplan, & Pierce, 1997; Clark, Kirisci, & Tarter, 1998; Kandel, Chen, Warner, Kessler, & Grant, 1997; Patton, Coffey, Carlin, Sawyer, & Wakefield, 2006; Volkow, 2004).

Preclinical studies have demonstrated that adolescent rats may be more sensitive to the anxiogenic effects of acute nicotine exposure (Elliott, Faraday, Phillips, & Grunberg, 2004). Likewise, in a study using adolescent rats given nicotine injections twice a day, the researchers found that exposure—during the adolescent period only—led to increased anxiety-like behaviors, depressive-like behaviors, and anhedonia, as measured by decreased sensitivity to sucrose reward later in life (Iniguez et al., 2009). This issue is especially concerning in adolescent smokers. Nicotine-induced c-fos expression, an immediate early gene commonly used as a marker of neuronal activity, was greater in the paraventricular nucleus of the hypothalamus (PVN) in adult exposed rats (Cao et al., 2007), while c-fos was increased in the ventral tegmental area

(VTA), nucleus accumbens (NAc) shell, and bed nucleus of the stria terminalis (BNST) of adolescent exposed rats, suggesting more brain areas become involved following nicotine administration in adolescent than in adult rats (Shram, Funk, Li, & Le, 2007). These data implicate adolescence as a vulnerable period of life in which nicotine consumption may precipitate stress and mood disorders, such as anxiety and depression (O'Loughlin, Kishchuk, DiFranza, Tremblay, & Paradis, 2002; Parrott, 1999; Volkow, 2004).

2. Adolescence: A vulnerable developmental period

Age has been shown to play a critical role in the initiation of addictive behaviors associated with nicotine and other addictive drugs (Spear, 2000, 2014). Adolescence is a vulnerable period of life for the induction of long-lasting effects of nicotine, primarily due to differences in brain maturity between adolescents and adults (Holliday & Gould, 2016; Slotkin, Bodwell, Ryde, & Seidler, 2008; Torres, Tejeda, Natividad, & O'Dell, 2008). Up to 80% of adult smokers first began using nicotine during adolescence (Lenney & Enderby, 2008; Warren et al., 2008). Gray matter increases dramatically during the adolescent period before slowing to adult rates at the end of adolescence. During adolescence, this peak in neuronal activity is thought to play a role in the increased likelihood to partake in novelty- and risk-taking behaviors, while the brain regions involved with self-control and inhibition undergo slower maturation than subcortical regions (Casey, Getz, & Galvan, 2008). These changes lead to impulsive behaviors and alterations in hedonic homeostasis, which promote nicotine use during adolescence and concomitant or subsequent alcohol use and thus nicotine-alcohol codependence.

2.1 Hedonic homeostasis and adolescent nicotine use

Hedonic equilibrium can be altered by life events, palatable foods, and addictive drugs (Koob, 2008a; Stojakovic, Espinosa, Farhad, & Lutfy, 2017). Dysregulated hedonic homeostasis results from chronic use of psychoactive substances, which usurp the brain's reward neurocircuitry (Koob, 2008a; Koob & Kreek, 2007; Koob & Le Moal, 2008a, 2008b; Koob & Volkow, 2010; Volkow et al., 2010; Volkow & Morales, 2015). As a result, it becomes more difficult to take pleasure in daily activities and normally pleasurable events. Besides, the dependent individuals experience a negative affective state when the subject cannot have access to the psychoactive substance (Madras, 2017). A combination of decreased reward and increased recruitment of the brain stress neurocircuitry has been theorized to cause dysregulation in hedonic homeostasis, leading to a negative emotional state in the absence of the drug, promoting drug-seeking and drug-taking behaviors (Koob, 2008a). Many studies on humans and rodents have demonstrated that abstinence from chronic nicotine use or exposure causes an anhedonic state, likely related to decreased reward value toward non-drug reinforcers (Epping-Jordan, Watkins, Koob, & Markou, 1998; Snuggs & Hajek, 2013).

With the consumption of nicotine during peri-adolescence, hedonic homeostasis is altered. Preclinical studies have demonstrated that adolescent rats demonstrated increased nicotine reward and higher nicotine self-administration levels while also experiencing lesser withdrawal symptoms than adult rats (Belluzzi, Lee, Oliff, & Leslie, 2004; Kota, Martin, Robinson, & Damaj, 2007; O'Dell, Bruijnzeel, Ghozland, Markou, & Koob, 2004; O'Dell, Torres, Natividad, & Tejeda, 2007). These data show an age-related disparity in the rewarding and aversive effects of nicotine, which may contribute to increased adolescent smoking levels. On the other hand, withdrawal-based addiction theories have shown that continued and escalated nicotine use during abstinence periods is driven by negative affective states (Aguirre, Madrid, & Leventhal, 2015; Doubeni, Reed, & Difranza, 2010). A study conducted using mice examining the effects of contingent and non-contingent adolescent nicotine exposure on saccharin and nicotine intake during adulthood found that prior non-contingent nicotine exposure increased nicotine but reduced saccharine self-administration using an oral operant conditioning paradigm (Cole et al., 2019). These data demonstrated that nicotine's reinforcing properties were increased in subjects with prior exposure to nicotine during adolescence (Adriani, Deroche-Gamonet, Le Moal, Laviola, & Piazza, 2006; Weaver et al., 2012). The increased hedonic value of nicotine in drug-treated mice demonstrates that early age nicotine exposure sensitizes reward neurocircuitry (Hilario, Turner, & Blendy, 2012).

Another study examined the relationship between hedonic capacity—an ability to derive pleasure in response to rewardable stimuli—to the onset and escalation of adolescent smoking (Audrain-McGovern et al., 2012; Bogdan & Pizzagalli, 2009; Meehl, 2001). Individuals with lower hedonic capacity demonstrated reduced responses to rewarding stimuli linked to interruptions of neural pathways related to reward and motivation. Thus, such subjects are at higher risk for relying on unnatural rewards, such as smoking tobacco, to experience pleasure (Nestler & Carlezon Jr., 2006;

Tremblay et al., 2005). A study conducted on high school students found that individuals with lower hedonic capacity were 2.5-fold more likely to initiate and escalate smoking behaviors (Audrain-McGovern et al., 2012). This altered responsiveness might be because a decrease in the activity of neural systems involved in reward-processing promotes drug intake during adolescence (Forbes & Dahl, 2005; Galvan, 2010). Lower mesocorticolimbic dopaminergic activity may indicate a decreased hedonic state and, therefore, a reduced phasic mesocorticolimbic dopamine response to natural reinforcers (Juckel et al., 2006). The lowered basal dopamine increases the sensitivity of the system to drugs, such as nicotine, that stimulate mesolimbic neurons (Barrett, Boileau, Okker, Pihl, & Dagher, 2004; Brody et al., 2004; Epping-Jordan et al., 1998). Thus, initial smoking behaviors to combat low hedonic capacity may eventually promote smoking behaviors.

2.2 Adolescent nicotine use and subsequent drug use

Research has shown that initial use of some drugs at an early age primes the brain to reinforce the use of other illicit substances later in life, leading to polydrug use (McCabe, Cranford, Morales, & Young, 2006). For example, early age nicotine exposure can promote the use of cocaine and other addictive substances (Levine et al., 2011; McQuown, Belluzzi, & Leslie, 2007; Orsini, Bonito-Oliva, Conversi, & Cabib, 2008). Indeed, 90.4% of cocaine users between the ages of 18–34 years in the US have also reported smoking cigarettes previously (Levine et al., 2011). Studies in rodents have also shown the impact of early nicotine use on the reinforcing and rewarding effects of cocaine later in life. For example, nicotine pretreatment has been reported to promote cocaine self-administration (McQuown et al., 2007), cocaine-induced conditioned place preference (Li et al., 2014), and sensitization (Kandel & Kandel, 2014). Moreover, the potency of cocaine reinforcement was increased by nicotine pretreatment, as evidenced by a left-ward shift of the cocaine dose-response curve in Rhesus monkeys (Mello & Newman, 2011).

It is noteworthy that the affective signs of nicotine withdrawal are reduced while nicotine reward is increased in adolescents compared to adult rats (Carcoba et al., 2014; Dickson, Miller, Rogers, Blaha, & Mittleman, 2014; Natividad, Buczynski, Parsons, Torres, & O'Dell, 2012; Natividad, Tejeda, Torres, & O'Dell, 2010; O'Dell et al., 2004, 2007), making this age group a vulnerable population for nicotine use and abuse. Moreover, nicotine exposure during this vulnerable life period alters cocaine-induced aversion (Hutchison & Riley, 2008) and reward (Kelley & Rowan, 2004; Schochet,

Kelley, & Landry, 2004). It has been reported that prior nicotine treatment during adolescence, but not adulthood, enhanced the rewarding effects of cocaine in rats (McQuown, Dao, Belluzzi, & Leslie, 2009; Mojica, Belluzzi, & Leslie, 2014). Interestingly, early work from our group showed that nicotine treatment during adulthood could reduce the rewarding effects of cocaine in male mice (Singh & Lutfy, 2017). Similar to the former findings, adult and adolescent rats pretreated with nicotine showed enhanced cocaine self-administration (Collins & Izenwasser, 2004; McMillen, Davis, Williams, & Soderstrom, 2005; McQuown et al., 2007, 2009), and application of a neurotensin analog, NT69L, was shown to lower rates of subsequent nicotine self-administration (Boules et al., 2011; Fredrickson, Boules, Yerbury, & Richelson, 2003). The former findings support the hypothesis that nicotine is more rewarding in adolescents than in adults. In another study, levels of dopamine were lower in adolescent than adult rats that underwent nicotine withdrawal (Natividad et al., 2010), showing that adolescents might be at a greater risk of initiating drug use and becoming addicted.

2.3 Adolescent nicotine use and subsequent alcohol use

Nicotine has been shown to disproportionally affect different age groups. In particular, the peri-adolescent ages are more vulnerable to the effects of nicotine. Previous research has shown that nicotine use is initiated during the adolescent period and often proceeds to alcohol use and abuse (Best et al., 2000; Torabi, Bailey, & Majd-Jabbari, 1993).

Preadolescence, the juvenile period of development that precedes adolescence, which includes ages between 10 and 12, is the period of prepubertal development, marked by changes in hormone levels, sexual maturity, cognition, among other changes (Gilmore & Meersand, 2014). It has long been established that nicotine exposure during the adolescent period may lead to further abuse of nicotine, alcohol, and other substances during adulthood due to the increased neural, psychosocial, and hormonal changes which increase vulnerability to risk-taking behaviors (Spear, 2000; Steinberg, 2004, 2008). Increased drug reinforcement in adulthood has been linked to nicotine exposure during, but not after, the adolescent period (Adriani et al., 2003; Larraga, Belluzzi, & Leslie, 2017). In particular, nicotine use during peri-adolescence has been strongly associated with pathological alcohol consumption (Grant, 1998; Harrison & McKee, 2011).

In addition, dopaminergic fibers projecting from the midbrain A9 and A10 origins to the striatum and NAc mature earlier than fiber projections to the medial prefrontal cortex, which usually happens around late adolescence (Benes, Taylor, & Cunningham, 2000; Ciliax et al., 1995; Naneix, Marchand, Di Scala, Pape, & Coutureau, 2012; Tarazi, Tomasini, & Baldessarini, 1998). Dopaminergic activity increases in early adolescence to mid-adolescence before decreasing to adult levels. Dopaminergic D1 and D2 receptors are found in the medial prefrontal cortex (mPFC), striatum, and NAc (Andersen, Dumont, & Teicher, 1997; Mathews, Waters, & McCormick, 2009; Robinson, Zitzman, Smith, & Spear, 2011). During the adolescent period, there is a peak in dopamine receptors in the striatum and NAc, while dopamine receptors peak in the mPFC during late adolescence (Wahlstrom, Collins, White, & Luciana, 2010). Research has shown earlier DA receptor maturation in the NAc and delayed maturation in the mPFC in response to stress (Burke & Miczek, 2014). Similarly, there was an increase in neuronal activation in the NAc in response to early-adolescent stress versus the mPFC for late-adolescent stress (Lyss, Andersen, LeBlanc, & Teicher, 1999). Proper development of central dopaminergic systems is thought to be linked to proper adult development (Andersen, 2003; Burke & Miczek, 2014; Spear, 2000; Weinberger, 1987). During withdrawal, there is a lesser decrease in dopamine levels in adolescents than in adults (Li, Mead, & Bevins, 2009; Natividad et al., 2010; Shram & Le, 2010).

Mesocorticolimbic dopaminergic neurons originating in the VTA and projecting to the medial prefrontal cortex and NAc play a crucial role in processing salient events (Everitt & Wolf, 2002; Schultz, 2002; Wise, 1996). Almost all drugs abused by humans can elicit an increase in extracellular dopamine in the NAc (Di Chiara & Imperato, 1988). Nicotine and alcohol can increase dopamine in this brain region, and the two drugs' effects are synergistic (Tizabi, Bai, Copeland, & Taylor, 2007). Nicotinic acetylcholine receptors are located in the VTA, NAc, and prefrontal cortex (Vidal, 1996). Stimulation of the nAChRs in the VTA increases dopaminergic neurons firing, while stimulation of prefrontal cortex receptors increases glutamatergic activity, similarly causing dopamine release (Mereu et al., 1987). Accumbal dopaminergic neuronal firing rate has been shown to increase after both acute and chronic nicotine use in adolescents, while chronic nicotine use decreases dopamine release in adults (Benwell & Balfour, 1992; Benwell, Balfour, & Lucchi, 1993; Imperato, Mulas, & Di Chiara, 1986; Vezina, Blanc, Glowinski, & Tassin, 1992). The mesolimbic dopaminergic system undergoes many changes during adolescence, influencing an individual's response toward nicotine and may subsequently play a role in adolescents' susceptibility to nicotine addiction (Badanich & Kirsteina, 2004).

One link between early nicotine use and later alcohol consumption stems from altered γ-amino-butyric acid (GABA) type A ($GABA_A$) receptors in the VTA, which have been shown in adult rats to be positively modified by ethanol as well as to play a role in ethanol reinforcement (Boyle, Segal, Smith, & Amit, 1993; Doyon et al., 2013; Rassnick, D'Amico, Riley, & Koob, 1993). In particular, the upregulation of nAChRs on GABAergic interneurons in the VTA contributes to increased GABAergic inhibition of dopaminergic neurons. This inhibition may play a role in withdrawal states by reducing tonic dopamine levels in the NAc. Further exposure to nicotine or alcohol during this withdrawal phase may elevate dopaminergic neurotransmission and increase the chances of relapse (Nashmi et al., 2007; Rahman, Zhang, Engleman, & Corrigall, 2004; Shiffman, 2006). A study examining how nicotine exposure during adolescence affects alcohol intake later in life found that nicotine exposure during adolescence—but not during adulthood—altered GABAergic signaling in the VTA and a long-term amplification of alcohol self-administration (Thomas et al., 2018). These changes in GABA signaling were dependent on activation of glucocorticoid receptors and were also linked to decreased dopaminergic responses to alcohol within the lateral VTA (Thomas et al., 2018).

2.4 Nicotine and alcohol co-abuse and codependence

Nicotine and alcohol are often used and abused concurrently, and an alcohol-nicotine comorbid disorder represents a more daunting public health and socioeconomic concern than either alone. Indeed, the prevalence of alcoholism is higher in smokers than in non-smokers (DiFranza & Guerrera, 1990; Miller & Gold, 1998). For example, 50–80% of alcohol-dependent individuals smoke regularly and consume more alcohol than nonsmoking alcoholics (Hurt et al., 1994; Pomerleau, Aubin, & Pomerleau, 1997; Romberger & Grant, 2004). Also, chronic nicotine use in alcoholic smokers is linked to a higher quantity and frequency of alcohol intake than nonsmoking or former-smoking alcoholics (York & Hirsch, 1995). A nicotine and alcohol comorbid disorder is also difficult to cease (Carmody, Brischetto, Matarazzo, O'Donnell, & Connor, 1985; DiFranza & Guerrera, 1990; Miller & Gold, 1998; Romberger & Grant, 2004). Indeed, while about 70% of smokers want to quit, only a small percentage (7%) of them succeed, which may be even worse in subjects with a comorbid disorder. Although smoking in the US has declined in recent years, there is virtually no reduction in concurrently dependent subjects (West, 2017). This may be because each drug serves as a precipitating factor in eliciting craving for the other drug in subjects with this comorbid disorder (Businelle et al., 2013; Lam et al., 2014; Lisha, Carmody, Humfleet, & Delucchi, 2014).

The interactions between nicotine and alcohol effects play a role in their codependence, and nAChR subunit combinations have been thought to play a key part in this relationship. Chronic nicotine treatment was shown to increase nicotinic binding, while chronic alcohol/ethanol treatment did not, indicating that cross-tolerance between alcohol and nicotine may not be attributed to alterations in receptor binding (Burch, de Fiebre, Marks, & Collins, 1988). Likewise, tolerance to ethanol may not be attributed to nAChRs upregulation but is due, at least in part, to ethanol metabolism (Collins, Wilkins, Slobe, Cao, & Bullock, 1996). Other studies have demonstrated that each drug increases the metabolism of the other, leading to cross-tolerance (Schoedel & Tyndale, 2003). Pretreatment with nicotine increases nicotine preference only in early adolescence, showing increased nicotine place preference compared to adult exposure (Kota, Robinson, & Imad Damaj, 2009). Nicotine administration in rats induced early escalation of 10% alcohol drinking in an operant conditioning paradigm (Leao et al., 2015). In this study, mecamylamine, a nAChR antagonist, decreased alcohol consumption in rats with prior nicotine experience (Leao et al., 2015). Additionally, intra-VTA mecamylamine administration blocked ethanol's dopamine-stimulating effect in the rat NAc (Ericson, Blomqvist, Engel, & Soderpalm, 1998). When it comes to withdrawal from both drugs, choline acetyltransferase activity increased with nicotine withdrawal but reversed after ethanol withdrawal; nAChRs were upregulated after withdrawal from both drugs (Ribeiro-Carvalho et al., 2009). When used concomitantly, nicotine has been shown to decrease alcohol's negative effects, while alcohol decreases negative affective states associated with nicotine withdrawal (Gulick & Gould, 2008).

Sex as a biological variable plays a role in nicotine use and abuse and modulates comorbid disorders. In general, females between 12 and 17 have higher cigarette smoking levels than males, while adult women have lower rates of successful and long-term cigarette cessation attempts than males (Becker & Hu, 2008; Bjornson et al., 1995; Pogun & Yararbas, 2009). In rodents, females have been shown to quickly relapse after drug extinction and consume and readily self-administer nicotine (Adriani et al., 2003; Adriani, Macri, Pacifici, & Laviola, 2002; Donny et al., 2000; Isiegas, Mague, & Blendy, 2009). Nicotine and alcohol have been reported to increase synaptic dopamine levels in the NAc, which is the reward's principal element (Corrigall, Coen, & Adamson, 1994; Larsson, Edstrom, Svensson, Soderpalm, & Engel, 2005). Both ethanol and nicotine target the reward pathway via cholinergic inputs to the VTA (Bito-Onon, Simms, Chatterjee, Holgate, & Bartlett, 2011; Hendrickson, Zhao-Shea, & Tapper, 2009; Imperato et al., 1986; Pidoplichko et al., 2004). Since the adolescent brain undergoes many changes, in particular, the cortical as well as limbic areas where there are high amounts of circuit remodeling, the developing brain is vulnerable to exogenous stimuli, which may subsequently alter their maturation (Laviola & Marco, 2011; Spear, 2000, 2014). A study examined how adolescent nicotine exposure in female rodents affected binge alcohol consumption and the subsequent changes in nAChRs in the VTA, NAc, cortical regions, and the dorsal tegmental area (Locker, Marks, Kamens, & Klein, 2016). The results demonstrated that nicotine consumption was linked to increased ethanol binge-like consumption. Besides, mice exposed to both nicotine and ethanol had increased nAChR density in the prefrontal cortex compared to mice only exposed to ethanol, possibly indicating that upregulation of nAChRs may be related to increased ethanol intake (Locker et al., 2016).

Some individuals may have been exposed to nicotine and alcohol during the prenatal stage, and its consequences appear during adolescence. For example, prenatal tobacco exposure has been associated with Attention Deficit Hyperactivity Disorder (Froehlich et al., 2009). Also, the effects of prenatal nicotine exposure on alcohol use and abuse carry into the adolescent period and adulthood, with 13–17% of women reporting smoking during pregnancy (Darville & Hahn, 2014; Kozlowski, Skinner, Kent, & Pope, 1989). It has been long established that drinking alcohol during pregnancy increases the likelihood that their children will abuse ethanol later in life, both among humans (Alati et al., 2006; Baer, Barr, Bookstein, Sampson, & Streissguth, 1998; Baer, Sampson, Barr, Connor, & Streissguth, 2003) as well as in animals (Spear & Molina, 2005). Similarly, the probability of alcohol abuse is also significantly increased in the offspring of mothers who smoke during pregnancy. A study looking at the relationship between smoking during pregnancy and the lifetime risk for alcohol use disorder in offspring found that offspring of mothers who smoked 20 or more cigarettes per day than controls were at a greater risk for developing an alcohol use disorder (Nomura, Gilman, & Buka, 2011). In addition to the increased likelihood of nicotine abuse, individuals exposed to nicotine during this gestational period also showed higher hospitalization rates due to substance abuse (Brennan, Hammen, Katz, & Le Brocque, 2002). As with nicotine use, gender-specific differences have also been found regarding the severity of alcohol consumption among male and female offspring, with different effects seen in both adolescent and adult rats (McMurray et al., 2008); however, this may be influenced by external variables such as gender-specific reactions to stressful environments. A longitudinal research study examining the effects of prenatal nicotine exposure on adolescent and adult substance abuse behaviors found that when controlling for socioeconomic status, other drug exposure, prenatal alcohol, and gender, individuals prenatally exposed to nicotine did not have an increased risk for substance abuse disorders, including alcohol-use (O'Brien & Hill, 2014). Thus, while some research has demonstrated that prenatal nicotine exposure has long-lasting effects on alcohol consumption extending into adolescence and adulthood, others have shown opposite results, indicating that more research needs to be done regarding the direct effects of prenatal nicotine exposure on the abuse of alcohol by offspring later in life.

3. Adolescent nicotine/alcohol exposure and alterations in neuropeptides

Nicotine exposure during the adolescent period has also been demonstrated to alter central neuropeptide activity. Several neuropeptide systems have been implicated in the addictive actions of nicotine and alcohol. Here we describe the role of some of the centrally-acting peptides involved in the stress response, consummatory behaviors, and emotion and motivation. In particular, we focus on corticotropin releasing hormone (CRH, also known as CRF), glucagon-like peptide 1 (GLP-1), neuropeptide Y (NPY), pituitary adenylyl cyclase activating polypeptides (PACAP), and opioid peptides.

3.1 Corticotropin releasing hormone (CRH, also known as CRF)

The hypothalamic-pituitary-adrenal (HPA) axis mediates stress responses and plays a functional role in stress-related increases in drug-seeking behavior (Goeders, 2002; Steiner & Wotjak, 2008). Following the stress circuit's activation, CRH is released by the PVN, reaching the anterior pituitary, stimulating adrenocorticotropic hormone (ACTH) release. ACTH then reaches the adrenal gland and stimulates the synthesis of glucocorticoids from the adrenal cortex, which via a negative feedback loop regulates the expression of proopiomelanocortin (POMC), and the release of ACTH, melanocortins, and beta-endorphin (Antoni, 1986; Herman et al., 2003; Keller-Wood & Dallman, 1984; Rivier, Brownstein, Spiess, Rivier, &

Vale, 1982). Adolescents have been shown to display a more exaggerated stress response, with larger amounts of glucocorticoid secreted in response to restraint stress, with a peak response in early to mid-adolescence (Foilb, Lui, & Romeo, 2011; Gomez, Houshyar, & Dallman, 2002). The recent rise in e-cigarettes has been shown to play a large role in this population, with about 21 and 5 percent of high schoolers and middle schoolers, respectively, reporting vaping in the past 30-days in 2019, amounting to an additional 1.5 million youth in the past year (Cavallo & Krishnan-Sarin, 2019; Gentzke et al., 2019).

Stress plays a major role in smoking behavior. Nicotine is an activator of the HPA axis and corticosteroid release in adults, raising the possibility that glucocorticoids may mediate some of nicotine's actions (Balfour, Khullar, & Longden, 1975; Caggiula et al., 1998; Cam & Bassett, 1983; Pauly, Ullman, & Collins, 1988). The HPA axis, the stress response, as well as the HPA axis responsiveness to nicotine continue to develop through adolescence, explaining differences in stress responsiveness between adult and adolescent individuals (Cruz, DeLucia, & Planeta, 2008; Matta, Valentine, & Sharp, 1997; McCormick & Ibrahim, 2007). In a study examining mounting a stress response in response to an intravenous injection of nicotine or saline in adolescent versus adult rats, the authors found no nicotine-induced corticosterone secretion in early- to mid-adolescent rats. In contrast, there was a significant release in adults. These data indicated that nicotine-induced HPA axis activation is limited in adolescents, suggesting that the adolescent brain and cholinergic system are still developing (Cao et al., 2010).

Another mechanism by which nicotine is thought to potentiate subsequent ethanol consumption is through the release of stress hormones (Clark, Lindgren, Brooks, Watson, & Little, 2001; Doyon et al., 2013; Le, Wang, Harding, Juzytsch, & Shaham, 2003; Truitt et al., 2015). As stated above, nicotine activates the HPA axis, resulting in glucocorticoid secretions from the adrenal gland. Glucocorticoids have been shown to lower ethanol-induced dopamine signaling by increasing GABAergic inhibition on dopaminergic neurons in the VTA (Doyon et al., 2013; Larsson & Engel, 2004; Le et al., 2003; Locker et al., 2016). The effect of this reduced dopaminergic signaling has been linked to greater susceptibility to alcohol use (Martinez et al., 2005). A study examining the effects of adolescent nicotine exposure on ethanol consumption as well as gene expression in mice found that nicotine exposure during adolescence not only increased ethanol intake and blood ethanol concentrations but also altered HPA axis responsiveness to stress, brain neuronal plasticity, as well as the expression of those genes involved in GABAergic and glutamatergic neurotransmission. These results demonstrated that adolescent nicotine exposure induces neuroadaptations of many systems that could play a role in increased self-administration of ethanol (Silva et al., 2018).

Additionally, nicotine exposure during the adolescent period has been shown to cause neuronal loss and increased nicotine binding in hippocampal and cortical regions (Chan, Hinds, Impey, & Storm, 1998; Stoltenburg-Didinger, 1994; Trauth, McCook, Seidler, & Slotkin, 2000; Xu, Seidler, Ali, Slikker Jr., & Slotkin, 2001). Decreases in cholinergic activity in the hippocampus suggest cholinergic apoptosis following nicotine exposure in adolescents (Trauth et al., 2000). When it comes to stress-related neuropeptide changes, CRF mRNA expression was shown to be increased in the PVN of adolescent, but not adult, subjects (Kupferschmidt, Funk, Erb, & Lê, 2010), while there was increased expression of c-fos in the PVN (Shram et al., 2007). These widespread changes demonstrate the broader effects of nicotine throughout the adolescent brain via the CRH system. Alcohol also acutely activates the HPA axis. Additionally, alcohol withdrawal is associated with increased anti-reward signaling, particularly through the CRH system (Koob, 2008a). Given that both drugs affect the stress neurocircuitry, this may be a potential target for developing medication to treat this comorbid disorder.

Nicotine consumption has been shown to have varying effects on anxiety, which also vary between adolescents and adults (Murray, 1991). CRF exerts anxiogenic effects by binding to CRF1 receptors, but anxiolytic effects by binding to CRF2 receptors. Areas in the brain with the most CRF include the PVN and amygdala, which are important in the HPA axis response to stress and aversive and negative affective states (Boorse & Denver, 2006; Smith & Vale, 2006). These brain subregions project to and influence the VTA and NAc, both of which are involved in reward processing (Rodaros, Caruana, Amir, & Stewart, 2007). Thus, CRF plays a significant role in inhibiting reward neurocircuitry as well as mediating negative affective states (Bruijnzeel et al., 2012; Grieder et al., 2014). CRF also plays an important role in nicotine dependence, especially during the adolescent period. Higher stress levels during this time are associated with a greater likelihood of starting smoking (Kutlu, Holliday, & Gould, 2016). Thus, activation of the CRF system during adolescence can contribute to smoking initiation (Brielmaier, McDonald, & Smith, 2012; Bruijnzeel et al., 2012). These data also provide evidence for targeting CRF for medication development. Still, one should also be aware of the negative impact of blocking the CRH receptor in initiating the stress response. CRH2 receptor agonists may be potential medications to elicit anti-anxiety effects in subjects undergoing withdrawal.

3.2 Neuropeptide Y (NPY)

The stimulation of neurogenesis in orexigenic peptides in adolescent rats prenatally exposed to nicotine is accompanied by increased alcohol consumption during the adolescent period (Chang, Karatayev, & Leibowitz, 2013). The impact of nicotine exposure on anxiety and depression in adolescent nicotine users are well known and largely linked to NPY and CRF. Both peptides are important in regulating depression and anxiety (Heim & Nemeroff, 2001; Redrobe, Dumont, & Quirion, 2002). NPY also has a role in food intake, cerebrocortical excitability, cardiovascular homeostasis, and emotional behavior integration (Clark, Kalra, Crowley, & Kalra, 1984; Dumont, Martel, Fournier, St-Pierre, & Quirion, 1992; Heilig & Widerlov, 1995; Pedrazzini et al., 1998; Woldbye, Greisen, Bolwig, Larsen, & Mikkelsen, 1996). NPY has also been implicated in depressive disorders, substance use disorders, neurological disorders, and anxiety disorders (Brothers & Wahlestedt, 2010; Tatemoto, Carlquist, & Mutt, 1982). NPY has been shown to have anxiolytic effects, while conversely, CRF has anxiogenic effects, as stated above (Heilig, Soderpalm, Engel, & Widerlov, 1989; Spina et al., 1996; Stogner & Holmes, 2000; Stout, Mortas, Owens, Nemeroff, & Moreau, 2000). In hypothalamic and extrahypothalamic brain regions, nicotine has a role in activating these systems; thus, adolescent nicotine exposure may alter NPY and/or CRF system activity, causing anxiety, depression, or other mood disorders (Li et al., 2000; Matta et al., 1997). In rats, the effects of CRF on cortical electroencephalogram were reduced in adult rats exposed to nicotine during adolescence (Slawecki & Ehlers, 2003). Another study examining the effect of adolescent nicotine exposure on depression and anxiety behavior found consistent changes in both the NPY and CRF systems (Slawecki, Thorsell, El Khoury, Mathé, & Ehlers, 2005). Increased frontal cortical and hypothalamic CRF and hypothalamic NPY levels by nicotine have been hypothesized to increase anxiety (Koob, 2008b). Similarly, acute stress exposure has also been shown to activate the hypothalamic and cortical NPY and CRF systems (Serra et al., 1999; Sweerts, Jarrott, & Lawrence, 2001). Thus, increased CRF and NPY levels suggest activation of the HPA axis (Carrasco & Van de Kar, 2003; Sainsbury et al., 1997).

NPY plays a role in the neurobiological responses to voluntary alcohol consumption as well. Rats selectively bred for alcohol preferences display differing levels of NPY in different brain regions (Ehlers et al., 1998; Hwang, Zhang, Ehlers, Lumeng, & Li, 1999). Levels of NPY in both knockout and transgenic mice are inversely related to resistance to alcohol intoxication and ethanol consumption. In contrast, injection of exogenous NPY into the brain has been shown to modulate rats' alcohol consumption (Kelley, Nannini, Bratt, & Hodge, 2001; Thiele, Marsh, Ste Marie, Bernstein, & Palmiter, 1998). An earlier study found that the Y1 receptors regulate voluntary alcohol consumption and alcohol intoxication effects in adult mice (Thiele, Koh, & Pedrazzini, 2002). NPY and Y1 knockout mice demonstrated increased ethanol consumption and decreased ethanol-induced sleep times, while NPY-overexpressing mice demonstrated low alcohol consumption and higher ethanol-induced sleep times (Thiele et al., 1998). Other studies have implicated the Y1 receptor in mediating the anxiolytic effects of NPY, and NPY knockout mice display high levels of anxiety-related behaviors (Heilig & Widerlov, 1995; Stewart, Gatto, Lumeng, Li, & Murphy, 1993; Thiele et al., 1998). In addition, NPY is thought to play a role in specific brain regions regarding voluntary ethanol consumption, as intracerebroventricular injections have been shown to have little effect on alcohol consumption, while infusions into the PVN increases alcohol consumption (Kelley et al., 2001; Slawecki, Betancourt, Walpole, & Ehlers, 2000; Thiele et al., 2002). Although a role for NPY in alcohol consumption has been established, its role in nicotine-alcohol comorbid disorders has not been characterized. Thus, NPY and its receptors could be a potential target to develop medications to treat nicotine-alcohol codependence.

3.3 Glucagon-like peptide-1 (GLP-1)

GLP-1 has been implicated in drug addiction (Brunchmann, Thomsen, & Fink-Jensen, 2019). GLP-1 is produced in the small intestine and brain and acts as both a neurotransmitter and hormone, with roles in regulating blood sugar, feeding behaviors, alcohol consumption, reward, and addiction (Drucker, Habener, & Holst, 2017; Holst, 2013; Jerlhag, 2018; Sorensen et al., 2015). In the brain, GLP-1 is produced by neurons in the nucleus tractus solitarius, which projects to the hypothalamus and mesocortical pathways (Alvarez, Roncero, Chowen, Thorens, & Blazquez, 1996; Baggio & Drucker, 2007; Holst, 2007; Merchenthaler, Lane, & Shughrue, 1999).

GLP-1 receptors have been found in the VTA and NAc, thus making it a potential target to treat substance use disorders and reward dysregulation (Alhadeff, Rupprecht, & Hayes, 2012; Cork et al., 2015; Goke, Larsen, Mikkelsen, & Sheikh, 1995; Heppner et al., 2015; Merchenthaler et al., 1999).

Studies targeting GLP-1 have found that activation of these receptors in the VTA and NAc shell and core reduces palatable food self-administration (Alhadeff et al., 2012). In addition, a GLP-1 analog, exendin-4 (Ex4), has been shown to suppress alcohol-induced locomotion, conditioned-place preference, accumbal dopamine release, and reduce alcohol-seeking and consummatory behaviors (Egecioglu, Engel, Jerlhag, & Engel, 2013). Another study investigating the effects of Ex4 on nicotine-induced behavior found lower nicotine-induced locomotor stimulation, accumbal dopamine release, conditioned place preference, as well as locomotor sensitization (Egecioglu et al., 2013). Additionally, GLP-1 causes excitation of medial habenular projections to the interpeduncular nucleus, and activation of its receptors has been shown to decrease nicotine intake and reward (Tuesta et al., 2017). Thus, GLP-1 has been shown to modulate satiety and reward-related behaviors associated with addictive drug intake, such as alcohol, nicotine, etc. (Egecioglu et al., 2013; Hernandez et al., 2018; Hernandez, O'Donovan, Ortinski, & Schmidt, 2019), raising the possibility that GLP-1 and its receptors may be a potential target to treat this comorbid disorder. Besides, there are no studies yet examining the role of GLP-1 in nicotine use and abuse during the adolescent period and subsequent alcohol or nicotinic use in adulthood.

3.4 Pituitary adenylyl cyclase activating polypeptide (PACAP)

PACAP has been implicated as a regulator of the stress response (Lutfy & Shankar, 2019). Previous findings have also shown the involvement of PACAP in alcohol's actions (Stojakovic et al., 2020). Alcohol intake in a two-bottle choice procedure was greater in mice lacking PACAP than wild-type mice (Tanaka et al., 2010). PACAP expression was increased in the paraventricular nucleus of the thalamus (PVT) in rats following alcohol self-administration (Gupta et al., 2018). Recent evidence suggests that PACAP may also be involved in the actions of nicotine (Stojakovic et al., 2020). For instance, we observed that nicotine-induced aversion was reduced in mice lacking PACAP compared to their wild-type controls (Tseng, Singh, Marquez, Hamid, & Lutfy, 2019). We also found that anxiety-like behaviors associated with nicotine withdrawal were reduced in these mice (Nega, Marquez, Hamid, Ahmad, & Lutfy, 2020). The PACAP/PAC1 receptor system may also play a functional role in alcohol and nicotine codependence. Indeed, an earlier study showed that PACAP reduces toxicity induced by alcohol, nicotine, and their combination (Manavalan et al., 2017). However, there is no evidence regarding the role of PACAP in the crosstalk between nicotine and alcohol as well as the impact of age and sex in the involvement of PACAP in the actions of nicotine and alcohol.

3.5 Opioid peptides

The endogenous opioid system has been implicated in the action of nicotine and alcohol (Drews & Zimmer, 2010; Xue & Domino, 2008). The endogenous opioid system consists of three different subtypes of opioid receptors, namely mu, delta, and kappa opioid receptors and three families of endogenous opioid peptides derived from their respective precursor molecules, namely pro-opiomelanocortin (POMC), prodynorphin (PDYN), and proenkephalin (PENK).

Early research using opioid antagonists has linked the endogenous opioid system to nicotine's effects (Balfour, 1982). Recent studies have focused on knockout mice without one of these receptors or peptides. It was found that the mu-opioid receptor is involved in nicotine dependence, reward, and antinociception (Berrendero et al., 2005; Berrendero, Kieffer, & Maldonado, 2002; Berrendero, Robledo, Trigo, Martin-Garcia, & Maldonado, 2010). Mice lacking beta-endorphins demonstrated attenuated anxiogenic and rewarding effects after acute nicotine treatment with no change in nicotine dependence, hypolocomotion, and antinociception. The difference in nicotine's physical dependence and rewarding properties between the two mouse lines suggests that mu opioid receptors mediate other effects of nicotine not involving beta-endorphin. Besides, these mice had a higher nociceptive threshold and spontaneous hyperlocomotion (Trigo, Zimmer, & Maldonado, 2009). In addition to its analgesic effects, beta-endorphins induce reward and play a major role in the actions of addictive drugs, particularly in the actions of cocaine (Roth-Deri, Green-Sadan, & Yadid, 2008). Nicotine exerts its rewarding effects via increasing the activity of the mesolimbic dopaminergic neurons through cholinergic inputs in the VTA, leading to increased dopamine in the NAc (Bassareo & Di Chiara, 1997; Di Chiara & Imperato, 1988). Mu receptor knockout mice expressed no behavioral sensitization and reduced conditioned place preference following repeated conditioning with nicotine (Berrendero et al., 2002; Yoo, Lee, Loh, Ho, & Jang, 2004). Additionally, acute exposure to nicotine did not increase levels of beta-endorphins in the NAc of rats, whereas amphetamines, cocaine, and alcohol did. The lack of a rise in extracellular endorphin levels in response to nicotine (Olive, Koenig, Nannini, & Hodge, 2001) demonstrated that the increase in beta-endorphin levels induced by alcohol, cocaine, and amphetamine is not due to stress of handling or injection. In humans, in contrast, elevated beta-endorphin levels were observed following nicotine exposure, and higher levels were seen in light smokers versus heavier smokers (Backon, 1989; del Arbol et al., 2000; Gilbert, Meliska, Williams, & Jensen, 1992; Meliska & Gilbert, 1991).

Nicotine and ethanol have been shown to regulate the release of opioid peptides in the brain (Drews & Zimmer, 2010), suggesting that opioids may contribute to their pharmacological effects. Research has demonstrated that acute nicotine injection causes the release of enkephalin as well as increases preproenkephalin mRNA transcription (Eiden, Giraud, Dave, Hotchkiss, & Affolter, 1984; Hollt & Horn, 1992; Houdi, Pierzchala, Marson, Palkovits, & Van Loon, 1991; Kiguchi et al., 2008; Livett, Dean, Whelan, Udenfriend, & Rossier, 1981; MacArthur, Iacangelo, Hsu, & Eiden, 1992). Large changes in beta-endorphin, enkephalin expression, and enkephalin release are associated with acute- and chronic nicotine exposure. Acute nicotine treatment has been shown to increase the expression and release of neuropeptides. In contrast, chronic nicotine exposure has been associated with adaptive changes in opioid peptide or receptor levels, with decreased mu-opioid receptor binding and density in response to chronic alcohol and nicotine intake (Galeote, Kieffer, Maldonado, & Berrendero, 2006; Marco et al., 2007; Scott et al., 2007; Wewers, Dhatt, Snively, & Tejwani, 1999). Therefore, the endogenous opioid system may be differentially affected by chronic versus acute exposure to nicotine and alcohol. The similarities in downstream opioid signaling between nicotine and alcohol may contribute to the comorbidities shared between the two drugs. Injection of an enkephalin analog into the hypothalamus of rats enhanced ethanol consumption (Barson et al., 2010; Clegg, Air, Woods, & Seeley, 2002; Naleid, Grace, Chimukangara, Billington, & Levine, 2007), while injection of receptor antagonists was shown to reduce ethanol consumption (Moorman & Aston-Jones, 2009).

Endomorphin-1 (EM-1) and endomorrphin-2 (EM-2), two additional endogenous opioid peptides, were discovered in 1997 and showed to have the highest affinity for the mu receptors and possess strong analgesic effects (Zadina, Hackler, Ge, & Kastin, 1997). EM-1 increased locomotor activity and the number of lever-pressing following direct injection into the VTA (Zangen, Ikemoto, Zadina, & Wise, 2002). Regarding the role of EM1 in the action of alcohol, an earlier study showed a steady decrease in EM-1 serum levels in mice exposed to chronic ethanol for 56 days (Banks, Kumar, & Morley, 2004). In addition, a derivative of EM-2, [N-allyl-Dmt(1)] endomorphin-2, which acts as a mu opioid receptor antagonist, reduces ethanol-induced GABA release in the hippocampus (Li et al., 2009). No evidence exists in the literature regarding the role of endogenous EM-1 or EM-2 or the effect of exogenous EM-1 and EM-2 on the actions of nicotine.

While FDA-approved pharmacotherapies are available for individuals with alcohol or nicotine addiction alone, there are no FDA-approved medications for treating the nicotine-alcohol codependence. In heavy drinkers who also smoke, limited treatment options are available. For those individuals who have this concurrent disorder, rates of smoking cessation are significantly less due to difficulties in maintaining abstinence for more than 1 month at a time (Kahler et al., 2009). Naltrexone is an opioid receptor antagonist with the highest affinity at the mu opioid receptor and is commonly used to treat opioid dependence and alcohol-use disorder (Singh & Saadabadi, 2021). In both clinical trials and human laboratory studies, naltrexone effectively lowers alcohol consumption (Hendershot, Wardell, Samokhvalov, & Rehm, 2017; Maisel, Blodgett, Wilbourne, Humphreys, & Finney, 2013). Recently, naltrexone was shown to reduce cue-induced withdrawal symptoms and lower smoking and alcohol consumption, thus playing a role in smoking cessation (King et al., 2012; King, Cao, Zhang, & O'Malley, 2013; Rohsenow et al., 2003). Additionally, naltrexone is more beneficial for heavy drinkers who wish to quit smoking by increasing smoking abstinence and lowering the urge to smoke (Fridberg, Cao, Grant, & King, 2014). Previous research has also shown that naltrexone, especially when used as a stand-alone drug, may not be effective toward smoking cessation (David, Lancaster, Stead, Evins, & Prochaska, 2013; King et al., 2006, 2012; Rohsenow et al., 2007). However, additional literature suggests that naltrexone may be beneficial as an adjunct therapeutic agent for smoking cessation in certain (i.e., women or heavy drinkers) groups of smokers (David et al., 2013; King et al., 2006, 2012; Rohsenow et al., 2007).

4. Conclusion

Adolescence is a vulnerable developmental age for the initiation of drug addiction. Nicotine and alcohol each are widely used at an early age, and the use of one can promote the use of the other and may lead to the development of nicotine-alcohol codependence. Indeed, previous research has shown that the prevalence of alcoholism is higher in smokers than in non-smokers. Despite the negative consequences of nicotine dependence, alcohol dependence, and codependence, many pharmacotherapeutic agents are unavailable to treat nicotine addiction or alcohol addiction. Notably, there is no pharmacotherapy to treat the alcohol-nicotine comorbid disorder, varenicline having a limited effect (Mackowick, Lynch, Weinberger, & George, 2012). Presently, the underlying mechanisms of the comorbid disorder are not fully understood. Thus, further research is needed to define key mechanisms that alter the crosstalk between nicotine and alcohol's effects. This information is essential because it is crucial to advance our understanding of this comorbid disorder's neurobiology and lay a foundation for developing novel pharmacotherapies to treat nicotine and alcohol codependence. Several neuropeptides, such as opioids, PACAP, NPY, CRH, etc., have been implicated in the rewarding actions of nicotine and alcohol. Regarding these neuropeptides, this review has highlighted the need for further studies to explore the role of these neuropeptides in nicotine-alcohol codependence.

Acknowledgments

We are grateful to Western University of Health Sciences for providing intramural funds to K.L. and a summer fellowship to G.C.

References

Adriani, W., Deroche-Gamonet, V., Le Moal, M., Laviola, G., & Piazza, P. V. (2006). Preexposure during or following adolescence differently affects nicotine-rewarding properties in adult rats. *Psychopharmacology, 184*(3–4), 382–390. https://doi.org/10.1007/s00213-005-0125-1.

Adriani, W., Macri, S., Pacifici, R., & Laviola, G. (2002). Peculiar vulnerability to nicotine oral self-administration in mice during early adolescence. *Neuropsychopharmacology, 27*(2), 212–224. https://doi.org/10.1016/S0893-133X(02)00295-6.

Adriani, W., Spijker, S., Deroche-Gamonet, V., Laviola, G., Le Moal, M., Smit, A. B., et al. (2003). Evidence for enhanced neurobehavioral vulnerability to nicotine during periadolescence in rats. *The Journal of Neuroscience, 23*(11), 4712–4716. https://www.ncbi.nlm.nih.gov/pubmed/12805310.

Aguirre, C. G., Madrid, J., & Leventhal, A. M. (2015). Tobacco withdrawal symptoms mediate motivation to reinstate smoking during abstinence. *Journal of Abnormal Psychology, 124*(3), 623–634. https://doi.org/10.1037/abn0000060.

Alati, R., Al Mamun, A., Williams, G. M., O'Callaghan, M., Najman, J. M., & Bor, W. (2006). In utero alcohol exposure and prediction of alcohol disorders in early adulthood: A birth cohort study. *Archives of General Psychiatry, 63*(9), 1009–1016. https://doi.org/10.1001/archpsyc.63.9.1009.

Alhadeff, A. L., Rupprecht, L. E., & Hayes, M. R. (2012). GLP-1 neurons in the nucleus of the solitary tract project directly to the ventral tegmental area and nucleus accumbens to control for food intake. *Endocrinology, 153*(2), 647–658. https://doi.org/10.1210/en.2011-1443.

Alvarez, E., Roncero, I., Chowen, J. A., Thorens, B., & Blazquez, E. (1996). Expression of the glucagon-like peptide-1 receptor gene in rat brain. *Journal of Neurochemistry, 66*(3), 920–927. https://doi.org/10.1046/j.1471-4159.1996.66030920.x.

Andersen, S. L. (2003). Trajectories of brain development: Point of vulnerability or window of opportunity? *Neuroscience and Biobehavioral Reviews, 27*(1–2), 3–18. https://doi.org/10.1016/s0149-7634(03)00005-8.

Andersen, S. L., Dumont, N. L., & Teicher, M. H. (1997). Developmental differences in dopamine synthesis inhibition by (+/−)-7-OH-DPAT. *Naunyn-Schmiedeberg's Archives of Pharmacology, 356*(2), 173–181. https://doi.org/10.1007/pl00005038.

Antoni, F. A. (1986). Hypothalamic control of adrenocorticotropin secretion: Advances since the discovery of 41-residue corticotropin-releasing factor. *Endocrine Reviews,* 7(4), 351–378. https://doi.org/10.1210/edrv-7-4-351.

Audrain-McGovern, J., Rodriguez, D., Leventhal, A. M., Cuevas, J., Rodgers, K., & Sass, J. (2012). Where is the pleasure in that? Low hedonic capacity predicts smoking onset and escalation. *Nicotine & Tobacco Research, 14*(10), 1187–1196. https://doi.org/10.1093/ntr/nts017.

Backon, J. (1989). Negative correlation of cigarette smoking and dysmenorrhea: Reduced prostaglandin synthesis due to beta-endorphin, nicotine, or acrolein antagonism. *Medical Hypotheses, 28*(3), 213–214. https://doi.org/10.1016/0306-9877(89)90054-6.

Badanich, K. A., & Kirsteina, C. L. (2004). Nicotine administration significantly alters accumbal dopamine in the adult but not in the adolescent rat. *Annals of the New York Academy of Sciences, 1021*, 410–417. https://doi.org/10.1196/annals.1308.054.

Baer, J. S., Barr, H. M., Bookstein, F. L., Sampson, P. D., & Streissguth, A. P. (1998). Prenatal alcohol exposure and family history of alcoholism in the etiology of adolescent alcohol problems. *Journal of Studies on Alcohol, 59*(5), 533–543. https://doi.org/10.15288/jsa.1998.59.533.

Baer, J. S., Sampson, P. D., Barr, H. M., Connor, P. D., & Streissguth, A. P. (2003). A 21-year longitudinal analysis of the effects of prenatal alcohol exposure on young adult drinking. *Archives of General Psychiatry, 60*(4), 377–385. https://doi.org/10.1001/archpsyc.60.4.377.

Baggio, L. L., & Drucker, D. J. (2007). Biology of incretins: GLP-1 and GIP. *Gastroenterology, 132*(6), 2131–2157. https://doi.org/10.1053/j.gastro.2007.03.054.

Balfour, D. J. (1982). The effects of nicotine on brain neurotransmitter systems. *Pharmacology & Therapeutics, 16*(2), 269–282. https://doi.org/10.1016/0163-7258(82)90058-4.

Balfour, D. J., Khullar, A. K., & Longden, A. (1975). Effects of nicotine on plasma corticosterone and brain amines in stressed and unstressed rats. *Pharmacology, Biochemistry, and Behavior, 3*(2), 179–184. https://doi.org/10.1016/0091-3057(75)90145-8.

Banks, W. A., Kumar, V. B., & Morley, J. E. (2004). Influence of ethanol dependence and methionine enkephalin antisense on serum endomorphin-1 and methionine enkephalin levels. *Alcoholism, Clinical and Experimental Research, 28*(5), 792–796. https://doi.org/10.1097/01.alc.0000125357.54776.41.

Barrett, S. P., Boileau, I., Okker, J., Pihl, R. O., & Dagher, A. (2004). The hedonic response to cigarette smoking is proportional to dopamine release in the human striatum as measured by positron emission tomography and [11C]raclopride. *Synapse, 54*(2), 65–71. https://doi.org/10.1002/syn.20066.

Barson, J. R., Carr, A. J., Soun, J. E., Sobhani, N. C., Rada, P., Leibowitz, S. F., et al. (2010). Opioids in the hypothalamic paraventricular nucleus stimulate ethanol intake. *Alcoholism, Clinical and Experimental Research, 34*(2), 214–222. https://doi.org/10.1111/j.1530-0277.2009.01084.x.

Bassareo, V., & Di Chiara, G. (1997). Differential influence of associative and nonassociative learning mechanisms on the responsiveness of prefrontal and accumbal dopamine transmission to food stimuli in rats fed ad libitum. *The Journal of Neuroscience*, *17*(2), 851–861. https://www.ncbi.nlm.nih.gov/pubmed/8987806.

Becker, J. B., & Hu, M. (2008). Sex differences in drug abuse. *Frontiers in Neuroendocrinology*, *29*(1), 36–47. https://doi.org/10.1016/j.yfrne.2007.07.003.

Belluzzi, J. D., Lee, A. G., Oliff, H. S., & Leslie, F. M. (2004). Age-dependent effects of nicotine on locomotor activity and conditioned place preference in rats. *Psychopharmacology*, *174*(3), 389–395. https://doi.org/10.1007/s00213-003-1758-6.

Benes, F. M., Taylor, J. B., & Cunningham, M. C. (2000). Convergence and plasticity of monoaminergic systems in the medial prefrontal cortex during the postnatal period: Implications for the development of psychopathology. *Cerebral Cortex*, *10*(10), 1014–1027. https://doi.org/10.1093/cercor/10.10.1014.

Benwell, M. E., & Balfour, D. J. (1992). The effects of acute and repeated nicotine treatment on nucleus accumbens dopamine and locomotor activity. *British Journal of Pharmacology*, *105*(4), 849–856. https://doi.org/10.1111/j.1476-5381.1992.tb09067.x.

Benwell, M. E., Balfour, D. J., & Lucchi, H. M. (1993). Influence of tetrodotoxin and calcium on changes in extracellular dopamine levels evoked by systemic nicotine. *Psychopharmacology*, *112*(4), 467–474. https://doi.org/10.1007/BF02244896.

Berrendero, F., Kieffer, B. L., & Maldonado, R. (2002). Attenuation of nicotine-induced antinociception, rewarding effects, and dependence in mu-opioid receptor knock-out mice. *The Journal of Neuroscience*, *22*(24), 10935–10940. https://www.ncbi.nlm.nih.gov/pubmed/12486188.

Berrendero, F., Mendizabal, V., Robledo, P., Galeote, L., Bilkei-Gorzo, A., Zimmer, A., et al. (2005). Nicotine-induced antinociception, rewarding effects, and physical dependence are decreased in mice lacking the preproenkephalin gene. *The Journal of Neuroscience*, *25*(5), 1103–1112. https://doi.org/10.1523/JNEUROSCI.3008-04.2005.

Berrendero, F., Robledo, P., Trigo, J. M., Martin-Garcia, E., & Maldonado, R. (2010). Neurobiological mechanisms involved in nicotine dependence and reward: Participation of the endogenous opioid system. *Neuroscience and Biobehavioral Reviews*, *35*(2), 220–231. https://doi.org/10.1016/j.neubiorev.2010.02.006.

Best, D., Rawaf, S., Rowley, J., Floyd, K., Manning, V., & Strang, J. (2000). Drinking and smoking as concurrent predictors of illicit drug use and positive drug attitudes in adolescents. *Drug and Alcohol Dependence*, *60*(3), 319–321. https://www.ncbi.nlm.nih.gov/pubmed/11053768.

Bito-Onon, J. J., Simms, J. A., Chatterjee, S., Holgate, J., & Bartlett, S. E. (2011). Varenicline, a partial agonist at neuronal nicotinic acetylcholine receptors, reduces nicotine-induced increases in 20% ethanol operant self-administration in Sprague-Dawley rats. *Addiction Biology*, *16*(3), 440–449. https://doi.org/10.1111/j.1369-1600.2010.00309.x.

Bjornson, W., Rand, C., Connett, J. E., Lindgren, P., Nides, M., Pope, F., et al. (1995). Gender differences in smoking cessation after 3 years in the lung health study. *American Journal of Public Health*, *85*(2), 223–230. https://doi.org/10.2105/ajph.85.2.223.

Bogdan, R., & Pizzagalli, D. A. (2009). The heritability of hedonic capacity and perceived stress: A twin study evaluation of candidate depressive phenotypes. *Psychological Medicine*, *39*(2), 211–218. https://doi.org/10.1017/S0033291708003619.

Boorse, G. C., & Denver, R. J. (2006). Widespread tissue distribution and diverse functions of corticotropin-releasing factor and related peptides. *General and Comparative Endocrinology*, *146*(1), 9–18. https://doi.org/10.1016/j.ygcen.2005.11.014.

Boules, M., Oliveros, A., Liang, Y., Williams, K., Shaw, A., Robinson, J., et al. (2011). A neurotensin analog, NT69L, attenuates intravenous nicotine self-administration in rats. *Neuropeptides*, *45*(1), 9–16. https://doi.org/10.1016/j.npep.2010.09.003.

Boyle, A. E., Segal, R., Smith, B. R., & Amit, Z. (1993). Bidirectional effects of GABAergic agonists and antagonists on maintenance of voluntary ethanol intake in rats. *Pharmacology, Biochemistry, and Behavior, 46*(1), 179–182. https://doi.org/10.1016/0091-3057(93)90338-t.

Brennan, P. A., Hammen, C., Katz, A. R., & Le Brocque, R. M. (2002). Maternal depression, paternal psychopathology, and adolescent diagnostic outcomes. *Journal of Consulting and Clinical Psychology, 70*(5), 1075–1085. https://doi.org/10.1037//0022-006x.70.5.1075.

Breslau, N., Novak, S. P., & Kessler, R. C. (2004). Psychiatric disorders and stages of smoking. *Biological Psychiatry, 55*(1), 69–76. https://doi.org/10.1016/s0006-3223(03)00317-2.

Breslau, N., Peterson, E. L., Schultz, L. R., Chilcoat, H. D., & Andreski, P. (1998). Major depression and stages of smoking. A longitudinal investigation. *Archives of General Psychiatry, 55*(2), 161–166. https://doi.org/10.1001/archpsyc.55.2.161.

Brielmaier, J., McDonald, C. G., & Smith, R. F. (2012). Effects of acute stress on acquisition of nicotine conditioned place preference in adolescent rats: A role for corticotropin-releasing factor 1 receptors. *Psychopharmacology, 219*(1), 73–82. https://doi.org/10.1007/s00213-011-2378-1.

Brody, A. L., Olmstead, R. E., London, E. D., Farahi, J., Meyer, J. H., Grossman, P., et al. (2004). Smoking-induced ventral striatum dopamine release. *The American Journal of Psychiatry, 161*(7), 1211–1218. https://doi.org/10.1176/appi.ajp.161.7.1211.

Brothers, S. P., & Wahlestedt, C. (2010). Therapeutic potential of neuropeptide Y (NPY) receptor ligands. *EMBO Molecular Medicine, 2*(11), 429–439. https://doi.org/10.1002/emmm.201000100.

Brown, R. A., Lewinsohn, P. M., Seeley, J. R., & Wagner, E. F. (1996). Cigarette smoking, major depression, and other psychiatric disorders among adolescents. *Journal of the American Academy of Child and Adolescent Psychiatry, 35*(12), 1602–1610. https://doi.org/10.1097/00004583-199612000-00011.

Bruijnzeel, A. W., Ford, J., Rogers, J. A., Scheick, S., Ji, Y., Bishnoi, M., et al. (2012). Blockade of CRF1 receptors in the central nucleus of the amygdala attenuates the dysphoria associated with nicotine withdrawal in rats. *Pharmacology, Biochemistry, and Behavior, 101*(1), 62–68. https://doi.org/10.1016/j.pbb.2011.12.001.

Brunchmann, A., Thomsen, M., & Fink-Jensen, A. (2019). The effect of glucagon-like peptide-1 (GLP-1) receptor agonists on substance use disorder (SUD)-related behavioural effects of drugs and alcohol: A systematic review. *Physiology & Behavior, 206*, 232–242. https://doi.org/10.1016/j.physbeh.2019.03.029.

Burch, J. B., de Fiebre, C. M., Marks, M. J., & Collins, A. C. (1988). Chronic ethanol or nicotine treatment results in partial cross-tolerance between these agents. *Psychopharmacology, 95*(4), 452–458. https://doi.org/10.1007/BF00172954.

Burke, A. R., & Miczek, K. A. (2014). Stress in adolescence and drugs of abuse in rodent models: Role of dopamine, CRF, and HPA axis. *Psychopharmacology, 231*(8), 1557–1580. https://doi.org/10.1007/s00213-013-3369-1.

Businelle, M. S., Lam, C. Y., Kendzor, D. E., Cofta-Woerpel, L., McClure, J. B., Cinciripini, P. M., et al. (2013). Alcohol consumption and urges to smoke among women during a smoking cessation attempt. *Experimental and Clinical Psychopharmacology, 21*(1), 29–37. https://doi.org/10.1037/a0031009.

Caggiula, A. R., Donny, E. C., Epstein, L. H., Sved, A. F., Knopf, S., Rose, C., et al. (1998). The role of corticosteroids in nicotine's physiological and behavioral effects. *Psychoneuroendocrinology, 23*(2), 143–159. https://doi.org/10.1016/s0306-4530(97)00078-4.

Cam, G. R., & Bassett, J. R. (1983). The effect of acute nicotine administration on plasma levels of the thyroid hormones and corticosterone in the rat. *Pharmacology, Biochemistry, and Behavior, 19*(3), 559–561. https://doi.org/10.1016/0091-3057(83)90135-1.

Cao, J., Belluzzi, J. D., Loughlin, S. E., Dao, J. M., Chen, Y., & Leslie, F. M. (2010). Locomotor and stress responses to nicotine differ in adolescent and adult rats. *Pharmacology, Biochemistry, and Behavior*, *96*(1), 82–90. https://doi.org/10.1016/j.pbb.2010.04.010.

Cao, J., Belluzzi, J. D., Loughlin, S. E., Keyler, D. E., Pentel, P. R., & Leslie, F. M. (2007). Acetaldehyde, a major constituent of tobacco smoke, enhances behavioral, endocrine, and neuronal responses to nicotine in adolescent and adult rats. *Neuropsychopharmacology*, *32*(9), 2025–2035. https://doi.org/10.1038/sj.npp.1301327.

Carcoba, L. M., Orfila, J. E., Natividad, L. A., Torres, O. V., Pipkin, J. A., Ferree, P. L., et al. (2014). Cholinergic transmission during nicotine withdrawal is influenced by age and pre-exposure to nicotine: Implications for teenage smoking. *Developmental Neuroscience*, *36*(3–4), 347–355. https://doi.org/10.1159/000360133.

Carmody, T. P., Brischetto, C. S., Matarazzo, J. D., O'Donnell, R. P., & Connor, W. E. (1985). Co-occurrent use of cigarettes, alcohol, and coffee in healthy, community-living men and women. *Health Psychology*, *4*(4), 323–335. http://www.ncbi.nlm.nih.gov/pubmed/4054078 (Research Support, U.S. Gov't, P.H.S.).

Carrasco, G. A., & Van de Kar, L. D. (2003). Neuroendocrine pharmacology of stress. *European Journal of Pharmacology*, *463*(1–3), 235–272. https://doi.org/10.1016/s0014-2999(03)01285-8.

Casey, B. J., Getz, S., & Galvan, A. (2008). The adolescent brain. *Developmental Review*, *28*(1), 62–77. https://doi.org/10.1016/j.dr.2007.08.003.

Cavallo, D. A., & Krishnan-Sarin, S. (2019). Nicotine use disorders in adolescents. *Pediatric Clinics of North America*, *66*(6), 1053–1062. https://doi.org/10.1016/j.pcl.2019.08.002.

Chan, G. C., Hinds, T. R., Impey, S., & Storm, D. R. (1998). Hippocampal neurotoxicity of Delta9-tetrahydrocannabinol. *The Journal of Neuroscience*, *18*(14), 5322–5332. https://www.ncbi.nlm.nih.gov/pubmed/9651215.

Chang, G. Q., Karatayev, O., & Leibowitz, S. F. (2013). Prenatal exposure to nicotine stimulates neurogenesis of orexigenic peptide-expressing neurons in hypothalamus and amygdala. *The Journal of Neuroscience*, *33*(34), 13600–13611. https://doi.org/10.1523/jneurosci.5835-12.2013.

Choi, W. S., Patten, C. A., Gillin, J. C., Kaplan, R. M., & Pierce, J. P. (1997). Cigarette smoking predicts development of depressive symptoms among U.S. adolescents. *Annals of Behavioral Medicine*, *19*(1), 42–50. https://doi.org/10.1007/BF02883426.

Ciliax, B. J., Heilman, C., Demchyshyn, L. L., Pristupa, Z. B., Ince, E., Hersch, S. M., et al. (1995). The dopamine transporter: Immunochemical characterization and localization in brain. *The Journal of Neuroscience*, *15*(3 Pt 1), 1714–1723. https://www.ncbi.nlm.nih.gov/pubmed/7534339.

Clark, J. T., Kalra, P. S., Crowley, W. R., & Kalra, S. P. (1984). Neuropeptide Y and human pancreatic polypeptide stimulate feeding behavior in rats. *Endocrinology*, *115*(1), 427–429. https://doi.org/10.1210/endo-115-1-427.

Clark, D. B., Kirisci, L., & Tarter, R. E. (1998). Adolescent versus adult onset and the development of substance use disorders in males. *Drug and Alcohol Dependence*, *49*(2), 115–121. https://doi.org/10.1016/s0376-8716(97)00154-3.

Clark, A., Lindgren, S., Brooks, S. P., Watson, W. P., & Little, H. J. (2001). Chronic infusion of nicotine can increase operant self-administration of alcohol. *Neuropharmacology*, *41*(1), 108–117. https://doi.org/10.1016/s0028-3908(01)00037-5.

Clegg, D. J., Air, E. L., Woods, S. C., & Seeley, R. J. (2002). Eating elicited by orexin-a, but not melanin-concentrating hormone, is opioid mediated. *Endocrinology*, *143*(8), 2995–3000. https://doi.org/10.1210/endo.143.8.8977.

Cole, R. D., Wolsh, C., Zimmerman, M., Harrington, E., Gould, T. J., & Parikh, V. (2019). Adolescent and adult nicotine exposure differentially impacts oral nicotine and oral saccharin self-administration in mice. *Behavioural Brain Research*, *359*, 836–844. https://doi.org/10.1016/j.bbr.2018.07.019.

Collins, S. L., & Izenwasser, S. (2004). Chronic nicotine differentially alters cocaine-induced locomotor activity in adolescent vs. adult male and female rats. *Neuropharmacology*, *46*(3), 349–362. https://doi.org/10.1016/j.neuropharm.2003.09.024.

Collins, A. C., Wilkins, L. H., Slobe, B. S., Cao, J. Z., & Bullock, A. E. (1996). Long-term ethanol and nicotine treatment elicit tolerance to ethanol. *Alcoholism, Clinical and Experimental Research*, *20*(6), 990–999. https://doi.org/10.1111/j.1530-0277.1996.tb01936.x.

Cork, S. C., Richards, J. E., Holt, M. K., Gribble, F. M., Reimann, F., & Trapp, S. (2015). Distribution and characterisation of glucagon-like peptide-1 receptor expressing cells in the mouse brain. *Molecular Metabolism*, *4*(10), 718–731. https://doi.org/10.1016/j.molmet.2015.07.008.

Corrigall, W. A., Coen, K. M., & Adamson, K. L. (1994). Self-administered nicotine activates the mesolimbic dopamine system through the ventral tegmental area. *Brain Research*, *653*(1–2), 278–284. http://www.ncbi.nlm.nih.gov/pubmed/7982062 (Research Support, Non-U.S. Gov't).

Cruz, F. C., DeLucia, R., & Planeta, C. S. (2008). Effects of chronic stress on nicotine-induced locomotor activity and corticosterone release in adult and adolescent rats. *Addiction Biology*, *13*(1), 63–69. https://doi.org/10.1111/j.1369-1600.2007.00080.x.

Dani, J. A., & Harris, R. A. (2005). Nicotine addiction and comorbidity with alcohol abuse and mental illness. *Nature Neuroscience*, *8*(11), 1465–1470. https://doi.org/10.1038/nn1580 (pii).

Darville, A., & Hahn, E. J. (2014). Hardcore smokers: What do we know? *Addictive Behaviors*, *39*(12), 1706–1712. https://doi.org/10.1016/j.addbeh.2014.07.020.

David, S. P., Lancaster, T., Stead, L. F., Evins, A. E., & Prochaska, J. J. (2013). Opioid antagonists for smoking cessation. *Cochrane Database of Systematic Reviews*, *6*, CD003086. https://doi.org/10.1002/14651858.CD003086.pub3.

del Arbol, J. L., Muñoz, J. R., Ojeda, L., Cascales, A. L., Irles, J. R., Miranda, M. T., et al. (2000). Plasma concentrations of beta-endorphin in smokers who consume different numbers of cigarettes per day. *Pharmacology, Biochemistry, and Behavior*, *67*(1), 25–28. https://doi.org/10.1016/s0091-3057(00)00291-4.

Di Chiara, G., & Imperato, A. (1988). Drugs abused by humans preferentially increase synaptic dopamine concentrations in the mesolimbic system of freely moving rats. *Proceedings of the National Academy of Sciences of the United States of America*, *85*(14), 5274–5278. http://www.ncbi.nlm.nih.gov/pubmed/2899326.

Dickson, P. E., Miller, M. M., Rogers, T. D., Blaha, C. D., & Mittleman, G. (2014). Effects of adolescent nicotine exposure and withdrawal on intravenous cocaine self-administration during adulthood in male C57BL/6J mice. *Addiction Biology*, *19*(1), 37–48. https://doi.org/10.1111/j.1369-1600.2012.00496.x.

DiFranza, J. R., & Guerrera, M. P. (1990). Alcoholism and smoking. *Journal of Studies on Alcohol*, *51*(2), 130–135. http://www.ncbi.nlm.nih.gov/pubmed/2308350.

Donny, E. C., Caggiula, A. R., Rose, C., Jacobs, K. S., Mielke, M. M., & Sved, A. F. (2000). Differential effects of response-contingent and response-independent nicotine in rats. *European Journal of Pharmacology*, *402*(3), 231–240. https://doi.org/10.1016/s0014-2999(00)00532-x.

Doubeni, C. A., Reed, G., & Difranza, J. R. (2010). Early course of nicotine dependence in adolescent smokers. *Pediatrics*, *125*(6), 1127–1133. https://doi.org/10.1542/peds.2009-0238.

Doyon, W. M., Dong, Y., Ostroumov, A., Thomas, A. M., Zhang, T. A., & Dani, J. A. (2013). Nicotine decreases ethanol-induced dopamine signaling and increases self-administration via stress hormones. *Neuron*, *79*(3), 530–540. https://doi.org/10.1016/j.neuron.2013.06.006.

Drews, E., & Zimmer, A. (2010). Modulation of alcohol and nicotine responses through the endogenous opioid system. *Progress in Neurobiology*, *90*(1), 1–15. https://doi.org/10.1016/j.pneurobio.2009.09.004.

Drucker, D. J., Habener, J. F., & Holst, J. J. (2017). Discovery, characterization, and clinical development of the glucagon-like peptides. *The Journal of Clinical Investigation, 127*(12), 4217–4227. https://doi.org/10.1172/JCI97233.

Dumont, Y., Martel, J. C., Fournier, A., St-Pierre, S., & Quirion, R. (1992). Neuropeptide Y and neuropeptide Y receptor subtypes in brain and peripheral tissues. *Progress in Neurobiology, 38*(2), 125–167. https://doi.org/10.1016/0301-0082(92)90038-g.

Egecioglu, E., Engel, J. A., Jerlhag, E., & Engel, J. A. (2013). The glucagon-like peptide 1 analogue Exendin-4 attenuates the nicotine-induced locomotor stimulation, accumbal dopamine release, conditioned place preference as well as the expression of locomotor sensitization in mice. *PLoS One, 8*(10), e77284.

Ehlers, C. L., Li, T. K., Lumeng, L., Hwang, B. H., Somes, C., Jimenez, P., et al. (1998). Neuropeptide Y levels in ethanol-naive alcohol-preferring and nonpreferring rats and in Wistar rats after ethanol exposure. *Alcoholism, Clinical and Experimental Research, 22*(8), 1778–1782. https://www.ncbi.nlm.nih.gov/pubmed/9835294.

Eiden, L. E., Giraud, P., Dave, J. R., Hotchkiss, A. J., & Affolter, H. U. (1984). Nicotinic receptor stimulation activates enkephalin release and biosynthesis in adrenal chromaffin cells. *Nature, 312*(5995), 661–663. https://doi.org/10.1038/312661a0.

Elliott, B. M., Faraday, M. M., Phillips, J. M., & Grunberg, N. E. (2004). Effects of nicotine on elevated plus maze and locomotor activity in male and female adolescent and adult rats. *Pharmacology, Biochemistry, and Behavior,* 77(1), 21–28. https://doi.org/10.1016/j.pbb.2003.09.016.

Epping-Jordan, M. P., Watkins, S. S., Koob, G. F., & Markou, A. (1998). Dramatic decreases in brain reward function during nicotine withdrawal. *Nature, 393*(6680), 76–79. https://doi.org/10.1038/30001.

Ericson, M., Blomqvist, O., Engel, J. A., & Soderpalm, B. (1998). Voluntary ethanol intake in the rat and the associated accumbal dopamine overflow are blocked by ventral tegmental mecamylamine. *European Journal of Pharmacology, 358*(3), 189–196. https://doi.org/10.1016/s0014-2999(98)00602-5.

Everitt, B. J., & Wolf, M. E. (2002). Psychomotor stimulant addiction: A neural systems perspective. *The Journal of Neuroscience, 22*(9), 3312–3320. https://doi.org/10.1523/JNEUROSCI.22-09-03312.2002 (pii).

Fiore, M. C., Schroeder, S. A., & Baker, T. B. (2014). Smoke, the chief killer—Strategies for targeting combustible tobacco use. *The New England Journal of Medicine, 370*(4), 297–299. https://doi.org/10.1056/NEJMp1314942.

Foilb, A. R., Lui, P., & Romeo, R. D. (2011). The transformation of hormonal stress responses throughout puberty and adolescence. *The Journal of Endocrinology, 210*(3), 391–398. https://doi.org/10.1530/JOE-11-0206.

Forbes, E. E., & Dahl, R. E. (2005). Neural systems of positive affect: Relevance to understanding child and adolescent depression? *Development and Psychopathology, 17*(3), 827–850. https://doi.org/10.1017/s095457940505039x.

Fredrickson, P., Boules, M., Yerbury, S., & Richelson, E. (2003). Novel neurotensin analog blocks the initiation and expression of nicotine-induced locomotor sensitization. *Brain Research, 979*(1–2), 245–248. https://doi.org/10.1016/s0006-8993(03)02895-6.

Fridberg, D. J., Cao, D., Grant, J. E., & King, A. C. (2014). Naltrexone improves quit rates, attenuates smoking urge, and reduces alcohol use in heavy drinking smokers attempting to quit smoking. *Alcoholism, Clinical and Experimental Research, 38*(10), 2622–2629. https://doi.org/10.1111/acer.12513.

Froehlich, T. E., Lanphear, B. P., Auinger, P., Hornung, R., Epstein, J. N., Braun, J., et al. (2009). Association of tobacco and lead exposures with attention-deficit/hyperactivity disorder. *Pediatrics, 124*(6), e1054–e1063. https://doi.org/10.1542/peds.2009-0738.

Galeote, L., Kieffer, B. L., Maldonado, R., & Berrendero, F. (2006). Mu-opioid receptors are involved in the tolerance to nicotine antinociception. *Journal of Neurochemistry, 97*(2), 416–423. https://doi.org/10.1111/j.1471-4159.2006.03751.x.

Galvan, A. (2010). Adolescent development of the reward system. *Frontiers in Human Neuroscience*, *4*, 6. https://doi.org/10.3389/neuro.09.006.2010.
Gentzke, A. S., Creamer, M., Cullen, K. A., Ambrose, B. K., Willis, G., Jamal, A., et al. (2019). Vital signs: Tobacco product use among middle and high school students—United States, 2011-2018. *MMWR. Morbidity and Mortality Weekly Report*, *68*(6), 157–164. https://doi.org/10.15585/mmwr.mm6806e1.
Gilbert, D. G., Meliska, C. J., Williams, C. L., & Jensen, R. A. (1992). Subjective correlates of cigarette-smoking-induced elevations of peripheral beta-endorphin and cortisol. *Psychopharmacology*, *106*(2), 275–281. https://doi.org/10.1007/BF02801984.
Gilmore, K., & Meersand, P. (2014). Normal child and adolescent development. In R. E. Hales, S. C. Yudofsky, & L. W. Roberts (Eds.), *The American Psychiatric Publishing textbook of psychiatry* (pp. 139–173). American Psychiatric Publishing Inc. https://doi.org/10.1176/appi.books.9781585625031.rh05.
Goeders, N. E. (2002). Stress and cocaine addiction. *The Journal of Pharmacology and Experimental Therapeutics*, *301*(3), 785–789. http://www.ncbi.nlm.nih.gov/pubmed/12023504 (Research Support, U.S. Gov't, P.H.S. Review).
Goke, R., Larsen, P. J., Mikkelsen, J. D., & Sheikh, S. P. (1995). Distribution of GLP-1 binding sites in the rat brain: Evidence that exendin-4 is a ligand of brain GLP-1 binding sites. *The European Journal of Neuroscience*, 7(11), 2294–2300. https://doi.org/10.1111/j.1460-9568.1995.tb00650.x.
Gomez, F., Houshyar, H., & Dallman, M. F. (2002). Marked regulatory shifts in gonadal, adrenal, and metabolic system responses to repeated restraint stress occur within a 3-week period in pubertal male rats. *Endocrinology*, *143*(8), 2852–2862. https://doi.org/10.1210/endo.143.8.8929.
Grant, B. F. (1998). Age at smoking onset and its association with alcohol consumption and DSM-IV alcohol abuse and dependence: Results from the National Longitudinal Alcohol Epidemiologic Survey. *Journal of Substance Abuse*, *10*(1), 59–73. https://doi.org/10.1016/s0899-3289(99)80141-2.
Grant, B. F., Hasin, D. S., Chou, S. P., Stinson, F. S., & Dawson, D. A. (2004). Nicotine dependence and psychiatric disorders in the United States: Results from the national epidemiologic survey on alcohol and related conditions. *Archives of General Psychiatry*, *61*(11), 1107–1115. https://doi.org/10.1001/archpsyc.61.11.1107.
Grieder, T. E., Herman, M. A., Contet, C., Tan, L. A., Vargas-Perez, H., Cohen, A., et al. (2014). VTA CRF neurons mediate the aversive effects of nicotine withdrawal and promote intake escalation. *Nature Neuroscience*, *17*(12), 1751–1758. https://doi.org/10.1038/nn.3872.
Gulick, D., & Gould, T. J. (2008). Interactive effects of ethanol and nicotine on learning in C57BL/6J mice depend on both dose and duration of treatment. *Psychopharmacology*, *196*(3), 483–495. https://doi.org/10.1007/s00213-007-0982-x.
Gupta, A., Gargiulo, A. T., Curtis, G. R., Badve, P. S., Pandey, S., & Barson, J. R. (2018). Pituitary adenylate cyclase-activating POLYPEPTIDE-27 (PACAP-27) in the thalamic paraventricular nucleus is stimulated by ethanol drinking. *Alcoholism, Clinical and Experimental Research*, *42*(9), 1650–1660. https://doi.org/10.1111/acer.13826.
Harrison, E. L., & McKee, S. A. (2011). Non-daily smoking predicts hazardous drinking and alcohol use disorders in young adults in a longitudinal U.S. sample. *Drug and Alcohol Dependence*, *118*(1), 78–82. https://doi.org/10.1016/j.drugalcdep.2011.02.022.
Heilig, M., Soderpalm, B., Engel, J. A., & Widerlov, E. (1989). Centrally administered neuropeptide Y (NPY) produces anxiolytic-like effects in animal anxiety models. *Psychopharmacology*, *98*(4), 524–529. https://doi.org/10.1007/BF00441953.
Heilig, M., & Widerlov, E. (1995). Neurobiology and clinical aspects of neuropeptide Y. *Critical Reviews in Neurobiology*, *9*(2–3), 115–136. https://www.ncbi.nlm.nih.gov/pubmed/8581979.

Heim, C., & Nemeroff, C. B. (2001). The role of childhood trauma in the neurobiology of mood and anxiety disorders: Preclinical and clinical studies. *Biological Psychiatry*, *49*(12), 1023–1039. https://doi.org/10.1016/s0006-3223(01)01157-x.

Hendershot, C. S., Wardell, J. D., Samokhvalov, A. V., & Rehm, J. (2017). Effects of naltrexone on alcohol self-administration and craving: Meta-analysis of human laboratory studies. *Addiction Biology*, *22*(6), 1515–1527. https://doi.org/10.1111/adb.12425.

Hendrickson, L. M., Zhao-Shea, R., & Tapper, A. R. (2009). Modulation of ethanol drinking-in-the-dark by mecamylamine and nicotinic acetylcholine receptor agonists in C57BL/6J mice. *Psychopharmacology*, *204*(4), 563–572. https://doi.org/10.1007/s00213-009-1488-5.

Heppner, K. M., Kirigiti, M., Secher, A., Paulsen, S. J., Buckingham, R., Pyke, C., et al. (2015). Expression and distribution of glucagon-like peptide-1 receptor mRNA, protein and binding in the male nonhuman primate (Macaca mulatta) brain. *Endocrinology*, *156*(1), 255–267. https://doi.org/10.1210/en.2014-1675.

Herman, J. P., Figueiredo, H., Mueller, N. K., Ulrich-Lai, Y., Ostrander, M. M., Choi, D. C., et al. (2003). Central mechanisms of stress integration: Hierarchical circuitry controlling hypothalamo-pituitary-adrenocortical responsiveness. *Frontiers in Neuroendocrinology*, *24*(3), 151–180. https://doi.org/10.1016/j.yfrne.2003.07.001.

Hernandez, N. S., Ige, K. Y., Mietlicki-Baase, E. G., Molina-Castro, G. C., Turner, C. A., Hayes, M. R., et al. (2018). Glucagon-like peptide-1 receptor activation in the ventral tegmental area attenuates cocaine seeking in rats. *Neuropsychopharmacology*, *43*(10), 2000–2008. https://doi.org/10.1038/s41386-018-0010-3.

Hernandez, N. S., O'Donovan, B., Ortinski, P. I., & Schmidt, H. D. (2019). Activation of glucagon-like peptide-1 receptors in the nucleus accumbens attenuates cocaine seeking in rats. *Addiction Biology*, *24*(2), 170–181. https://doi.org/10.1111/adb.12583.

Hilario, M. R., Turner, J. R., & Blendy, J. A. (2012). Reward sensitization: Effects of repeated nicotine exposure and withdrawal in mice. *Neuropsychopharmacology*, *37*(12), 2661–2670. https://doi.org/10.1038/npp.2012.130.

Holliday, E., & Gould, T. J. (2016). Nicotine, adolescence, and stress: A review of how stress can modulate the negative consequences of adolescent nicotine abuse. *Neuroscience and Biobehavioral Reviews*, *65*, 173–184. https://doi.org/10.1016/j.neubiorev.2016.04.003.

Hollt, V., & Horn, G. (1992). Effect of nicotine on mRNA levels encoding opioid peptides, vasopressin and alpha 3 nicotinic receptor subunit in the rat. *The Clinical Investigator*, *70*(3–4), 224–231. https://doi.org/10.1007/BF00184655.

Holst, J. J. (2007). The physiology of glucagon-like peptide 1. *Physiological Reviews*, *87*(4), 1409–1439. https://doi.org/10.1152/physrev.00034.2006.

Holst, J. J. (2013). Incretin hormones and the satiation signal. *International Journal of Obesity*, *37*(9), 1161–1168. https://doi.org/10.1038/ijo.2012.208.

Houdi, A. A., Pierzchala, K., Marson, L., Palkovits, M., & Van Loon, G. R. (1991). Nicotine-induced alteration in Tyr-Gly-Gly and met-enkephalin in discrete brain nuclei reflects altered enkephalin neuron activity. *Peptides*, *12*(1), 161–166. https://doi.org/10.1016/0196-9781(91)90183-p.

Hurt, R. D., Eberman, K. M., Croghan, I. T., Offord, K. P., Davis, L. J., Jr., Morse, R. M., et al. (1994). Nicotine dependence treatment during inpatient treatment for other addictions: A prospective intervention trial. *Alcoholism, Clinical and Experimental Research*, *18*(4), 867–872.

Hutchison, M. A., & Riley, A. L. (2008). Adolescent exposure to nicotine alters the aversive effects of cocaine in adult rats. *Neurotoxicology and Teratology*, *30*(5), 404–411. https://doi.org/10.1016/j.ntt.2008.04.004.

Hwang, B. H., Zhang, J. K., Ehlers, C. L., Lumeng, L., & Li, T. K. (1999). Innate differences of neuropeptide Y (NPY) in hypothalamic nuclei and central nucleus of the amygdala between selectively bred rats with high and low alcohol preference. *Alcoholism,*

Clinical and Experimental Research, *23*(6), 1023–1030. https://www.ncbi.nlm.nih.gov/pubmed/10397286.
Imperato, A., Mulas, A., & Di Chiara, G. (1986). Nicotine preferentially stimulates dopamine release in the limbic system of freely moving rats. *European Journal of Pharmacology*, *132*(2–3), 337–338. https://doi.org/10.1016/0014-2999(86)90629-1.
Iniguez, S. D., Warren, B. L., Parise, E. M., Alcantara, L. F., Schuh, B., Maffeo, M. L., et al. (2009). Nicotine exposure during adolescence induces a depression-like state in adulthood. *Neuropsychopharmacology*, *34*(6), 1609–1624. https://doi.org/10.1038/npp.2008.220.
Isiegas, C., Mague, S. D., & Blendy, J. A. (2009). Sex differences in response to nicotine in C57Bl/6:129SvEv mice. *Nicotine & Tobacco Research*, *11*(7), 851–858. https://doi.org/10.1093/ntr/ntp076.
Jerlhag, E. (2018). GLP-1 signaling and alcohol-mediated behaviors; preclinical and clinical evidence. *Neuropharmacology*, *136*(Pt B), 343–349. https://doi.org/10.1016/j.neuropharm.2018.01.013.
John, U., Meyer, C., Rumpf, H. J., & Hapke, U. (2004). Depressive disorders are related to nicotine dependence in the population but do not necessarily hamper smoking cessation. *The Journal of Clinical Psychiatry*, *65*(2), 169–176. https://doi.org/10.4088/jcp.v65n0205.
Juckel, G., Schlagenhauf, F., Koslowski, M., Wüstenberg, T., Villringer, A., Knutson, B., et al. (2006). Dysfunction of ventral striatal reward prediction in schizophrenia. *NeuroImage*, *29*(2), 409–416. https://doi.org/10.1016/j.neuroimage.2005.07.051.
Kahler, C. W., Borland, R., Hyland, A., McKee, S. A., Thompson, M. E., & Cummings, K. M. (2009). Alcohol consumption and quitting smoking in the international Tobacco Control (ITC) Four Country Survey. *Drug and Alcohol Dependence*, *100*(3), 214–220. https://doi.org/10.1016/j.drugalcdep.2008.10.006.
Kandel, D., Chen, K., Warner, L. A., Kessler, R. C., & Grant, B. (1997). Prevalence and demographic correlates of symptoms of last year dependence on alcohol, nicotine, marijuana and cocaine in the U.S. population. *Drug and Alcohol Dependence*, *44*(1), 11–29. https://doi.org/10.1016/s0376-8716(96)01315-4.
Kandel, D. B., & Kandel, E. R. (2014). A molecular basis for nicotine as a gateway drug. *The New England Journal of Medicine*, *371*(21), 2038–2039. https://doi.org/10.1056/NEJMc1411785.
Keller-Wood, M. E., & Dallman, M. F. (1984). Corticosteroid inhibition of ACTH secretion. *Endocrine Reviews*, *5*(1), 1–24. https://doi.org/10.1210/edrv-5-1-1.
Kelley, S. P., Nannini, M. A., Bratt, A. M., & Hodge, C. W. (2001). Neuropeptide-Y in the paraventricular nucleus increases ethanol self-administration. *Peptides*, *22*(3), 515–522. https://doi.org/10.1016/s0196-9781(01)00361-8.
Kelley, B. M., & Rowan, J. D. (2004). Long-term, low-level adolescent nicotine exposure produces dose-dependent changes in cocaine sensitivity and reward in adult mice. *International Journal of Developmental Neuroscience*, *22*(5–6), 339–348. https://doi.org/10.1016/j.ijdevneu.2004.04.002.
Kiguchi, N., Maeda, T., Tsuruga, M., Yamamoto, A., Yamamoto, C., Ozaki, M., et al. (2008). Involvement of spinal Met-enkephalin in nicotine-induced antinociception in mice. *Brain Research*, *1189*, 70–77. https://doi.org/10.1016/j.brainres.2007.10.086.
King, A. C., Cao, D., O'Malley, S. S., Kranzler, H. R., Cai, X., deWit, H., et al. (2012). Effects of naltrexone on smoking cessation outcomes and weight gain in nicotine-dependent men and women. *Journal of Clinical Psychopharmacology*, *32*(5), 630–636. https://doi.org/10.1097/JCP.0b013e3182676956.
King, A. C., Cao, D., Zhang, L., & O'Malley, S. S. (2013). Naltrexone reduction of long-term smoking cessation weight gain in women but not men: A randomized controlled trial. *Biological Psychiatry*, *73*(9), 924–930. https://doi.org/10.1016/j.biopsych.2012.09.025.

King, A., de Wit, H., Riley, R. C., Cao, D., Niaura, R., & Hatsukami, D. (2006). Efficacy of naltrexone in smoking cessation: A preliminary study and an examination of sex differences. *Nicotine & Tobacco Research*, *8*(5), 671–682. https://doi.org/10.1080/14622200600789767.

Klungsoyr, O., Nygard, J. F., Sorensen, T., & Sandanger, I. (2006). Cigarette smoking and incidence of first depressive episode: An 11-year, population-based follow-up study. *American Journal of Epidemiology*, *163*(5), 421–432. https://doi.org/10.1093/aje/kwj058.

Koob, G. F. (2008a). Hedonic homeostatic dysregulation as a driver of drug-seeking behavior. *Drug Discovery Today: Disease Models*, *5*(4), 207–215. https://doi.org/10.1016/j.ddmod.2009.04.002.

Koob, G. F. (2008b). A role for brain stress systems in addiction. *Neuron*, *59*(1), 11–34. https://doi.org/10.1016/j.neuron.2008.06.012.

Koob, G., & Kreek, M. J. (2007). Stress, dysregulation of drug reward pathways, and the transition to drug dependence. *The American Journal of Psychiatry*, *164*(8), 1149–1159. https://doi.org/10.1176/appi.ajp.2007.05030503 (Research Support, N.I.H., Extramural Review).

Koob, G. F., & Le Moal, M. (2008a). Addiction and the brain antireward system. *Annual Review of Psychology*, *59*, 29–53. https://doi.org/10.1146/annurev.psych.59.103006.093548 (review).

Koob, G. F., & Le Moal, M. (2008b). Review. Neurobiological mechanisms for opponent motivational processes in addiction. *Philosophical Transactions of the Royal Society of London. Series B, Biological Sciences*, *363*(1507), 3113–3123. https://doi.org/10.1098/rstb.2008.0094 (Research Support, N.I.H., Extramural Research Support, Non-U.S. Gov't Review).

Koob, G. F., & Volkow, N. D. (2010). Neurocircuitry of addiction. *Neuropsychopharmacology*, *35*(1), 217–238. https://doi.org/10.1038/npp.2009.110 (Research Support, N.I.H., Extramural Review).

Kota, D., Martin, B. R., Robinson, S. E., & Damaj, M. I. (2007). Nicotine dependence and reward differ between adolescent and adult male mice. *The Journal of Pharmacology and Experimental Therapeutics*, *322*(1), 399–407. https://doi.org/10.1124/jpet.107.121616.

Kota, D., Robinson, S. E., & Imad Damaj, M. (2009). Enhanced nicotine reward in adulthood after exposure to nicotine during early adolescence in mice. *Biochemical Pharmacology*, *78*(7), 873–879. https://doi.org/10.1016/j.bcp.2009.06.099.

Kozlowski, L. T., Skinner, W., Kent, C., & Pope, M. A. (1989). Prospects for smoking treatment in individuals seeking treatment for alcohol and other drug problems. *Addictive Behaviors*, *14*(3), 273–278. https://doi.org/10.1016/0306-4603(89)90058-0.

Kupferschmidt, D. A., Funk, D., Erb, S., & Lê, A. D. (2010). Age-related effects of acute nicotine on behavioural and neuronal measures of anxiety. *Behavioural Brain Research*, *213*(2), 288–292. https://doi.org/10.1016/j.bbr.2010.05.022.

Kutlu, M. G., Holliday, E., & Gould, T. J. (2016). High-affinity alpha4beta2 nicotinic receptors mediate the impairing effects of acute nicotine on contextual fear extinction. *Neurobiology of Learning and Memory*, *128*, 17–22. https://doi.org/10.1016/j.nlm.2015.11.021.

Lam, C. Y., Businelle, M. S., Aigner, C. J., McClure, J. B., Cofta-Woerpel, L., Cinciripini, P. M., et al. (2014). Individual and combined effects of multiple high-risk triggers on postcessation smoking urge and lapse. *Nicotine & Tobacco Research*, *16*(5), 569–575. https://doi.org/10.1093/ntr/ntt190.

Larraga, A., Belluzzi, J. D., & Leslie, F. M. (2017). Nicotine increases alcohol intake in adolescent male rats. *Frontiers in Behavioral Neuroscience*, *11*, 25. https://doi.org/10.3389/fnbeh.2017.00025.

Larsson, A., Edstrom, L., Svensson, L., Soderpalm, B., & Engel, J. A. (2005). Voluntary ethanol intake increases extracellular acetylcholine levels in the ventral tegmental area in the rat. *Alcohol and Alcoholism*, *40*(5), 349–358. https://doi.org/10.1093/alcalc/agh180.

Larsson, A., & Engel, J. A. (2004). Neurochemical and behavioral studies on ethanol and nicotine interactions. *Neuroscience and Biobehavioral Reviews*, *27*(8), 713–720. https://doi.org/10.1016/j.neubiorev.2003.11.010.

Laviola, G., & Marco, E. M. (2011). Passing the knife edge in adolescence: Brain pruning and specification of individual lines of development. *Neuroscience and Biobehavioral Reviews*, *35*(8), 1631–1633. https://doi.org/10.1016/j.neubiorev.2011.05.011.

Le, A. D., Wang, A., Harding, S., Juzytsch, W., & Shaham, Y. (2003). Nicotine increases alcohol self-administration and reinstates alcohol seeking in rats. *Psychopharmacology*, *168*(1–2), 216–221. https://doi.org/10.1007/s00213-002-1330-9.

Leao, R. M., Cruz, F. C., Vendruscolo, L. F., de Guglielmo, G., Logrip, M. L., Planeta, C. S., et al. (2015). Chronic nicotine activates stress/reward-related brain regions and facilitates the transition to compulsive alcohol drinking. *The Journal of Neuroscience*, *35*(15), 6241–6253. https://doi.org/10.1523/JNEUROSCI.3302-14.2015.

Lenney, W., & Enderby, B. (2008). "blowing in the wind": A review of teenage smoking. *Archives of Disease in Childhood*, *93*(1), 72–75. https://doi.org/10.1136/adc.2006.109702.

Leonard, S., Adler, L. E., Benhammou, K., Berger, R., Breese, C. R., Drebing, C., et al. (2001). Smoking and mental illness. *Pharmacology, Biochemistry, and Behavior*, *70*(4), 561–570. https://doi.org/10.1016/s0091-3057(01)00677-3.

Levine, A., Huang, Y., Drisaldi, B., Griffin, E. A., Jr., Pollak, D. D., Xu, S., et al. (2011). Molecular mechanism for a gateway drug: Epigenetic changes initiated by nicotine prime gene expression by cocaine. *Science Translational Medicine*, *3*(107), 107ra109. https://doi.org/10.1126/scitranslmed.3003062.

Li, H., Bu, Q., Chen, B., Shao, X., Hu, Z., Deng, P., et al. (2014). Mechanisms of metabonomic for a gateway drug: Nicotine priming enhances behavioral response to cocaine with modification in energy metabolism and neurotransmitter level. *PLoS One*, *9*(1), e87040. https://doi.org/10.1371/journal.pone.0087040.

Li, M. D., Kane, J. K., Parker, S. L., McAllen, K., Matta, S. G., & Sharp, B. M. (2000). Nicotine administration enhances NPY expression in the rat hypothalamus. *Brain Research*, *867*(1–2), 157–164. https://doi.org/10.1016/s0006-8993(00)02283-6.

Li, M., Mead, A., & Bevins, R. A. (2009). Individual differences in responses to nicotine: Tracking changes from adolescence to adulthood. *Acta Pharmacologica Sinica*, *30*(6), 868–878. https://doi.org/10.1038/aps.2009.55.

Li, Q., Okada, Y., Marczak, E., Wilson, W. A., Lazarus, L. H., & Swartzwelder, H. S. (2009). The novel micro-opioid receptor antagonist, [N-allyl-Dmt(1)]endomorphin-2, attenuates the enhancement of GABAergic neurotransmission by ethanol. *Alcohol and Alcoholism*, *44*(1), 13–19. https://doi.org/10.1093/alcalc/agn085.

Lisha, N. E., Carmody, T. P., Humfleet, G. L., & Delucchi, K. L. (2014). Reciprocal effects of alcohol and nicotine in smoking cessation treatment studies. *Addictive Behaviors*, *39*(3), 637–643. https://doi.org/10.1016/j.addbeh.2013.11.018.

Livett, B. G., Dean, D. M., Whelan, L. G., Udenfriend, S., & Rossier, J. (1981). Co-release of enkephalin and catecholamines from cultured adrenal chromaffin cells. *Nature*, *289*(5795), 317–319. https://doi.org/10.1038/289317a0.

Locker, A. R., Marks, M. J., Kamens, H. M., & Klein, L. C. (2016). Exposure to nicotine increases nicotinic acetylcholine receptor density in the reward pathway and binge ethanol consumption in C57BL/6J adolescent female mice. *Brain Research Bulletin*, *123*, 13–22. https://doi.org/10.1016/j.brainresbull.2015.09.009.

Lutfy, K., & Shankar, G. (2019). Emerging evidence for the role of pituitary adenylate cyclase-activating peptide in neuropsychiatric disorders. *Progress in Molecular Biology and Translational Science*, *167*, 143–157. https://doi.org/10.1016/bs.pmbts.2019.06.009.

Lyss, P. J., Andersen, S. L., LeBlanc, C. J., & Teicher, M. H. (1999). Degree of neuronal activation following FG-7142 changes across regions during development. *Brain*

Research. Developmental Brain Research, 116(2), 201–203. https://doi.org/10.1016/s0165-3806(99)00069-3.
MacArthur, L., Iacangelo, A. L., Hsu, C. M., & Eiden, L. E. (1992). Enkephalin biosynthesis is coupled to secretory activity via transcription of the proenkephalin A gene. *Journal of Physiology, Paris, 86*(1–3), 89–98. https://doi.org/10.1016/s0928-4257(05)80012-9.
Mackowick, K. M., Lynch, M. J., Weinberger, A. H., & George, T. P. (2012). Treatment of tobacco dependence in people with mental health and addictive disorders. *Current Psychiatry Reports, 14*(5), 478–485. https://doi.org/10.1007/s11920-012-0299-2.
Madras, B. K. (2017). The growing problem of new psychoactive substances (NPS). *Current Topics in Behavioral Neurosciences, 32*, 1–18. https://doi.org/10.1007/7854_2016_34.
Maisel, N. C., Blodgett, J. C., Wilbourne, P. L., Humphreys, K., & Finney, J. W. (2013). Meta-analysis of naltrexone and acamprosate for treating alcohol use disorders: When are these medications most helpful? *Addiction, 108*(2), 275–293. https://doi.org/10.1111/j.1360-0443.2012.04054.x.
Manavalan, S., Getachew, B., Manaye, K. F., Khundmiri, S. J., Csoka, A. B., McKinley, R., et al. (2017). PACAP protects against ethanol and nicotine toxicity in SH-SY5Y cells: Implications for drinking-smoking co-morbidity. *Neurotoxicity Research, 32*(1), 8–13. https://doi.org/10.1007/s12640-017-9727-8.
Marco, E. M., Granstrem, O., Moreno, E., Llorente, R., Adriani, W., Laviola, G., et al. (2007). Subchronic nicotine exposure in adolescence induces long-term effects on hippocampal and striatal cannabinoid-CB1 and mu-opioid receptors in rats. *European Journal of Pharmacology, 557*(1), 37–43. https://doi.org/10.1016/j.ejphar.2006.11.013.
Markou, A., Kosten, T. R., & Koob, G. F. (1998). Neurobiological similarities in depression and drug dependence: A self-medication hypothesis. *Neuropsychopharmacology, 18*(3), 135–174. https://doi.org/10.1016/S0893-133X(97)00113-9.
Martinez, D., Gil, R., Slifstein, M., Hwang, D. R., Huang, Y., Perez, A., et al. (2005). Alcohol dependence is associated with blunted dopamine transmission in the ventral striatum. *Biological Psychiatry, 58*(10), 779–786. https://doi.org/10.1016/j.biopsych.2005.04.044.
Mathews, I. Z., Waters, P., & McCormick, C. M. (2009). Changes in hyporesponsiveness to acute amphetamine and age differences in tyrosine hydroxylase immunoreactivity in the brain over adolescence in male and female rats. *Developmental Psychobiology, 51*(5), 417–428. https://doi.org/10.1002/dev.20381.
Matta, S. G., Valentine, J. D., & Sharp, B. M. (1997). Nicotinic activation of CRH neurons in extrahypothalamic regions of the rat brain. *Endocrine, 7*(2), 245–253. https://doi.org/10.1007/BF02778147.
McCabe, S. E., Cranford, J. A., Morales, M., & Young, A. (2006). Simultaneous and concurrent polydrug use of alcohol and prescription drugs: Prevalence, correlates, and consequences. *Journal of Studies on Alcohol, 67*(4), 529–537.
McCormick, C. M., & Ibrahim, F. N. (2007). Locomotor activity to nicotine and Fos immunoreactivity in the paraventricular nucleus of the hypothalamus in adolescent socially-stressed rats. *Pharmacology, Biochemistry, and Behavior, 86*(1), 92–102. https://doi.org/10.1016/j.pbb.2006.12.012.
McGinnis, J. M., & Foege, W. H. (1999). Mortality and morbidity attributable to use of addictive substances in the United States. *Proceedings of the Association of American Physicians, 111*(2), 109–118 (review) http://www.ncbi.nlm.nih.gov/pubmed/10220805.
McMillen, B. A., Davis, B. J., Williams, H. L., & Soderstrom, K. (2005). Periadolescent nicotine exposure causes heterologous sensitization to cocaine reinforcement. *European Journal of Pharmacology, 509*(2–3), 161–164. https://doi.org/10.1016/j.ejphar.2005.01.002.
McMurray, M. S., Williams, S. K., Jarrett, T. M., Cox, E. T., Fay, E. E., Overstreet, D. H., et al. (2008). Gestational ethanol and nicotine exposure: Effects on maternal behavior,

oxytocin, and offspring ethanol intake in the rat. *Neurotoxicology and Teratology*, *30*(6), 475–486. https://doi.org/10.1016/j.ntt.2008.07.001.

McQuown, S. C., Belluzzi, J. D., & Leslie, F. M. (2007). Low dose nicotine treatment during early adolescence increases subsequent cocaine reward. *Neurotoxicology and Teratology*, *29*(1), 66–73. https://doi.org/10.1016/j.ntt.2006.10.012.

McQuown, S. C., Dao, J. M., Belluzzi, J. D., & Leslie, F. M. (2009). Age-dependent effects of low-dose nicotine treatment on cocaine-induced behavioral plasticity in rats. *Psychopharmacology*, *207*(1), 143–152. https://doi.org/10.1007/s00213-009-1642-0.

Meehl, P. E. (2001). Primary and secondary hypohedonia. *Journal of Abnormal Psychology*, *110*(1), 188–193. https://doi.org/10.1037//0021-843x.110.1.188.

Meliska, C. J., & Gilbert, D. G. (1991). Hormonal and subjective effects of smoking the first five cigarettes of the day: A comparison in males and females. *Pharmacology, Biochemistry, and Behavior*, *40*(2), 229–235. https://doi.org/10.1016/0091-3057(91)90544-c.

Mello, N. K., & Newman, J. L. (2011). Discriminative and reinforcing stimulus effects of nicotine, cocaine, and cocaine + nicotine combinations in rhesus monkeys. *Experimental and Clinical Psychopharmacology*, *19*(3), 203–214. https://doi.org/10.1037/a0023373.

Merchenthaler, I., Lane, M., & Shughrue, P. (1999). Distribution of pre-pro-glucagon and glucagon-like peptide-1 receptor messenger RNAs in the rat central nervous system. *The Journal of Comparative Neurology*, *403*(2), 261–280. https://doi.org/10.1002/(sici)1096-9861(19990111)403:2<261::aid-cne8>3.0.co;2-5.

Mereu, G., Yoon, K. W., Boi, V., Gessa, G. L., Naes, L., & Westfall, T. C. (1987). Preferential stimulation of ventral tegmental area dopaminergic neurons by nicotine. *European Journal of Pharmacology*, *141*(3), 395–399. https://doi.org/10.1016/0014-2999(87)90556-5.

Miller, N. S., & Gold, M. S. (1998). Comorbid cigarette and alcohol addiction: Epidemiology and treatment. *Journal of Addictive Diseases*, *17*(1), 55–66. http://www.ncbi.nlm.nih.gov/pubmed/9549603.

Mojica, C. Y., Belluzzi, J. D., & Leslie, F. M. (2014). Age-dependent alterations in reward-seeking behavior after brief nicotine exposure. *Psychopharmacology*, *231*(8), 1763–1773. https://doi.org/10.1007/s00213-013-3266-7.

Moorman, D. E., & Aston-Jones, G. (2009). Orexin-1 receptor antagonism decreases ethanol consumption and preference selectively in high-ethanol--preferring Sprague—Dawley rats. *Alcohol*, *43*(5), 379–386. https://doi.org/10.1016/j.alcohol.2009.07.002.

Murray, J. B. (1991). Nicotine as a psychoactive drug. *The Journal of Psychology*, *125*(1), 5–25. https://doi.org/10.1080/00223980.1991.10543265.

Murray, C. J., & Lopez, A. D. (1997a). Alternative projections of mortality and disability by cause 1990-2020: Global burden of disease study [Research Support, non-U.S. Gov't]. *Lancet*, *349*(9064), 1498–1504. https://doi.org/10.1016/S0140-6736(96)07492-2.

Murray, C. J., & Lopez, A. D. (1997b). Global mortality, disability, and the contribution of risk factors: Global Burden of Disease Study. *Lancet*, *349*(9063), 1436–1442. https://doi.org/10.1016/S0140-6736(96)07495-8 (Research Support, Non-U.S. Gov't).

Murray, C. J., & Lopez, A. D. (1997c). Mortality by cause for eight regions of the world: Global Burden of Disease Study. *Lancet*, *349*(9061), 1269–1276. https://doi.org/10.1016/S0140-6736(96)07493-4 (Research Support, non-U.S. Gov't).

Myers, M. G., & Brown, S. A. (1994). Smoking and health in substance-abusing adolescents: A two-year follow-up. *Pediatrics*, *93*(4), 561–566. https://www.ncbi.nlm.nih.gov/pubmed/8134209.

Myers, M. G., & Kelly, J. F. (2006). Cigarette smoking among adolescents with alcohol and other drug use problems. *Alcohol Research & Health*, *29*(3), 221–227. https://www.ncbi.nlm.nih.gov/pubmed/17373413.

Naleid, A. M., Grace, M. K., Chimukangara, M., Billington, C. J., & Levine, A. S. (2007). Paraventricular opioids alter intake of high-fat but not high-sucrose diet depending on diet preference in a binge model of feeding. *American Journal of Physiology. Regulatory, Integrative and Comparative Physiology*, *293*(1), R99–105. https://doi.org/10.1152/ajpregu.00675.2006.

Naneix, F., Marchand, A. R., Di Scala, G., Pape, J. R., & Coutureau, E. (2012). Parallel maturation of goal-directed behavior and dopaminergic systems during adolescence. *The Journal of Neuroscience*, *32*(46), 16223–16232. https://doi.org/10.1523/JNEUROSCI.3080-12.2012.

Nashmi, R., Xiao, C., Deshpande, P., McKinney, S., Grady, S. R., Whiteaker, P., et al. (2007). Chronic nicotine cell specifically upregulates functional alpha 4* nicotinic receptors: Basis for both tolerance in midbrain and enhanced long-term potentiation in perforant path. *The Journal of Neuroscience*, *27*(31), 8202–8218. https://doi.org/10.1523/JNEUROSCI.2199-07.2007.

Natividad, L. A., Buczynski, M. W., Parsons, L. H., Torres, O. V., & O'Dell, L. E. (2012). Adolescent rats are resistant to adaptations in excitatory and inhibitory mechanisms that modulate mesolimbic dopamine during nicotine withdrawal. *Journal of Neurochemistry*, *123*(4), 578–588. https://doi.org/10.1111/j.1471-4159.2012.07926.x.

Natividad, L. A., Tejeda, H. A., Torres, O. V., & O'Dell, L. E. (2010). Nicotine withdrawal produces a decrease in extracellular levels of dopamine in the nucleus accumbens that is lower in adolescent versus adult male rats. *Synapse*, *64*(2), 136–145. https://doi.org/10.1002/syn.20713.

Nega, S., Marquez, P., Hamid, A., Ahmad, S. M., & Lutfy, K. (2020). The role of pituitary adenylyl cyclase activating polypeptide in affective signs of nicotine withdrawal. *Journal of Neuroscience Research*, *98*(8), 1549–1560. https://doi.org/10.1002/jnr.24649.

Nestler, E. J., & Carlezon, W. A., Jr. (2006). The mesolimbic dopamine reward circuit in depression. *Biological Psychiatry*, *59*(12), 1151–1159. https://doi.org/10.1016/j.biopsych.2005.09.018 (Research Support, N.I.H., Extramural Review).

Nomura, Y., Gilman, S. E., & Buka, S. L. (2011). Maternal smoking during pregnancy and risk of alcohol use disorders among adult offspring. *Journal of Studies on Alcohol and Drugs*, *72*(2), 199–209. https://doi.org/10.15288/jsad.2011.72.199.

O'Brien, J. W., & Hill, S. Y. (2014). Effects of prenatal alcohol and cigarette exposure on offspring substance use in multiplex, alcohol-dependent families. *Alcoholism, Clinical and Experimental Research*, *38*(12), 2952–2961. https://doi.org/10.1111/acer.12569.

O'Dell, L. E., Bruijnzeel, A. W., Ghozland, S., Markou, A., & Koob, G. F. (2004). Nicotine withdrawal in adolescent and adult rats. *Annals of the New York Academy of Sciences*, *1021*, 167–174. https://doi.org/10.1196/annals.1308.022.

O'Dell, L. E., Torres, O. V., Natividad, L. A., & Tejeda, H. A. (2007). Adolescent nicotine exposure produces less affective measures of withdrawal relative to adult nicotine exposure in male rats. *Neurotoxicology and Teratology*, *29*(1), 17–22. https://doi.org/10.1016/j.ntt.2006.11.003.

Olive, M. F., Koenig, H. N., Nannini, M. A., & Hodge, C. W. (2001). Stimulation of endorphin neurotransmission in the nucleus accumbens by ethanol, cocaine, and amphetamine. *The Journal of Neuroscience*, *21*(23), Rc184. https://doi.org/10.1523/JNEUROSCI.21-23-j0002.2001.

O'Loughlin, J., Kishchuk, N., DiFranza, J., Tremblay, M., & Paradis, G. (2002). The hardest thing is the habit: A qualitative investigation of adolescent smokers' experience of nicotine dependence. *Nicotine & Tobacco Research*, *4*(2), 201–209. https://doi.org/10.1080/14622200210123000.

Orsini, C., Bonito-Oliva, A., Conversi, D., & Cabib, S. (2008). Genetic liability increases propensity to prime-induced reinstatement of conditioned place preference in mice exposed to low cocaine. *Psychopharmacology*, *198*(2), 287–296. https://doi.org/10.1007/s00213-008-1137-4.

Parrott, A. C. (1999). Does cigarette smoking cause stress? *The American Psychologist, 54*(10), 817–820. https://doi.org/10.1037//0003-066x.54.10.817.

Patton, G. C., Coffey, C., Carlin, J. B., Sawyer, S. M., & Wakefield, M. (2006). Teen smokers reach their mid twenties. *The Journal of Adolescent Health, 39*(2), 214–220. https://doi.org/10.1016/j.jadohealth.2005.11.027.

Pauly, J. R., Ullman, E. A., & Collins, A. C. (1988). Adrenocortical hormone regulation of nicotine sensitivity in mice. *Physiology & Behavior, 44*(1), 109–116. https://doi.org/10.1016/0031-9384(88)90353-8.

Pedrazzini, T., Seydoux, J., Kunstner, P., Aubert, J. F., Grouzmann, E., Beermann, F., et al. (1998). Cardiovascular response, feeding behavior and locomotor activity in mice lacking the NPY Y1 receptor. *Nature Medicine, 4*(6), 722–726. https://doi.org/10.1038/nm0698-722.

Pidoplichko, V. I., Noguchi, J., Areola, O. O., Liang, Y., Peterson, J., Zhang, T., et al. (2004). Nicotinic cholinergic synaptic mechanisms in the ventral tegmental area contribute to nicotine addiction. *Learning & Memory, 11*(1), 60–69. https://doi.org/10.1101/lm.70004.

Pogun, S., & Yararbas, G. (2009). Sex differences in nicotine action. In J. E. Henningfield, E. D. London, & S. Pogun (Eds.), *Vol. 192. Nicotine psychopharmacology. Handbook of experimental pharmacology*. Berlin, Heidelberg: Springer. https://doi.org/10.1007/978-3-540-69248-5_10.

Pomerleau, C. S., Aubin, H. J., & Pomerleau, O. F. (1997). Self-reported alcohol use patterns in a sample of male and female heavy smokers. *Journal of Addictive Diseases, 16*(3), 19–24. https://doi.org/10.1300/J069v16n03_02.

Rahman, S., Zhang, J., Engleman, E. A., & Corrigall, W. A. (2004). Neuroadaptive changes in the mesoaccumbens dopamine system after chronic nicotine self-administration: A microdialysis study. *Neuroscience, 129*(2), 415–424. https://doi.org/10.1016/j.neuroscience.2004.08.010.

Rassnick, S., D'Amico, E., Riley, E., & Koob, G. F. (1993). GABA antagonist and benzodiazepine partial inverse agonist reduce motivated responding for ethanol. *Alcoholism, Clinical and Experimental Research, 17*(1), 124–130. https://doi.org/10.1111/j.1530-0277.1993.tb00736.x.

Redrobe, J. P., Dumont, Y., & Quirion, R. (2002). Neuropeptide Y (NPY) and depression: From animal studies to the human condition. *Life Sciences, 71*(25), 2921–2937. https://doi.org/10.1016/s0024-3205(02)02159-8.

Ribeiro-Carvalho, A., Lima, C. S., Medeiros, A. H., Siqueira, N. R., Filgueiras, C. C., Manhaes, A. C., et al. (2009). Combined exposure to nicotine and ethanol in adolescent mice: Effects on the central cholinergic systems during short and long term withdrawal. *Neuroscience, 162*(4), 1174–1186. https://doi.org/10.1016/j.neuroscience.2009.05.032.

Rivier, C., Brownstein, M., Spiess, J., Rivier, J., & Vale, W. (1982). In vivo corticotropin-releasing factor-induced secretion of adrenocorticotropin, beta-endorphin, and corticosterone. *Endocrinology, 110*(1), 272–278. https://doi.org/10.1210/endo-110-1-272.

Robbins, A. S., Manson, J. E., Lee, I. M., Satterfield, S., & Hennekens, C. H. (1994). Cigarette smoking and stroke in a cohort of U.S. male physicians. *Annals of Internal Medicine, 120*(6), 458–462. http://www.ncbi.nlm.nih.gov/pubmed/8311368 (Clinical Trial Randomized Controlled Trial Research Support, U.S. Gov't, P.H.S.).

Robinson, D. L., Zitzman, D. L., Smith, K. J., & Spear, L. P. (2011). Fast dopamine release events in the nucleus accumbens of early adolescent rats. *Neuroscience, 176*, 296–307. https://doi.org/10.1016/j.neuroscience.2010.12.016.

Rodaros, D., Caruana, D. A., Amir, S., & Stewart, J. (2007). Corticotropin-releasing factor projections from limbic forebrain and paraventricular nucleus of the hypothalamus to the region of the ventral tegmental area. *Neuroscience, 150*(1), 8–13. https://doi.org/10.1016/j.neuroscience.2007.09.043.

Rohsenow, D. J., Monti, P. M., Colby, S. M., Gulliver, S. B., Swift, R. M., & Abrams, D. B. (2003). Naltrexone treatment for alcoholics: Effect on cigarette smoking rates. *Nicotine & Tobacco Research*, *5*(2), 231–236. https://doi.org/10.1080/1462220031000073298.

Rohsenow, D. J., Monti, P. M., Hutchison, K. E., Swift, R. M., MacKinnon, S. V., Sirota, A. D., et al. (2007). High-dose transdermal nicotine and naltrexone: Effects on nicotine withdrawal, urges, smoking, and effects of smoking. *Experimental and Clinical Psychopharmacology*, *15*(1), 81–92. https://doi.org/10.1037/1064-1297.15.1.81.

Romberger, D. J., & Grant, K. (2004). Alcohol consumption and smoking status: The role of smoking cessation. *Biomedicine & Pharmacotherapy*, *58*(2), 77–83. https://doi.org/10.1016/j.biopha.2003.12.002.

Roth-Deri, I., Green-Sadan, T., & Yadid, G. (2008). Beta-endorphin and drug-induced reward and reinforcement. *Progress in Neurobiology*, *86*(1), 1–21. https://doi.org/10.1016/j.pneurobio.2008.06.003.

Saha, S. P., Bhalla, D. K., Whayne, T. F., Jr., & Gairola, C. (2007). Cigarette smoke and adverse health effects: An overview of research trends and future needs. *International Journal of Angiology*, *16*(3), 77–83. http://www.ncbi.nlm.nih.gov/pubmed/22477297.

Sainsbury, A., Rohner-Jeanrenaud, F., Cusin, I., Zakrzewska, K. E., Halban, P. A., Gaillard, R. C., et al. (1997). Chronic central neuropeptide Y infusion in normal rats: Status of the hypothalamo-pituitary-adrenal axis, and vagal mediation of hyperinsulinaemia. *Diabetologia*, *40*(11), 1269–1277. https://doi.org/10.1007/s001250050820.

Schochet, T. L., Kelley, A. E., & Landry, C. F. (2004). Differential behavioral effects of nicotine exposure in adolescent and adult rats. *Psychopharmacology*, *175*(3), 265–273. https://doi.org/10.1007/s00213-004-1831-9.

Schoedel, K. A., & Tyndale, R. F. (2003). Induction of nicotine-metabolizing CYP2B1 by ethanol and ethanol-metabolizing CYP2E1 by nicotine: Summary and implications. *Biochimica et Biophysica Acta*, *1619*(3), 283–290. https://doi.org/10.1016/s0304-4165(02)00487-7.

Schultz, W. (2002). Getting formal with dopamine and reward. *Neuron*, *36*(2), 241–263. https://doi.org/10.1016/s0896-6273(02)00967-4.

Scott, D. J., Domino, E. F., Heitzeg, M. M., Koeppe, R. A., Ni, L., Guthrie, S., et al. (2007). Smoking modulation of mu-opioid and dopamine D2 receptor-mediated neurotransmission in humans. *Neuropsychopharmacology*, *32*(2), 450–457. https://doi.org/10.1038/sj.npp.1301238.

Serra, M., Concas, A., Mostallino, M. C., Chessa, M. F., Stomati, M., Petraglia, F., et al. (1999). Antagonism by pivagabine of stress-induced changes in GABAA receptor function and corticotropin-releasing factor concentrations in rat brain. *Psychoneuroendocrinology*, *24*(3), 269–284. https://doi.org/10.1016/s0306-4530(98)00049-3.

Shiffman, S. (2006). Reflections on smoking relapse research. *Drug and Alcohol Review*, *25*(1), 15–20. https://doi.org/10.1080/09595230500459479.

Shram, M. J., Funk, D., Li, Z., & Le, A. D. (2007). Acute nicotine enhances c-fos mRNA expression differentially in reward-related substrates of adolescent and adult rat brain. *Neuroscience Letters*, *418*(3), 286–291. https://doi.org/10.1016/j.neulet.2007.03.034.

Shram, M. J., & Le, A. D. (2010). Adolescent male Wistar rats are more responsive than adult rats to the conditioned rewarding effects of intravenously administered nicotine in the place conditioning procedure. *Behavioural Brain Research*, *206*(2), 240–244. https://doi.org/10.1016/j.bbr.2009.09.018.

Silva, C. P., Horton, W. J., Caruso, M. J., Sebastian, A., Klein, L. C., Albert, I., et al. (2018). The influence of adolescent nicotine exposure on ethanol intake and brain gene expression. *PLoS One*, *13*(6), e0198935. https://doi.org/10.1371/journal.pone.0198935.

Singh, P. K., & Lutfy, K. (2017). Nicotine pretreatment reduced cocaine-induced CPP and its reinstatement in a sex- and dose-related manner in adult C57BL/6J mice. *Pharmacology, Biochemistry, and Behavior*, *159*, 84–89. https://doi.org/10.1016/j.pbb.2017.07.010.

Singh, D., & Saadabadi, A. (2021). Naltrexone. In *StatPearls* StatPearls Publishing. https://www.ncbi.nlm.nih.gov/pubmed/30521232.
Slawecki, C. J., Betancourt, M., Walpole, T., & Ehlers, C. L. (2000). Increases in sucrose consumption, but not ethanol consumption, following ICV NPY administration. *Pharmacology, Biochemistry, and Behavior, 66*(3), 591–594. https://doi.org/10.1016/s0091-3057(00)00215-x.
Slawecki, C. J., & Ehlers, C. L. (2003). The effects of corticotropin-releasing factor on the cortical EEG are reduced following adolescent nicotine exposure. *Neuropeptides, 37*(1), 66–73. https://doi.org/10.1016/s0143-4179(03)00006-4.
Slawecki, C. J., Thorsell, A. K., El Khoury, A., Mathé, A. A., & Ehlers, C. L. (2005). Increased CRF-like and NPY-like immunoreactivity in adult rats exposed to nicotine during adolescence: Relation to anxiety-like and depressive-like behavior. *Neuropeptides, 39*(4), 369–377. https://doi.org/10.1016/j.npep.2005.06.002.
Slotkin, T. A., Bodwell, B. E., Ryde, I. T., & Seidler, F. J. (2008). Adolescent nicotine treatment changes the response of acetylcholine systems to subsequent nicotine administration in adulthood. *Brain Research Bulletin, 76*(1–2), 152–165. https://doi.org/10.1016/j.brainresbull.2007.12.009.
Smith, S. M., & Vale, W. W. (2006). The role of the hypothalamic-pituitary-adrenal axis in neuroendocrine responses to stress. *Dialogues in Clinical Neuroscience, 8*(4), 383–395. https://www.ncbi.nlm.nih.gov/pubmed/17290797.
Snuggs, S., & Hajek, P. (2013). Responsiveness to reward following cessation of smoking. *Psychopharmacology, 225*(4), 869–873. https://doi.org/10.1007/s00213-012-2874-y.
Sorensen, G., Reddy, I. A., Weikop, P., Graham, D. L., Stanwood, G. D., Wortwein, G., et al. (2015). The glucagon-like peptide 1 (GLP-1) receptor agonist exendin-4 reduces cocaine self-administration in mice. *Physiology & Behavior, 149*, 262–268. https://doi.org/10.1016/j.physbeh.2015.06.013.
Spear, L. P. (2000). The adolescent brain and age-related behavioral manifestations. *Neuroscience and Biobehavioral Reviews, 24*(4), 417–463. https://doi.org/10.1016/s0149-7634(00)00014-2.
Spear, L. P. (2014). Adolescents and alcohol: Acute sensitivities, enhanced intake, and later consequences. *Neurotoxicology and Teratology, 41*, 51–59. https://doi.org/10.1016/j.ntt.2013.11.006.
Spear, N. E., & Molina, J. C. (2005). Fetal or infantile exposure to ethanol promotes ethanol ingestion in adolescence and adulthood: A theoretical review. *Alcoholism, Clinical and Experimental Research, 29*(6), 909–929. https://doi.org/10.1097/01.alc.0000171046.78556.66.
Spina, M., Merlo-Pich, E., Chan, R. K., Basso, A. M., Rivier, J., Vale, W., et al. (1996). Appetite-suppressing effects of urocortin, a CRF-related neuropeptide. *Science, 273*(5281), 1561–1564. https://doi.org/10.1126/science.273.5281.1561.
Steinberg, L. (2004). Risk taking in adolescence: What changes, and why? *Annals of the New York Academy of Sciences, 1021*, 51–58. https://doi.org/10.1196/annals.1308.005.
Steinberg, L. (2008). A social neuroscience perspective on adolescent risk-taking. *Developmental Review, 28*(1), 78–106. https://doi.org/10.1016/j.dr.2007.08.002.
Steiner, M. A., & Wotjak, C. T. (2008). Role of the endocannabinoid system in regulation of the hypothalamic-pituitary-adrenocortical axis. *Progress in Brain Research, 170*, 397–432. https://doi.org/10.1016/S0079-6123(08)00433-0.
Stewart, R. B., Gatto, G. J., Lumeng, L., Li, T. K., & Murphy, J. M. (1993). Comparison of alcohol-preferring (P) and nonpreferring (NP) rats on tests of anxiety and for the anxiolytic effects of ethanol. *Alcohol, 10*(1), 1–10. https://doi.org/10.1016/0741-8329(93)90046-q.
Stogner, K. A., & Holmes, P. V. (2000). Neuropeptide-Y exerts antidepressant-like effects in the forced swim test in rats. *European Journal of Pharmacology, 387*(2), R9–10. https://doi.org/10.1016/s0014-2999(99)00800-6.

Stojakovic, A., Ahmad, S. M., Malhotra, S., Afzal, Z., Ahmed, M., & Lutfy, K. (2020). The role of pituitary adenylyl cyclase-activating polypeptide in the motivational effects of addictive drugs. *Neuropharmacology*, *171*, 108109. https://doi.org/10.1016/j.neuropharm.2020.108109.

Stojakovic, A., Espinosa, E. P., Farhad, O. T., & Lutfy, K. (2017). Effects of nicotine on homeostatic and hedonic components of food intake. *The Journal of Endocrinology*, *235*(1), R13–R31. https://doi.org/10.1530/JOE-17-0166.

Stoltenburg-Didinger, G. (1994). Neuropathology of the hippocampus and its susceptibility to neurotoxic insult. *Neurotoxicology*, *15*(3), 445–450. https://www.ncbi.nlm.nih.gov/pubmed/7854577.

Stout, S. C., Mortas, P., Owens, M. J., Nemeroff, C. B., & Moreau, J. (2000). Increased corticotropin-releasing factor concentrations in the bed nucleus of the stria terminalis of anhedonic rats. *European Journal of Pharmacology*, *401*(1), 39–46. https://doi.org/10.1016/s0014-2999(00)00412-x.

Sweerts, B. W., Jarrott, B., & Lawrence, A. J. (2001). The effect of acute and chronic restraint on the central expression of prepro-neuropeptide Y mRNA in normotensive and hypertensive rats. *Journal of Neuroendocrinology*, *13*(7), 608–617. https://doi.org/10.1046/j.1365-2826.2001.00674.x.

Tanaka, K., Kunishige-Yamamoto, A., Hashimoto, H., Shintani, N., Hayata, A., & Baba, A. (2010). Increased ethanol preference and serotonin 1A receptor-dependent attenuation of ethanol-induced hypothermia in PACAP-deficient mice. *Biochemical and Biophysical Research Communications*, *391*(1), 773–777. https://doi.org/10.1016/j.bbrc.2009.11.136.

Tarazi, F. I., Tomasini, E. C., & Baldessarini, R. J. (1998). Postnatal development of dopamine and serotonin transporters in rat caudate-putamen and nucleus accumbens septi. *Neuroscience Letters*, *254*(1), 21–24. https://doi.org/10.1016/s0304-3940(98)00644-2.

Tatemoto, K., Carlquist, M., & Mutt, V. (1982). Neuropeptide Y--a novel brain peptide with structural similarities to peptide YY and pancreatic polypeptide. *Nature*, *296*(5858), 659–660. https://doi.org/10.1038/296659a0.

Thiele, T. E., Koh, M. T., & Pedrazzini, T. (2002). Voluntary alcohol consumption is controlled via the neuropeptide Y Y1 receptor. *The Journal of Neuroscience*, *22*(3), RC208. https://www.ncbi.nlm.nih.gov/pubmed/11826154.

Thiele, T. E., Marsh, D. J., Ste Marie, L., Bernstein, I. L., & Palmiter, R. D. (1998). Ethanol consumption and resistance are inversely related to neuropeptide Y levels. *Nature*, *396*(6709), 366–369. https://doi.org/10.1038/24614.

Thomas, A. M., Ostroumov, A., Kimmey, B. A., Taormina, M. B., Holden, W. M., Kim, K., et al. (2018). Adolescent nicotine exposure alters GABAA receptor Signaling in the ventral tegmental area and increases adult ethanol self-administration. *Cell Reports*, *23*(1), 68–77. https://doi.org/10.1016/j.celrep.2018.03.030.

Tizabi, Y., Bai, L., Copeland, R. L., Jr., & Taylor, R. E. (2007). Combined effects of systemic alcohol and nicotine on dopamine release in the nucleus accumbens shell. *Alcohol and Alcoholism*, *42*(5), 413–416. https://doi.org/10.1093/alcalc/agm057.

Torabi, M. R., Bailey, W. J., & Majd-Jabbari, M. (1993). Cigarette smoking as a predictor of alcohol and other drug use by children and adolescents: Evidence of the "gateway drug effect". *The Journal of School Health*, *63*(7), 302–306. https://www.ncbi.nlm.nih.gov/pubmed/8246462.

Torres, O. V., Tejeda, H. A., Natividad, L. A., & O'Dell, L. E. (2008). Enhanced vulnerability to the rewarding effects of nicotine during the adolescent period of development. *Pharmacology, Biochemistry, and Behavior*, *90*(4), 658–663. https://doi.org/10.1016/j.pbb.2008.05.009.

Trauth, J. A., McCook, E. C., Seidler, F. J., & Slotkin, T. A. (2000). Modeling adolescent nicotine exposure: Effects on cholinergic systems in rat brain regions. *Brain Research*, *873*(1), 18–25. https://doi.org/10.1016/s0006-8993(00)02465-3.

Tremblay, L. K., Naranjo, C. A., Graham, S. J., Herrmann, N., Mayberg, H. S., Hevenor, S., et al. (2005). Functional neuroanatomical substrates of altered reward processing in major depressive disorder revealed by a dopaminergic probe. *Archives of General Psychiatry*, *62*(11), 1228–1236. https://doi.org/10.1001/archpsyc.62.11.1228.

Trigo, J. M., Zimmer, A., & Maldonado, R. (2009). Nicotine anxiogenic and rewarding effects are decreased in mice lacking beta-endorphin. *Neuropharmacology*, *56*(8), 1147–1153. https://doi.org/10.1016/j.neuropharm.2009.03.013.

Truitt, W. A., Hauser, S. R., Deehan, G. A., Jr., Toalston, J. E., Wilden, J. A., Bell, R. L., et al. (2015). Ethanol and nicotine interaction within the posterior ventral tegmental area in male and female alcohol-preferring rats: Evidence of synergy and differential gene activation in the nucleus accumbens shell. *Psychopharmacology*, *232*(3), 639–649. https://doi.org/10.1007/s00213-014-3702-3.

Tseng, A., Singh, P., Marquez, P., Hamid, A., & Lutfy, K. (2019). The role of endogenous pituitary adenylyl cyclase activating polypeptide (PACAP) in nicotine self-administration, reward and aversion. *Pharmacology, Biochemistry, and Behavior*, *181*, 46–52. https://doi.org/10.1016/j.pbb.2019.04.007.

Tuesta, L. M., Chen, Z., Duncan, A., Fowler, C. D., Ishikawa, M., Lee, B. R., et al. (2017). GLP-1 acts on habenular avoidance circuits to control nicotine intake. *Nature Neuroscience*, *20*(5), 708–716. https://doi.org/10.1038/nn.4540.

Vezina, P., Blanc, G., Glowinski, J., & Tassin, J. P. (1992). Nicotine and morphine differentially activate brain dopamine in prefrontocortical and subcortical terminal fields: Effects of acute and repeated injections. *The Journal of Pharmacology and Experimental Therapeutics*, *261*(2), 484–490. https://www.ncbi.nlm.nih.gov/pubmed/1578364.

Vidal, C. (1996). Nicotinic receptors in the brain. Molecular biology, function, and therapeutics. *Molecular and Chemical Neuropathology*, *28*(1–3), 3–11. https://doi.org/10.1007/BF02815199.

Volkow, N. D. (2004). The reality of comorbidity: Depression and drug abuse. *Biological Psychiatry*, *56*(10), 714–717. https://doi.org/10.1016/j.biopsych.2004.07.007.

Volkow, N. D., & Morales, M. (2015). The brain on drugs: From reward to addiction. *Cell*, *162*(4), 712–725. https://doi.org/10.1016/j.cell.2015.07.046.

Volkow, N. D., Wang, G. J., Fowler, J. S., Tomasi, D., Telang, F., & Baler, R. (2010). Addiction: Decreased reward sensitivity and increased expectation sensitivity conspire to overwhelm the brain's control circuit. *BioEssays*, *32*(9), 748–755 (review) https://doi.org/10.1002/bies.201000042.

Wahlstrom, D., Collins, P., White, T., & Luciana, M. (2010). Developmental changes in dopamine neurotransmission in adolescence: Behavioral implications and issues in assessment. *Brain and Cognition*, *72*(1), 146–159. https://doi.org/10.1016/j.bandc.2009.10.013.

Warren, C. W., Jones, N. R., Peruga, A., Chauvin, J., Baptiste, J. P., Costa de Silva, V., et al. (2008). Global youth tobacco surveillance, 2000-2007. *MMWR Surveillance Summaries*, *57*(1), 1–28. https://www.ncbi.nlm.nih.gov/pubmed/18219269.

Weaver, M. T., Geier, C. F., Levin, M. E., Caggiula, A. R., Sved, A. F., & Donny, E. C. (2012). Adolescent exposure to nicotine results in reinforcement enhancement but does not affect adult responding in rats. *Drug and Alcohol Dependence*, *125*(3), 307–312. https://doi.org/10.1016/j.drugalcdep.2012.03.006.

Weinberger, D. R. (1987). Implications of normal brain development for the pathogenesis of schizophrenia. *Archives of General Psychiatry*, *44*(7), 660–669. https://doi.org/10.1001/archpsyc.1987.01800190080012.

West, R. (2017). Tobacco smoking: Health impact, prevalence, correlates and interventions. *Psychology & Health*, *32*(8), 1018–1036. https://doi.org/10.1080/08870446.2017.1325890.

Wewers, M. E., Dhatt, R. K., Snively, T. A., & Tejwani, G. A. (1999). The effect of chronic administration of nicotine on antinociception, opioid receptor binding and met-enkelphalin levels in rats. *Brain Research*, *822*(1–2), 107–113. https://doi.org/10.1016/s0006-8993(99)01095-1.

Wise, R. A. (1996). Neurobiology of addiction. *Current Opinion in Neurobiology*, *6*(2), 243–251. https://doi.org/10.1016/s0959-4388(96)80079-1.

Woldbye, D. P., Greisen, M. H., Bolwig, T. G., Larsen, P. J., & Mikkelsen, J. D. (1996). Prolonged induction of c-fos in neuropeptide Y- and somatostatin-immunoreactive neurons of the rat dentate gyrus after electroconvulsive stimulation. *Brain Research*, *720*(1–2), 111–119. https://doi.org/10.1016/0006-8993(96)00158-8.

Xu, Z., Seidler, F. J., Ali, S. F., Slikker, W., Jr., & Slotkin, T. A. (2001). Fetal and adolescent nicotine administration: Effects on CNS serotonergic systems. *Brain Research*, *914*(1–2), 166–178. https://doi.org/10.1016/s0006-8993(01)02797-4.

Xue, Y., & Domino, E. F. (2008). Tobacco/nicotine and endogenous brain opioids. *Progress in Neuro-Psychopharmacology & Biological Psychiatry*, *32*(5), 1131–1138. https://doi.org/10.1016/j.pnpbp.2007.12.012.

Yoo, J. H., Lee, S. Y., Loh, H. H., Ho, I. K., & Jang, C. G. (2004). Loss of nicotine-induced behavioral sensitization in micro-opioid receptor knockout mice. *Synapse*, *51*(4), 219–223. https://doi.org/10.1002/syn.10303.

York, J. L., & Hirsch, J. A. (1995). Drinking patterns and health status in smoking and nonsmoking alcoholics. *Alcoholism, Clinical and Experimental Research*, *19*(3), 666–673.

Zadina, J. E., Hackler, L., Ge, L. J., & Kastin, A. J. (1997). A potent and selective endogenous agonist for the mu-opiate receptor. *Nature*, *386*(6624), 499–502. https://doi.org/10.1038/386499a0.

Zangen, A., Ikemoto, S., Zadina, J. E., & Wise, R. A. (2002). Rewarding and psychomotor stimulant effects of endomorphin-1: Anteroposterior differences within the ventral tegmental area and lack of effect in nucleus accumbens. *The Journal of Neuroscience*, *22*(16), 7225–7233. https://doi.org/20026722.

CHAPTER THREE

Cannabis exposure during adolescence: A uniquely sensitive period for neurobiological effects

K.Z. Peters[a,b,*], N.E. Zlebnik[a], and J.F. Cheer[a,c,d,*]
[a]Department of Anatomy and Neurobiology, University of Maryland School of Medicine, Baltimore, MD, United States
[b]Sussex Neuroscience, School of Psychology, University of Sussex, Falmer, United Kingdom
[c]Department of Psychiatry, University of Maryland School of Medicine, Baltimore, MD, United States
[d]Program in Neuroscience, University of Maryland School of Medicine, Baltimore, MD, United States
*Corresponding authors: e-mail address: katezarapeters@gmail.com; k.peters@sussex.ac.uk; jcheer@som.umaryland.edu

Contents

1. Introduction 96
2. Cannabinoids and the eCB system 98
   2.1 Cannabis—A primer 99
3. The endogenous cannabinoid system across the life span: Adolescence as a uniquely vulnerable period 100
4. Interactions between eCBs and dopamine: Effects on late maturing systems 103
   4.1 Maturation of the dopamine system during the adolescent sensitive period 103
   4.2 Cannabinoid modulation of dopaminergic neurotransmission and reward-related behaviors: Indirect mechanisms in adults 105
5. Adolescent THC exposure, effects on subcortical dopamine via alterations in prefrontal GABA circuits 107
   5.1 Development of the prefrontal cortex, importance of GABA changes during adolescence 107
   5.2 Adolescent CB1 receptor agonist exposure effects on prefrontal circuits and behavior 108
6. Conclusions 110
References 110

Abstract

Adolescence is a crucial developmental period where neural circuits are refined and the brain is especially vulnerable to external insults. The endocannabinoid (eCB) system undergoes changes during adolescence which affect the way in which it modulates the development of other systems, in particular dopamine circuits, which show protracted development into adolescence. Given the rise of cannabis use by adolescents and young people, as well as variants containing increasingly higher concentrations of THC, it is now crucial to understand the unique effects of adolescent exposure to

International Review of Neurobiology, Volume 161
ISSN 0074-7742
https://doi.org/10.1016/bs.irn.2021.07.002

cannabis on the developing brain and it might shape future adult vulnerabilities to conditions such as psychosis, schizophrenia, addiction and more. Here we discuss the development of the eCB system across the lifespan, how CB1 receptors modulate dopamine release and potential neurobiological and behavioral effects of adolescent THC exposure on the developing brain such as alterations in excitatory/inhibitory balance during this developmental period.

Abbreviations

2-AG	2-arachidonoylglycerol
AEA	N-arachidonoylethanolamine
CB1R	cannabinoid type 1 receptor
CB2R	cannabinoid type 2 receptor
CBD	cannabidiol
CIN	cholinergic interneuron
CNS	central nervous system
COMT	catechol-O-methyltransferase
DAGL	diacylglycerol lipase
DSE	depolarization-induced suppression of excitation
eCB	endocannabinoid
FAAH	fatty acid amidohydrolase
GABA	gamma aminobutyric acid
GPCR	G protein coupled receptor
MAGL	monoacylglycerol lipase
mPFC	medial prefrontal cortex
NAc	nucleus accumbens
NAPE	N-arachidonoyl phosphatidyl ethanol
PFC	prefrontal cortex
PND	postnatal day
THC	Δ^9- tetrahydrocannabinol
VTA	ventral tegmental area

1. Introduction

Cannabis is one of the most commonly used drugs worldwide, and its recent legalization in the US, Canada, and other countries has resulted in greater access and less perceived risk of harm for many individuals, including adolescents (World Drug Report, 2019). Across North America from 2007 to 2017, past-year cannabis use among people aged 15 and older increased significantly from 42 to 57 million people (World Drug Report, 2019), and among US adolescents (e.g., 12–17 year-olds), as many as 20% reported using cannabis (Miech et al., 2018). Record increases in regular cannabis use among adolescents (Miech et al., 2018; Miech, Patrick, O'Malley, Johnston, & Bachman, 2020) coincide with troubling reports that the transition from casual use to the development of a cannabis use disorder is more rapid among adolescents than young adults (e.g., 18–25 year-olds) (Volkow, Han, Einstein, & Compton, 2021).

A significant concern is that regular cannabis use during adolescence may interfere with crucial developmental processes occurring during this time period. Distributed brain regions, such as the prefrontal cortex (PFC) (Caballero, Granberg, & Tseng, 2016), nucleus accumbens (NAc) (Cuesta et al., 2020), and ventral tegmental area (VTA) (Manitt et al., 2011) and their neurocircuitry are still undergoing maturation to support the emergence of adult cognition and affective processing. Adolescence is characterized by behavioral changes such as increased risk taking, licit and illicit drug experimentation, and rapid cognitive development in humans. In rodents, which are used as models for understanding development and the neurobiological basis of psychiatric disorders, adolescence [postnatal day (PND) 30–50] is accompanied by increased play behaviors, exploration and impulsivity (Spear, 2000). These behavioral changes are related to the development of the aforementioned brain areas and neurotransmitter systems involved in motivation, emotion, and cognition. Exogenous insults such as cannabis use may have especially profound effects during this uniquely sensitive time period.

The endocannabinoid (eCB) system, which mediates the effects of cannabis via its principal psychoactive ingredient Δ^9- tetrahydrocannabinol or THC (see "Cannabis" primer), undergoes significant refinement while also participating in regulation of other neurotransmitter systems, including those within cortico-limbic regions (Zlebnik & Cheer, 2016). Given the role of eCBs in integrating chemical signals from diverse neurotransmitter systems (e.g., GABA, glutamate, dopamine, acetylcholine) with changes in neuronal excitability, activation by exogenous cannabinoids may have widespread effects on adolescent-specific ontogenetic processes across neurosignaling domains. Thus, repeated adolescent cannabis exposure has the potential for significant long-term effects on excitatory and inhibitory circuits across the brain, affecting neurobehavioral outcomes in adulthood. Indeed, a growing body of human and animal studies suggests that adolescent exposure to cannabis is associated with increased risk for various neuropsychiatric disorders later in life, such as schizophrenia/psychosis (Andréasson, Allebeck, Engström, & Rydberg, 1987; Bossong & Niesink, 2010; Laviolette & Grace, 2006; Renard, Krebs, Le Pen, & Jay, 2014; Stefanis et al., 2004), depression

(Degenhardt, Hall, & Lynskey, 2003; Rubino, Zamberletti, & Parolaro, 2012), anxiety (Degenhardt et al., 2013; O'Shea, Singh, McGregor, & Mallet, 2004), and substance abuse (Laviolette & Grace, 2006).

In this review, we outline the development of the eCB system during adolescence and discuss the influence of exogenous cannabinoids, via cannabinoid receptors, on dopamine neurons in the VTA and terminals in the PFC and NAc. The focus will be on dopamine transmission particularly because of its role in multiple psychiatric conditions and findings that dopamine pathways, especially those in the cortex, are still developing in adolescence. We synthesize recent evidence of the unique vulnerabilities of the adolescent brain to exogenous cannabinoids, predominantly THC, found in cannabis. Finally, we consider specific effects of adolescent THC exposure on frontal circuits and how these may have deleterious effects on the downstream, subcortical dopamine system as well as implications for symptoms of psychiatric disorders, such as psychosis and addiction.

2. Cannabinoids and the eCB system

The eCB system consists of several G-protein coupled receptors (GPCRs) and endogenous lipid molecules, termed eCBs, which are ligands of these receptors as well as their biosynthetic and metabolic enzymes. The eCB system was characterized following the isolation of Δ9-tetrahydrocannabinol (THC)—the main psychoactive compound found in cannabis (Gaoni & Mechoulam, 1964). THC was found to bind to two types of GPCRs, the cannabinoid type 1 (CB1R) (Devane, Dysarz, Johnson, Melvin, & Howlett, 1988) and cannabinoid type 2 receptor (CB2R) (Munro, Thomas, & Abu-Shaar, 1993). CB2 receptors are mostly expressed in immune cells, although they have been found in some neurons and glia (Atwood & Mackie, 2010; Zhang et al., 2014), whereas CB1 receptors are found throughout the CNS (Herkenham et al., 1990). The CB1 receptor is the most abundant GPCR in the brain (Mechoulam & Parker, 2013) and is highly expressed in reward and cognitive processing regions of the brain, including the basal ganglia, substantia nigra, and hippocampus (Herkenham et al., 1990). Both classes of these cannabinoid receptors are coupled to inhibitory Gi/o proteins and promote inhibition of adenylyl cyclase as well as several voltage-dependent calcium channels (Howlett et al., 2002). In addition, via several signaling mechanisms, they can activate potassium channels and MAP kinase (Bidaut-Russell, Devane, & Howlett, 1990; Henry & Chavkin, 1995; Hoffman & Lupica, 2000; Twitchell, Brown, & Mackie, 1997). Overall, the result of CB1 or CB2 receptor activation is a reduced probability of transmitter release from neurons expressing these receptors (Howlett et al., 2002).

There are two main endogenous ligands of cannabinoid receptors, the eCBs 2-arachidonoylglycerol (2-AG) (Mechoulam et al., 1995; Sugiura et al., 1995) and N-arachidonoylethanolamine (AEA or anandamide). These have different synthetic and metabolic pathways (Howlett & Mukhopadhyay, 2000). 2-AG is predominantly synthesized from 2-arachidonoyl-containing phospholipids such as diacylglycerol lipase (DAGL), and anandamide is produced from N-arachidonoyl phosphatidyl ethanol (NAPE) (Cravatt et al., 1996; Lu & Mackie, 2016). They are primarily metabolized by two enzymes: monoacylglycerol lipase (MAGL) in the case of 2-AG and fatty acid amidohydrolase (FAAH) for anandamide (Cravatt et al., 1996; Lu & Mackie, 2016).

Unlike classical neurotransmitters, eCBs are not stored in vesicles; instead they are synthesized and released *de novo* in times of sustained neuronal activity (Freund, Katona, & Piomelli, 2003). Once synthesized, eCBs traverse the plasma membrane and retrogradely activate cannabinoid CB1 receptors on presynaptic terminals (Wilson & Nicoll, 2001). ECBs function as retrograde messengers to filter and select afferent inputs via this negative feedback mechanism, depending on their level of activation, and this may be especially important in regulating synaptic plasticity (Alger, 2002).

2.1 Cannabis—A primer

BOX 1

Hundreds of phytocannabinoids have been isolated from the cannabis plant, and the two major active constituents identified are THC (Gaoni & Mechoulam, 1964) and cannabidiol (CBD) (Mechoulam & Shvo, 1963). THC is the prominent psychoactive phytocannabinoid and mediates the rewarding properties of cannabis (Huestis et al., 2001); CBD, in contrast, does not have reinforcing effects and thus has low abuse potential (Katsidoni, Anagnostou, & Panagis, 2013; Parker, Burton, Sorge, Yakiwchuk, & Mechoulam, 2004). Despite these differences in psychoactivity, THC and CBD are produced by co-dominant alleles of the same gene locus and segregate according to basic Mendelian inheritance, yielding

Continued

BOX 1—CONT'D

three cannabis chemovariants in a 1:2:1 ratio of pure THC, mixed THC/CBD, and pure CBD (de Meijer et al., 2003). Chemovariants with approximate 1:1 ratio of THC to CBD often reduce the unwanted central actions of THC (Russo & Guy, 2006) and temper abuse liability (Schoedel et al., 2011). In contrast, chemovariants containing predominantly THC are popular for recreational purposes, and these strains have been cultivated to maximize THC content, with a twofold increase over the last 20 years (Chandra et al., 2019; ElSohly et al., 2000). For example, in Europe from 2006 to 2016, THC content of cannabis flower increased from 5% to 10%, and THC content of cannabis resin increased from 8% to 17% (Freeman et al., 2019). Along with widespread legalization of cannabis in the US and Canada, several new high potency routes of administration have become accessible, including "dabbing" (Raber, Elzinga, & Kaplan, 2015; Stogner & Miller, 2015) and vaping (Ben Taleb et al., 2020; Budney, Sargent, & Lee, 2015; Knapp et al., 2019; Kowitt et al., 2019) of THC wax or oil. These concentrated products boast nearly 70% THC content (Raber et al., 2015; Stogner & Miller, 2015; World Drug Report, 2019). Such potent cannabis and cannabis-derived products may predispose individuals to cannabis use disorder as well as associated psychiatric co-morbidity (Lowe, Sasiadek, Coles, & George, 2019), and more research is required to characterize the long-term consequences of these products on fundamental neurodevelopmental processes.

3. The endogenous cannabinoid system across the life span: Adolescence as a uniquely vulnerable period

During adolescence, the brain undergoes synaptic over-expression, subsequent pruning, and other processes of refinement, particularly in areas that are late to mature, such as targets of the dopamine system like the PFC, NAc, and other limbic structures. CB1 receptors, along with anandamide and 2-AG, are already expressed during early growth of the central nervous system (CNS) and are involved in nearly all phases of development: progenitor differentiation, neuronal migration and proliferation, axonal guidance, fasciculation, etc. (Berghuis et al., 2007; Harkany et al., 2007; Harkany, Keimpema, Barabás, & Mulder, 2008; Maccarrone, Guzmán, Mackie, Doherty, & Harkany, 2014). By adulthood, the CB1 receptor will be the most highly expressed GPCR in the brain, enriched in brain regions important for cognition and reward processing (Peters, Oleson, & Cheer, 2021; Zlebnik & Cheer, 2016). However, throughout adolescence, the eCB system is itself undergoing transformation. In humans (Mato, Del Olmo, & Pazos, 2003), nonhuman primates (Eggan, Mizoguchi, Stoyak, & Lewis, 2010), and rodents (Dalton & Zavitsanou, 2010; Ellgren et al., 2008; Rodríguez de Fonseca, Ramos, Bonnin, & Fernández-Ruiz, 1993; Van Waes, Beverley, Siman, Tseng, & Steiner, 2012; Viveros et al., 2012), CB1 receptor expression increases significantly from the juvenile period to adulthood (Fig. 1A), achieving its highest levels early in adolescence (Berrendero et al., 1999; Van Waes et al., 2012), before diminishing somewhat and remaining stable throughout adulthood (Rodríguez de Fonseca et al., 1993). Anandamide and 2-AG levels also fluctuate during adolescent development. Anandamide levels gradually rise and then persist at stable levels into adulthood, while 2-AG levels sharply decline during early to mid-adolescence before increasing again prior to the onset of adulthood (Berrendero et al., 1999; Ellgren et al., 2008; Lee, Hill, Hillard, & Gorzalka, 2013).

Spatially- and temporally-specific patterns of CB1 receptor and eCB expression over adolescence are linked to functional maturation of distinct brain regions (Dalton & Zavitsanou, 2010; Eggan et al., 2010; Ellgren et al., 2008; Mato et al., 2003; Rodríguez de Fonseca et al., 1993). The mesocorticolimbic dopamine system is comprised of dopamine cells originating in the VTA of the midbrain that project rostrally to the NAc in the ventral forebrain and to the PFC (Wise & Bozarth, 1985). Throughout adulthood, CB1 receptor density is sparse in the VTA, and much of the CB1 receptor expression is limited to GABAergic and glutamatergic afferents (Herkenham et al., 1991; Mailleux & Vanderhaeghen, 1992; Tsou, Brown, Sañudo-Peña, Mackie, & Walker, 1998). Similar to brain-wide patterns of expression, CB1 receptor density increases in the ventral mesencephalon until PND 40 and then marginally decreases into adulthood at PND 70 (Rodríguez de Fonseca et al., 1993). In corticostriatal pathways, distinct inverse patterns of expression differentiate cortical vs striatal regions (Van Waes et al., 2012). In sensorimotor circuits, cortical regions (e.g., primary motor, somatosensory) have lower CB1 receptor expression and send projections to striatal regions (e.g., dorsolateral striatum) with higher expression; however, in limbic circuits, cortical regions (e.g., PFC) have higher CB1 receptor expression and project to striatal regions (e.g., NAc) with lower expression. Within the PFC, CB1 receptor expression decreases from early or mid-adolescence until adulthood (Ellgren et al., 2008; Heng, Beverley, Steiner, & Tseng, 2011; Van Waes et al., 2012), with the greatest decreases observed in the cingulate, prelimbic, and infralimbic regions (Ellgren et al., 2008; Heng et al., 2011). Similarly, dynamic changes in expression occur throughout

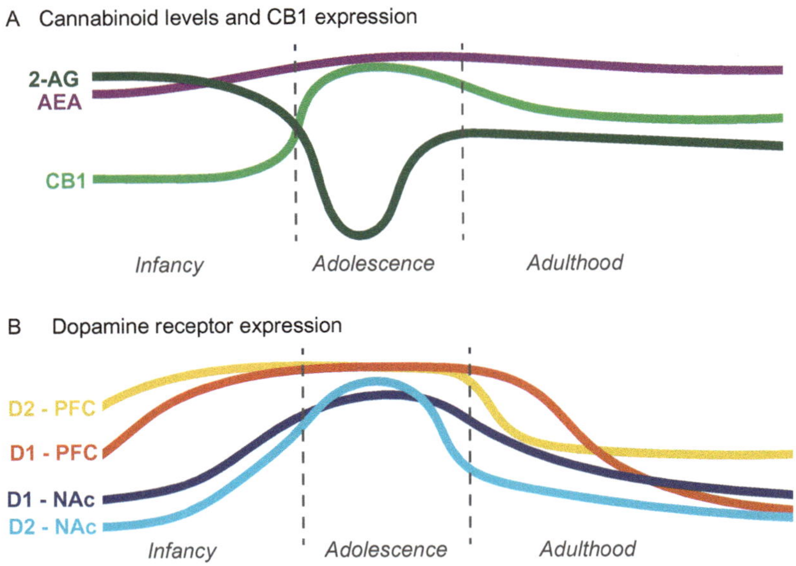

Fig. 1 Relative expression levels of receptors and ligands in the eCB and dopamine systems. Top panel (A): Schematic representation of the developmental time course of expression of cannabinoid receptors (CB1 light green) and the eCBs 2-arachidonoylglycerol (2-AG dark green) and anandamide (AEA purple). There is a rapid increase in CB1 receptor expression during adolescence, reaching maximal density early in adolescence (Berrendero, Sepe, Ramos, Di Marzo, & Fernández-Ruiz, 1999). Expression later decreases to sustained, adult levels (Rodríguez de Fonseca et al., 1993). ECBs show different developmental patterns of expression. AEA expression shows a slow but sustained increase from infancy through adolescence reaching maximal levels which are mostly maintained in adulthood, conversely, 2-AG shows a different pattern with a decrease in expression during adolescence before a later stabilization in adulthood (Bara, Ferland, Rompala, Szutorisz, & Hurd, 2021; Ellgren et al., 2008). Bottom panel (B): Schematic representation of relative expression of dopamine receptors in rodents across time. Dopamine D1 and D2 receptors are shown relative to their maximal expression levels, in the prefrontal cortex (PFC: D1 orange, D2 yellow) and nucleus accumbens (NAc: D1 dark blue, D2 light blue). Within the PFC, D1 and D2 receptors steadily increase during infancy. During adolescence, vast increases in dopamine innervation and dopamine concentration are accompanied a peak in both D1 and D2 receptor expression (Andersen, Thompson, Rutstein, Hostetter, & Teicher, 2000; Larsen & Luna, 2018) which later decreases to adult levels. Within subcortical regions such as the NAc, D1 and D2 receptors show a more rapid change from relatively low levels in childhood to maximal density during adolescence (Andersen et al., 2000; Reichelt, 2016). Dopamine cell firing and dopaminergic innervation is highest during adolescence (McCutcheon & Marinelli, 2009; Naneix, Marchand, Di Scala, Pape, & Coutureau, 2012).

adolescence within the NAc: increases in CB1 receptor density in the NAc shell (Ellgren et al., 2008) are accompanied by decreases in the NAc core (Ellgren et al., 2008; Van Waes et al., 2012).

Across the PFC and NAc, there are also attendant developmental changes in anandamide and 2-AG levels. Anandamide increases within the PFC from early to late adolescence (Ellgren et al., 2008). However, within the NAc, anandamide levels peak during mid-adolescence before declining somewhat during late adolescence and into adulthood. In contrast, 2-AG levels decrease from early to late adolescence in the PFC and NAc. Importantly, significant physiological alterations in synaptic function accompany time-dependent changes in CB1 receptor expression during the adolescent period. Depolarization-induced suppression of excitation or DSE is a well-established electrophysiological response to probe eCB regulation of synaptic plasticity (Kreitzer & Regehr, 2002; Lovinger, Davis, & Costa, 2010). In the adult PFC, CB1 receptor-dependent DSE is markedly diminished compared to that measured in the adolescent PFC (Heng et al., 2011). This indicates that functional changes in synaptic plasticity are concomitant with ontogenetic changes in CB1 receptor expression and may have significant impact on circuit dynamics within brain regions still undergoing remodeling and refinement during adolescence.

4. Interactions between eCBs and dopamine: Effects on late maturing systems

4.1 Maturation of the dopamine system during the adolescent sensitive period

Dopamine has critical roles in learning, reward prediction error and decision making. It is also fundamentally important in responding to drugs of abuse and cues which predict their availability (Ahmed, 2018; Berke, 2018; Berridge, 2012; Dayan, 2009; Schultz, 2016; Watabe-Uchida, Eshel, & Uchida, 2017). Alterations in dopamine signaling are implicated in multiple psychiatric conditions including psychosis, schizophrenia, and addiction (Grace & Gomes, 2019; Solinas, Belujon, Fernagut, Jaber, & Thiriet, 2019; Volkow, Wise, & Baler, 2017; Wise & Robble, 2020); which have a peri-adolescent onset in humans (Paus, Keshavan, & Giedd, 2008). The dopamine system is late to mature compared to other neurotransmitters, and its development continues throughout adolescence in both humans and rodents (Carta, Carlsson, Munoz, Kirik, & Bjorklund, 2008; Dwyer & Leslie, 2016; Hoops & Flores, 2017). This protracted development leaves dopamine circuits vulnerable to exogenous insults and may represent a uniquely sensitive period for the effects of drugs such as cannabis on the adolescent brain.

During adolescence, there is a steady increase in dopamine tissue content in cortical and subcortical regions, including vast increases in COMT and dopaminergic inputs to the PFC in particular (Bossong & Niesink, 2010; Spear, 2000; Tunbridge et al., 2007). This is accompanied by an over-expression of D1 and D2 receptors, which is particularly marked in the PFC. During late adolescence, when dopamine neurons exhibit their highest levels of spontaneous firing (McCutcheon & Marinelli, 2009), there is substantial pruning (Juraska & Markham, 2004) and downregulation of D1 and D2 receptor expression which then stabilizes to adult levels (Fig. 1B). Cortical dopamine, in particular the mesocortical pathway, terminating in the PFC, is considerably slower to mature than subcortical regions, such as the mesolimbic pathway, terminating in the NAc. Dopamine fiber density in the PFC shows a continued increase through adolescence, whereas these fibers are mostly mature in subcortical structures of the mesolimbic and nigrostriatal pathways. It is the PFC, and especially the anterior portion of the medial PFC, which are the latest to develop (Naneix et al., 2012; Voorn, Kalsbeek, Jorritsma-Byham, & Groenewegen, 1988). Axonal guidance systems, such as Dcc-Netrin-1 guidance cue system, are important mediators of final maturation of the mesocortical dopamine system during this time (Manitt et al., 2011).

In parallel to the late development of dopamine afferents to the PFC is the emergence of complex decision making. The adult ability to select and execute actions in a goal-directed manner, based on predicted consequences, is associated with this late maturation (Naneix et al., 2012). Immature "causal reasoning" is a frontal-dependent behavior and a characteristic of adolescence in humans. For example, adolescents often display reckless behavior, typified by increased risk-taking behavior, lack of inhibition, heavy drinking and drug taking, as well as impulsivity—all associated with immature cortical circuits (Laviola, Macrì, Morley-Fletcher, & Adriani, 2003; Spear, 2000). This has been demonstrated in rats using goal-directed behavioral tasks such as contingency degradation. Rats trained to lever press for two kinds of rewards (pellets and liquid sucrose) learn to associate the action of pressing with reward delivery. Then, if one of these rewards is delivered non-contingently without the need for lever pressing, adult rats correctly identify that the reward occurs without their action and reduce responding. Adolescents conversely continue to press the lever, not appearing to learn that the action-outcome contingency has been degraded and fail to adapt responding when action and outcomes are no longer causally related (Naneix et al., 2012). Importantly, adolescents do not show deficits in Pavlovian instrumental transfer task, which are subserved by more mature regions such as the NAc (Corbit, Muir, & Balleine, 2001) and amygdala (Balleine & Killcross, 2006). These data suggest the slow maturing PFC dopamine system is responsible for adolescent deficits in causal reasoning (Naneix et al., 2012). In Section 4.2 we discuss what is currently known about cannabinoid interactions with the adult dopamine system, and in Sections 5.1 and 5.2 we then consider specific effects of *adolescent* exposure to cannabinoid agonists on the developing brain.

4.2 Cannabinoid modulation of dopaminergic neurotransmission and reward-related behaviors: Indirect mechanisms in adults

There is evidence in rodents that CB1R agonism or antagonism can have effects on the dopamine system (Covey, Bunner, Schuweiler, Cheer, & Garris, 2016; Oleson et al., 2012; Wang, Treadway, Covey, Cheer, & Lupica, 2015). However, these effects occur through indirect mechanisms, including reduced activity in GABA and glutamate neurons, as dopamine cells do not express CB1Rs. Dopamine neurons contain the molecular machinery needed to synthesize and release eCBs, performing *de novo* synthesis in times of heightened neuronal activity (Peters, Oleson, & Cheer, 2021).

Systemic administration of CB1R agonists (such as WIN55,212-2 or THC) increase tonic and phasic dopamine release in terminal regions such as the NAc and PFC (Cheer, 2004; Gessa, Melis, Muntoni, & Diana, 1998; Peters, Oleson, & Cheer, 2021). Conversely, blockade of the CB1 receptor and disrupting eCB signaling using antagonists or inverse agonist, such as rimonabant, has the opposite effect leading to reduced dopamine release (Cheer, 2004). In particular, dopamine burst activity and the related rapid increase of transmitter release in response to drugs of abuse as well as cues which predict their availability, are markedly reduced in the presence of CB1 receptor antagonists (Cheer et al., 2007; Oleson et al., 2012). Furthermore, this reduced dopamine is accompanied by decreased reward seeking in several paradigms (Cheer et al., 2007; Oleson et al., 2012). The endogenous cannabinoid system plays a role in reward-related behaviors and modulates dopamine transmission involved in these responses.

ECBs and exogenously applied cannabinoids such as THC affect dopamine transmission via a number of indirect mechanisms (Fig. 2). One proposed mechanism is disinhibition via actions at GABA neurons in the VTA

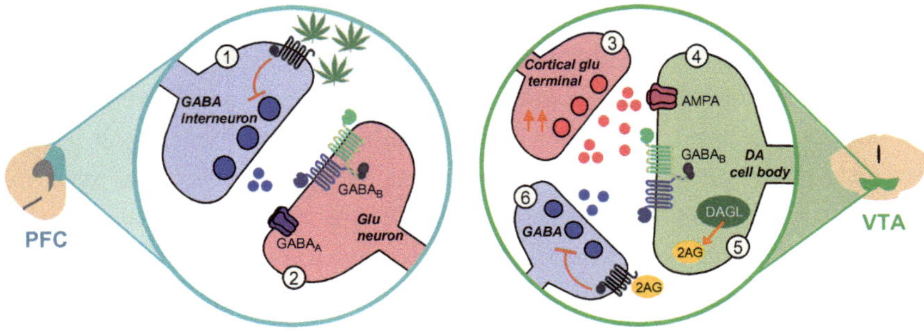

Fig. 2 Adolescent THC exposure effects on cortical GABA neurons and downstream consequences for the dopamine system (1) Exposure to THC during adolescence may preferentially effect GABA interneurons in the prefrontal cortex (PFC), agonism of CB1 receptors in GABA interneurons has the effect of reducing the probability of GABA release due to the Gi/o protein coupling of CB1 receptors (2) a consequent disinhibition of cortical pyramidal neurons which these GABA cells impinge upon occurs, in addition reduced GABA function in the PFC may trigger other changes in network activity such as enhanced gamma oscillatory activity (Renard et al., 2017) (3) excitatory cortical output is increased as a result and more glutamate is released from terminals in subcortical areas such as the ventral tegmental area (VTA) (4) dopamine cell bodies in the VTA receive increased glutamatergic drive and increase firing (5) increased activity in dopamine neurons has the parallel effect of engaging eCB synthesis and release (6) in addition to the effects of increased cortical drive, impinging GABA neurons in the VTA are further disinhibited by mobilized eCBs such as 2-AG. The net result of these proposed mechanisms is a dramatic increase in dopamine output and a hyperdopaminergic state in subcortical dopamine neurons which may manifest as cognitive deficits typical of schizophrenia (Renard, Rushlow, & Laviolette, 2018).

(Peters, Oleson, & Cheer, 2021). Here, at the cell body level, under standard conditions dopamine neurons receive considerable inhibitory GABA inputs (Paladini & Tepper, 2016). When such GABA inhibition is removed or attenuated, there is a resulting increased dopamine release within terminals in the NAc and this has been shown to promote reward-seeking (Nieh et al., 2016; Stamatakis et al., 2013). ECB action at presynaptic CB1 receptors suppresses GABA-mediated inhibition (Wilson & Nicoll, 2001), and systemic CB1 receptor antagonists reduce NAc dopamine release in response to various types of reinforcers and drugs of abuse in rodents (Covey et al., 2016; Oleson et al., 2012; Wang et al., 2015). Adolescent THC exposure, in particular, leads to long term changes in these mechanisms, including sensitization of VTA cell firing and increases in excitability (Renard et al., 2017). In addition, adolescent exposure can reduce expression of CB1 receptors on glutamatergic inputs to the VTA which normally play a role in modulating excitation and dopamine output (Bara et al., 2021; Kruse, Cao, Viray, Stella, & Clark, 2019; Peters, Cheer, & Tonini, 2021).

There are also several powerful modulators of dopamine release that occur at the terminal level, which influence dopamine output independent of cell body firing (Cachope et al., 2012; Nolan et al., 2020; Threlfell et al., 2012). One example is cholinergic interneurons (CIN) driven release within the NAc. CIN-driven dopamine release can be attenuated by CB1 receptor agonists or by increasing tissue levels of the eCB 2-AG pharmacologically—effects that are prevented by CB1 receptor antagonists (Mateo et al., 2017). However, CINs—like dopamine neurons—do not express CB1 receptors. The observed effects are hypothesized to result from cortical inputs impinging upon CINs. Cortical glutamate afferents synapse onto CINs in the NAc and increase CIN activity (Kosillo, Zhang, Threlfell, & Cragg, 2016). This in turn drives dopamine release via nicotinic acetylcholine receptors (Kosillo et al., 2016; Peters, Oleson, & Cheer, 2021; Threlfell et al., 2012; Threlfell & Cragg, 2011) CB1 receptors located on these cortical *glutamate* inputs may modulate CIN-driven release by reducing excitatory drive onto CINs (Peters, Oleson, & Cheer, 2021). Importantly, the actions of CB1 receptor agonist depend upon which neuronal populations receptors are expressed in and when they are activated. The exact mechanisms of eCB modulation of dopamine and other transmitters is still not fully characterized. However, it is highly likely that manipulation of the eCB system in adolescence will have an enhanced effect on these mechanisms due to the high levels of CB1R expression during adolescence and the uniquely developing state of multiple neural systems which may be altered substantially by excessive stimulation of the CB1 receptor.

5. Adolescent THC exposure, effects on subcortical dopamine via alterations in prefrontal GABA circuits

5.1 Development of the prefrontal cortex, importance of GABA changes during adolescence

The prefrontal cortex undergoes substantial refinement and remodeling during adolescence (Bossong & Niesink, 2010; Chambers, Taylor, & Potenza, 2003; Gogtay et al., 2004) with extensive pruning of glutamatergic excitatory synapses, decreased excitatory input (Spear, 2000) and profound anatomical and functional reorganization of local GABA circuits (Tseng & O'Donnell, 2006). Changes to local GABA interneurons in the PFC are especially important for the maturation of the PFC network (Tseng, Chambers, & Lipska, 2009) and the maturation of cognitive abilities, such as working memory, decision making, and impulse control (O'Donnell, 2011; Tseng et al., 2009; Uhlhaas & Singer, 2011). Within GABA interneurons, there are a number of structural and functional changes observed. For example, there is a 50% increase in parvalbumin expression in GABA interneurons across adolescence and a parallel decrease in calretinin expression (Caballero, Flores-Barrera, Cass, & Tseng, 2014). This shift towards PV expression represents functional changes in fast-spiking interneurons, which exert critical inhibitory influences on cortical output (Bartos & Elgueta, 2012) and are essential for fine-tuning pyramidal cell firing (Rao, Williams, & Goldman-Rakic, 2000). Altered or disrupted prefrontal development during the vulnerable period of adolescence can precipitate the onset of several psychiatric disorders including schizophrenia, drug abuse, and other mood disorders (Chambers et al., 2003; Paus et al., 2008). Changes to GABAergic circuits within the PFC may be especially relevant to the adolescent onset of cognitive deficits seen in schizophrenia (Hoftman & Lewis, 2011; Uhlhaas & Singer, 2011). It is crucially important to uncover how exogenous insults such as adolescent cannabis exposure may alter such developing neurocircuits, as well as possible downstream effects.

5.2 Adolescent CB1 receptor agonist exposure effects on prefrontal circuits and behavior

CB1 receptors play an essential role in adolescent maturation processes within the PFC. The critical role of CB1 receptors in synaptic plasticity is well known (Bara et al., 2021; Zlebnik & Cheer, 2016) and their expression and activation in the late developing PFC regulates excitatory/inhibitory balance (Cass et al., 2014; Ellgren et al., 2008; Heng et al., 2011) as well as normal eCB-mediated long-term depression in layer 2/3 of the medial PFC (mPFC) specifically (Bara et al., 2021; Cuccurazzu et al., 2018). Cannabis use in younger people is linked to later psychiatric diseases. In particular, cannabis use in adolescence, especially before the age of 17, increases the risk of schizophrenia and psychosis (Andréasson et al., 1987; Renard et al., 2014; Stefanis et al., 2004), bipolar disorder, and addiction (Lagerberg et al., 2011; Lucatch, Coles, Hill, & George, 2018). In mice and rats, chronic CB1 receptor agonist exposure during adolescence leads to a myriad of cognitive changes, many of which rely on the PFC. Such deficits include long-lasting increases in anxiety in adulthood (O'Shea et al., 2004), alterations to cortical network activity which is essential for normal cognitive functioning, short- and long-term spatial working memory deficits (Renard, Krebs, Jay, & Le Pen, 2013), impaired sensorimotor gating, and reduced motivation/anhedonia (Schneider & Koch, 2003) as well as altered social behavior, social play and self-grooming (Schneider, Schömig, & Leweke, 2008); many of these general impairments in information processing are characteristic of cognitive symptoms seen in psychoses and schizophrenia, in both humans and animals models (Gold & Harvey, 1993; O'Carroll, 2000). Importantly, these impairments are age-specific, with no such effects seen when cannabinoid agonists are administered in adulthood, suggesting CB1 receptor stimulation in this sensitive time-window may perturb the development of normal cognitive functions.

The mechanisms by which CB1 receptor agonists lead to such adverse effects, specifically when given during adolescence, are not fully understood. However, there is emerging evidence that GABA transmission in the PFC may be a candidate for the primary site of disruption (Cass et al., 2014; Renard et al., 2018) (Fig. 2). CB1 receptors are expressed on cortical GABA neurons, which target the cell bodies of pyramidal neurons (Lewis & González-Burgos, 2008). Activation of these receptors by eCBs or by exogenous compounds, such as THC in cannabis, suppresses GABA release through the inhibitory actions of the CB1R, resulting in a decrease in overall inhibition within the PFC (Katona et al., 1999). Reduced cortical inhibition during adolescent development when these circuits are still being formed may have profound consequences, such as increasing glutamate and dopamine release downstream (Pistis et al., 2002). An overstimulation of CB1 receptors by repeated, chronic cannabis use may lead to prolonged imbalance in PFC GABAergic function. Hyperexcitability in the cortex may trigger downstream consequences, particularly dopaminergic adaptations in subcortical areas, namely a hyperdopaminergic state in the mesolimbic system. This could manifest as cognitive disruptions seen in schizophrenia, such as altered pre-pulse inhibition (i.e., sensorimotor gating) (Enomoto, Tse, & Floresco, 2011; Nakazawa et al., 2012; Yee et al., 2005). Cortical dysfunction, subcortical hyperdopaminergic function, and the behavioral outcomes discussed here are cardinal features of schizophrenia and psychoses in humans and animal models (Carlsson, Hansson, Waters, & Carlsson, 1997; O'Carroll, 2000). In addition, disrupted GABAergic modulation of subcortical dopamine release has been recently reported in animal models of schizophrenia (Ferdinand, Peters, Yavas, & Young, 2021) and there is evidence that activating $GABA_A$ receptors in the PFC, in particular, can restore behavioral and dopaminergic abnormalities observed following adolescent THC exposure (Renard et al., 2017). GABAergic functionality is unaffected when cannabinoid receptor agonists are administered in adulthood (Cass et al., 2014), suggesting a specific effect of adolescent exposure. There is now emerging evidence in support of prefrontal GABA disruption as a key mechanism for some of the reported deleterious effects of adolescent cannabis exposure. Further studies to characterize the long term neurobiological and behavioral effects of adolescent cannabis exposure will be crucial in our understanding of the unique effects this drug can have when used in young people.

6. Conclusions

Adolescence is a dynamic and sensitive period of development, where multiple brain structures and neural systems are still maturing. Heightened CB1R expression during this time allows for vast influence of the eCB system throughout the brain. There are critical roles for eCBs in the development of long-range targets of the mesocorticolimbic dopamine pathway, including establishing appropriate excitatory/inhibitory balance in the PFC, which may set the trajectory for adult cognitive development. Any exogenous insult that results in supraphysiological stimulation of CB1 receptors—including regular cannabis use—may have enduring harmful effects and alter the course of the developing adolescent brain in ways we still do not fully understand. Therefore, escalation in the prevalence of adolescent cannabis use and the emergence of high potency cannabis products poses a significant risk for adverse developmental effects. In particular, adolescence is a time when cognitive and reward-related psychiatric disorders begin to emerge (Vosberg, Leyton, & Flores, 2020) and the impact of adolescent THC exposure on PFC function are linked to an elevated risk of psychosis and schizophrenia-like symptoms (Renard et al., 2017; Rubino et al., 2012). These findings highlight the critical importance of continuing research to characterize the precise cellular, molecular, and behavioral mechanisms related to adolescent cannabis exposure and associated neuropsychiatric risk.

References

Ahmed, S. H. (2018). Trying to make sense of rodents' drug choice behavior. *Progress in Neuro-Psychopharmacology & Biological Psychiatry*, *87*(Part A), 3–10. https://doi.org/10.1016/j.pnpbp.2017.09.027.

Alger, B. E. (2002). Retrograde signaling in the regulation of synaptic transmission: Focus on endocannabinoids. *Progress in Neurobiology*, *68*(4), 247–286. https://doi.org/10.1016/s0301-0082(02)00080-1.

Andersen, S. L., Thompson, A. T., Rutstein, M., Hostetter, J. C., & Teicher, M. H. (2000). Dopamine receptor pruning in prefrontal cortex during the periadolescent period in rats. *Synapse*, *37*(2), 167–169. https://doi.org/10.1002/1098-2396(200008).

Andréasson, S., Allebeck, P., Engström, A., & Rydberg, U. (1987). Cannabis and schizophrenia. A longitudinal study of Swedish conscripts. *Lancet, 2*(8574), 1483–1486. https://doi.org/10.1016/s0140-6736(87)92620-1.

Atwood, B. K., & Mackie, K. (2010). CB2: A cannabinoid receptor with an identity crisis. *British Journal of Pharmacology, 160*(3), 467–479. https://doi.org/10.1111/j.1476-5381.2010.00729.x.

Balleine, B. W., & Killcross, S. (2006). Parallel incentive processing: An integrated view of amygdala function. *Trends in Neurosciences, 29*(5), 272–279. https://doi.org/10.1016/j.tins.2006.03.002.

Bara, A., Ferland, J.-M. N., Rompala, G., Szutorisz, H., & Hurd, Y. L. (2021). Cannabis and synaptic reprogramming of the developing brain. *Nature Reviews Neuroscience, 22*, 423–438. https://doi.org/10.1038/s41583-021-00465-5.

Bartos, M., & Elgueta, C. (2012). Functional characteristics of parvalbumin- and cholecystokinin-expressing basket cells. *The Journal of Physiology, 590*(4), 669–681. https://doi.org/10.1113/jphysiol.2011.226175.

Ben Taleb, Z., Kalan, M. E., Bahelah, R., Boateng, G. O., Rahman, M., & Alshbool, F. Z. (2020). Vaping while high: Factors associated with vaping marijuana among youth in the United States. *Drug and Alcohol Dependence, 217*, 108290. https://doi.org/10.1016/j.drugalcdep.2020.108290.

Berghuis, P., Rajnicek, A. M., Morozov, Y. M., Ross, R. A., Mulder, J., Urbán, G. M., et al. (2007). Hardwiring the brain: Endocannabinoids shape neuronal connectivity. *Science, 316*(5828), 1212–1216. https://doi.org/10.1126/science.1137406.

Berke, J. D. (2018). What does dopamine mean? *Nature Neuroscience, 21*(6), 787–793. https://doi.org/10.1038/s41593-018-0152-y.

Berrendero, F., Sepe, N., Ramos, J. A., Di Marzo, V., & Fernández-Ruiz, J. J. (1999). Analysis of cannabinoid receptor binding and mRNA expression and endogenous cannabinoid contents in the developing rat brain during late gestation and early postnatal period. *Synapse, 33*(3), 181–191. https://doi.org/10.1002/(SICI)1098-2396(19990901).

Berridge, K. C. (2012). From prediction error to incentive salience: Mesolimbic computation of reward motivation. *European Journal of Neuroscience, 35*(7), 1124–1143. https://doi.org/10.1111/j.1460-9568.2012.07990.x.

Bidaut-Russell, M., Devane, W. A., & Howlett, A. C. (1990). Cannabinoid receptors and modulation of cyclic AMP accumulation in the rat brain. *Journal of Neurochemistry, 55*(1), 21–26. https://doi.org/10.1111/j.1471-4159.1990.tb08815.x.

Bossong, M. G., & Niesink, R. J. M. (2010). Adolescent brain maturation, the endogenous cannabinoid system and the neurobiology of cannabis-induced schizophrenia. *Progress in Neurobiology, 92*(3), 370–385. https://doi.org/10.1016/j.pneurobio.2010.06.010.

Budney, A. J., Sargent, J. D., & Lee, D. C. (2015). Vaping cannabis (marijuana): Parallel concerns to e-cigs? *Addiction, 110*(11), 1699–1704. https://doi.org/10.1111/add.13036.

Caballero, A., Flores-Barrera, E., Cass, D. K., & Tseng, K. Y. (2014). Differential regulation of parvalbumin and calretinin interneurons in the prefrontal cortex during adolescence. *Brain Structure & Function, 219*(1). https://doi.org/10.1007/s00429-013-0508-8.

Caballero, A., Granberg, R., & Tseng, K. Y. (2016). Mechanisms contributing to prefrontal cortex maturation during adolescence. *Neuroscience and Biobehavioral Reviews, 70*, 4–12. https://doi.org/10.1016/j.neubiorev.2016.05.013.

Cachope, R., Mateo, Y., Mathur, B. N., Irving, J., Wang, H.-L., Morales, M., et al. (2012). Selective activation of cholinergic interneurons enhances accumbal phasic dopamine release: Setting the tone for reward processing. *Cell Reports, 2*(1), 33–41. https://doi.org/10.1016/j.celrep.2012.05.011.

Carlsson, A., Hansson, L. O., Waters, N., & Carlsson, M. L. (1997). Neurotransmitter aberrations in schizophrenia: New perspectives and therapeutic implications. *Life Sciences, 61*(2), 75–94. https://doi.org/10.1016/s0024-3205(97)00228-2.

Carta, M., Carlsson, T., Munoz, A., Kirik, D., & Bjorklund, A. (2008). Serotonin–dopamine interaction in the induction and maintenance of L-DOPA-induced dyskinesias. In *Vol. 172. Progress in brain research* (pp. 465–478). Elsevier. https://doi.org/10.1016/S0079-6123(08)00922-9.

Cass, D. K., Flores-Barrera, E., Thomases, D. R., Vital, W. F., Caballero, A., & Tseng, K. Y. (2014). CB1 cannabinoid receptor stimulation during adolescence impairs the maturation of GABA function in the adult rat prefrontal cortex. *Molecular Psychiatry*, *19*(5), 536–543. https://doi.org/10.1038/mp.2014.14.

Chambers, R. A., Taylor, J. R., & Potenza, M. N. (2003). Developmental neurocircuitry of motivation in adolescence: A critical period of addiction vulnerability. *American Journal of Psychiatry*, *160*(6), 1041–1052.

Chandra, S., Radwan, M. M., Majumdar, C. G., Church, J. C., Freeman, T. P., & ElSohly, M. A. (2019). New trends in cannabis potency in USA and Europe during the last decade (2008–2017). *European Archives of Psychiatry and Clinical Neuroscience*, *269*(1), 5–15. https://doi.org/10.1007/s00406-019-00983-5.

Cheer, J. F. (2004). Cannabinoids enhance subsecond dopamine release in the nucleus accumbens of awake rats. *Journal of Neuroscience*, *24*(18), 4393–4400. https://doi.org/10.1523/JNEUROSCI.0529-04.2004.

Cheer, J. F., Aragona, B. J., Heien, M. L. A. V., Seipel, A. T., Carelli, R. M., & Wightman, R. M. (2007). Coordinated accumbal dopamine release and neural activity drive goal-directed behavior. *Neuron*, *54*(2), 237–244. https://doi.org/10.1016/j.neuron.2007.03.021.

Corbit, L. H., Muir, J. L., & Balleine, B. W. (2001). The role of the nucleus accumbens in instrumental conditioning: Evidence of a functional dissociation between accumbens core and shell. *Journal of Neuroscience*, *21*(9), 3251–3260.

Covey, D. P., Bunner, K. D., Schuweiler, D. R., Cheer, J. F., & Garris, P. A. (2016). Amphetamine elevates nucleus accumbens dopamine via an action potential-dependent mechanism that is modulated by endocannabinoids. *European Journal of Neuroscience*, *43*(12), 1661–1673. https://doi.org/10.1111/ejn.13248.

Cravatt, B. F., Giang, D. K., Mayfield, S. P., Boger, D. L., Lerner, R. A., & Gilula, N. B. (1996). Molecular characterization of an enzyme that degrades neuromodulatory fatty-acid amides. *Nature*, *384*(6604), 83–87. https://doi.org/10.1038/384083a0.

Cuccurazzu, B., Zamberletti, E., Nazzaro, C., Prini, P., Trusel, M., Grilli, M., et al. (2018). Adult cellular neuroadaptations induced by adolescent THC exposure in female rats are rescued by enhancing anandamide signaling. *International Journal of Neuropsychopharmacology*, *21*(11), 1014–1024. https://doi.org/10.1093/ijnp/pyy057.

Cuesta, S., Nouel, D., Reynolds, L. M., Morgunova, A., Torres-Berrío, A., White, A., et al. (2020). Dopamine axon targeting in the nucleus accumbens in adolescence requires netrin-1. *Frontiers in Cell and Developmental Biology*, *8*, 487. https://doi.org/10.3389/fcell.2020.00487.

Dalton, V. S., & Zavitsanou, K. (2010). Cannabinoid effects on CB1 receptor density in the adolescent brain: An autoradiographic study using the synthetic cannabinoid HU210. *Synapse*, *64*(11), 845–854. https://doi.org/10.1002/syn.20801.

Dayan, P. (2009). Dopamine, reinforcement learning, and addiction. *Pharmacopsychiatry*, *42*(Suppl. 1), S56–S65. https://doi.org/10.1055/s-0028-1124107.

de Meijer, E. P. M., Bagatta, M., Carboni, A., Crucitti, P., Moliterni, V. M. C., Ranalli, P., et al. (2003). The inheritance of chemical phenotype in *Cannabis sativa* L. *Genetics*, *163*(1), 335–346.

Degenhardt, L., Coffey, C., Romaniuk, H., Swift, W., Carlin, J. B., Hall, W. D., et al. (2013). The persistence of the association between adolescent cannabis use and common mental disorders into young adulthood. *Addiction*, *108*(1), 124–133. https://doi.org/10.1111/j.1360-0443.2012.04015.x.

Degenhardt, L., Hall, W., & Lynskey, M. (2003). Exploring the association between cannabis use and depression. *Addiction*, *98*(11), 1493–1504. https://doi.org/10.1046/j.1360-0443.2003.00437.x.

Devane, W. A., Dysarz, F. A., Johnson, M. R., Melvin, L. S., & Howlett, A. C. (1988). Determination and characterization of a cannabinoid receptor in rat brain. *Molecular Pharmacology*, *34*(5), 605–613.

Dwyer, J. B., & Leslie, F. M. (2016). Adolescent maturation of dopamine D1 and D2 receptor function and interactions in rodents. *PLoS One*, *11*(1), e0146966. https://doi.org/10.1371/journal.pone.0146966.

Eggan, S. M., Mizoguchi, Y., Stoyak, S. R., & Lewis, D. A. (2010). Development of cannabinoid 1 receptor protein and messenger RNA in monkey dorsolateral prefrontal cortex. *Cerebral Cortex*, *20*(5), 1164–1174. https://doi.org/10.1093/cercor/bhp179.

Ellgren, M., Artmann, A., Tkalych, O., Gupta, A., Hansen, H. S., Hansen, S. H., et al. (2008). Dynamic changes of the endogenous cannabinoid and opioid mesocorticolimbic systems during adolescence: THC effects. *European Neuropsychopharmacology*, *18*(11), 826–834. https://doi.org/10.1016/j.euroneuro.2008.06.009.

ElSohly, M. A., Ross, S. A., Mehmedic, Z., Arafat, R., Yi, B., & Banahan, B. F. (2000). Potency trends of delta9-THC and other cannabinoids in confiscated marijuana from 1980-1997. *Journal of Forensic Sciences*, *45*(1), 24–30.

Enomoto, T., Tse, M. T., & Floresco, S. B. (2011). Reducing prefrontal gamma-aminobutyric acid activity induces cognitive, behavioral, and dopaminergic abnormalities that resemble schizophrenia. *Biological Psychiatry*, *69*(5), 432–441. https://doi.org/10.1016/j.biopsych.2010.09.038.

Ferdinand, J.-M., Peters, K. Z., Yavas, E., & Young, A. M. J. (2021). Modulation of stimulated dopamine release in rat nucleus accumbens shell by GABA in vitro: Effect of sub-chronic phencyclidine pretreatment. *Journal of Neuroscience Research*, *99*(7), 1885–1901. https://doi.org/10.1002/jnr.24843.

Freeman, T. P., Groshkova, T., Cunningham, A., Sedefov, R., Griffiths, P., & Lynskey, M. T. (2019). Increasing potency and price of cannabis in Europe, 2006-16. *Addiction*, *114*(6), 1015–1023. https://doi.org/10.1111/add.14525.

Freund, T. F., Katona, I., & Piomelli, D. (2003). Role of endogenous cannabinoids in synaptic signaling. *Physiological Reviews*, *83*(3), 1017–1066. https://doi.org/10.1152/physrev.00004.2003.

Gaoni, Y., & Mechoulam, R. (1964). Isolation, structure, and partial synthesis of an active constituent of hashish. *Journal of the American Chemical Society*, *86*(8), 1646–1647. https://doi.org/10.1021/ja01062a046.

Gessa, G. L., Melis, M., Muntoni, A. L., & Diana, M. (1998). Cannabinoids activate mesolimbic dopamine neurons by an action on cannabinoid CB1 receptors. *European Journal of Pharmacology*, *341*(1), 39–44. https://doi.org/10.1016/s0014-2999(97)01442-8.

Gogtay, N., Giedd, J. N., Lusk, L., Hayashi, K. M., Greenstein, D., Vaituzis, A. C., et al. (2004). Dynamic mapping of human cortical development during childhood through early adulthood. *Proceedings of the National Academy of Sciences of the United States of America*, *101*(21), 8174–8179. https://doi.org/10.1073/pnas.0402680101.

Gold, J. M., & Harvey, P. D. (1993). Cognitive deficits in schizophrenia. *Psychiatric Clinics*, *16*(2), 295–312. https://doi.org/10.1016/S0193-953X(18)30175-8.

Grace, A. A., & Gomes, F. V. (2019). The circuitry of dopamine system regulation and its disruption in schizophrenia: Insights into treatment and prevention. *Schizophrenia Bulletin*, *45*(1), 148–157. https://doi.org/10.1093/schbul/sbx199.

Harkany, T., Guzmán, M., Galve-Roperh, I., Berghuis, P., Devi, L. A., & Mackie, K. (2007). The emerging functions of endocannabinoid signaling during CNS development. *Trends in Pharmacological Sciences*, *28*(2), 83–92. https://doi.org/10.1016/j.tips.2006.12.004.

Harkany, T., Keimpema, E., Barabás, K., & Mulder, J. (2008). Endocannabinoid functions controlling neuronal specification during brain development. *Molecular and Cellular Endocrinology*, *286*(1–2 Suppl. 1), S84–S90. https://doi.org/10.1016/j.mce.2008.02.011.

Heng, L., Beverley, J. A., Steiner, H., & Tseng, K. Y. (2011). Differential developmental trajectories for CB1 cannabinoid receptor expression in limbic/associative and sensorimotor cortical areas. *Synapse (New York, N.Y.)*, *65*(4), 278–286. https://doi.org/10.1002/syn.20844.

Henry, D. J., & Chavkin, C. (1995). Activation of inwardly rectifying potassium channels (GIRK1) by co-expressed rat brain cannabinoid receptors in Xenopus oocytes. *Neuroscience Letters*, *186*(2–3), 91–94. https://doi.org/10.1016/0304-3940(95)11289-9.

Herkenham, M., Lynn, A. B., Johnson, M. R., Melvin, L. S., de Costa, B. R., & Rice, K. C. (1991). Characterization and localization of cannabinoid receptors in rat brain: A quantitative in vitro autoradiographic study. *The Journal of Neuroscience: The Official Journal of the Society for Neuroscience*, *11*(2), 563–583.

Herkenham, M., Lynn, A. B., Little, M. D., Johnson, M. R., Melvin, L. S., de Costa, B. R., et al. (1990). Cannabinoid receptor localization in brain. *Proceedings of the National Academy of Sciences of the United States of America*, *87*(5), 1932–1936. https://doi.org/10.1073/pnas.87.5.1932.

Hoffman, A. F., & Lupica, C. R. (2000). Mechanisms of cannabinoid inhibition of GABA(A) synaptic transmission in the hippocampus. *The Journal of Neuroscience: The Official Journal of the Society for Neuroscience*, *20*(7), 2470–2479.

Hoftman, G. D., & Lewis, D. A. (2011). Postnatal developmental trajectories of neural circuits in the primate prefrontal cortex: Identifying sensitive periods for vulnerability to schizophrenia. *Schizophrenia Bulletin*, *37*(3), 493–503. https://doi.org/10.1093/schbul/sbr029.

Hoops, D., & Flores, C. (2017). Making dopamine connections in adolescence. *Trends in Neurosciences*, *40*(12), 709–719. https://doi.org/10.1016/j.tins.2017.09.004.

Howlett, A. C., Barth, F., Bonner, T. I., Cabral, G., Casellas, P., Devane, W. A., et al. (2002). International union of pharmacology. XXVII. Classification of cannabinoid receptors. *Pharmacological Reviews*, *54*(2), 161–202. https://doi.org/10.1124/pr.54.2.161.

Howlett, A. C., & Mukhopadhyay, S. (2000). Cellular signal transduction by anandamide and 2-arachidonoylglycerol. *Chemistry and Physics of Lipids*, *108*(1–2), 53–70. https://doi.org/10.1016/s0009-3084(00)00187-0.

Huestis, M. A., Gorelick, D. A., Heishman, S. J., Preston, K. L., Nelson, R. A., Moolchan, E. T., et al. (2001). Blockade of effects of smoked marijuana by the CB1-selective cannabinoid receptor antagonist SR141716. *Archives of General Psychiatry*, *58*(4), 322–328. https://doi.org/10.1001/archpsyc.58.4.322.

Juraska, J. M., & Markham, J. A. (2004). The cellular basis for volume changes in the rat cortex during puberty: White and gray matter. *Annals of the New York Academy of Sciences*, *1021*(1), 431–435. https://doi.org/10.1196/annals.1308.058.

Katona, I., Sperlágh, B., Sı́k, A., Käfalvi, A., Vizi, E. S., Mackie, K., et al. (1999). Presynaptically located CB1 cannabinoid receptors regulate GABA release from axon terminals of specific hippocampal interneurons. *The Journal of Neuroscience*, *19*(11), 4544–4558. https://doi.org/10.1523/JNEUROSCI.19-11-04544.1999.

Katsidoni, V., Anagnostou, I., & Panagis, G. (2013). Cannabidiol inhibits the reward-facilitating effect of morphine: Involvement of 5-HT1A receptors in the dorsal raphe nucleus. *Addiction Biology*, *18*(2), 286–296. https://doi.org/10.1111/j.1369-1600.2012.00483.x.

Knapp, A. A., Lee, D. C., Borodovsky, J. T., Auty, S. G., Gabrielli, J., & Budney, A. J. (2019). Emerging trends in cannabis administration among adolescent cannabis users. *The Journal of Adolescent Health: Official Publication of the Society for Adolescent Medicine*, *64*(4), 487–493. https://doi.org/10.1016/j.jadohealth.2018.07.012.

Kosillo, P., Zhang, Y.-F., Threlfell, S., & Cragg, S. J. (2016). Cortical control of striatal dopamine transmission via striatal cholinergic interneurons. *Cerebral Cortex (New York, N.Y.: 1991)*, *26*(11), 4160–4169. https://doi.org/10.1093/cercor/bhw252.
Kowitt, S. D., Osman, A., Meernik, C., Zarkin, G. A., Ranney, L. M., Martin, J., et al. (2019). Vaping cannabis among adolescents: Prevalence and associations with tobacco use from a cross-sectional study in the USA. *BMJ Open*, *9*(6), e028535. https://doi.org/10.1136/bmjopen-2018-028535.
Kreitzer, A. C., & Regehr, W. G. (2002). Retrograde signaling by endocannabinoids. *Current Opinion in Neurobiology*, *12*(3), 324–330. https://doi.org/10.1016/s0959-4388(02)00328-8.
Kruse, L. C., Cao, J. K., Viray, K., Stella, N., & Clark, J. J. (2019). Voluntary oral consumption of Δ9-tetrahydrocannabinol by adolescent rats impairs reward-predictive cue behaviors in adulthood. *Neuropsychopharmacology: Official Publication of the American College of Neuropsychopharmacology*, *44*(8), 1406–1414. https://doi.org/10.1038/s41386-019-0387-7.
Lagerberg, T. V., Sundet, K., Aminoff, S. R., Berg, A. O., Ringen, P. A., Andreassen, O. A., et al. (2011). Excessive cannabis use is associated with earlier age at onset in bipolar disorder. *European Archives of Psychiatry and Clinical Neuroscience*, *261*(6), 397–405. https://doi.org/10.1007/s00406-011-0188-4.
Larsen, B., & Luna, B. (2018). Adolescence as a neurobiological critical period for the development of higher-order cognition. *Neuroscience & Biobehavioral Reviews*, *94*, 179–195. https://doi.org/10.1016/j.neubiorev.2018.09.005.
Laviola, G., Macrì, S., Morley-Fletcher, S., & Adriani, W. (2003). Risk-taking behavior in adolescent mice: Psychobiological determinants and early epigenetic influence. *Neuroscience and Biobehavioral Reviews*, *27*(1–2), 19–31. https://doi.org/10.1016/s0149-7634(03)00006-x.
Laviolette, S. R., & Grace, A. A. (2006). The roles of cannabinoid and dopamine receptor systems in neural emotional learning circuits: Implications for schizophrenia and addiction. *Cellular and Molecular Life Sciences: CMLS*, *63*(14), 1597–1613. https://doi.org/10.1007/s00018-006-6027-5.
Lee, T. T.-Y., Hill, M. N., Hillard, C. J., & Gorzalka, B. B. (2013). Temporal changes in *N*-acylethanolamine content and metabolism throughout the peri-adolescent period. *Synapse*, *67*(1), 4–10. https://doi.org/10.1002/syn.21609.
Lewis, D. A., & González-Burgos, G. (2008). Neuroplasticity of neocortical circuits in schizophrenia. *Neuropsychopharmacology: Official Publication of the American College of Neuropsychopharmacology*, *33*(1), 141–165. https://doi.org/10.1038/sj.npp.1301563.
Lovinger, D. M., Davis, M. I., & Costa, R. M. (2010). Endocannabinoid signaling in the striatum. In H. Steiner, & K. Y. Tseng (Eds.), *Handbook of basal ganglia structure and function* (pp. 167–186). London: Academic Press.
Lowe, D. J. E., Sasiadek, J. D., Coles, A. S., & George, T. P. (2019). Cannabis and mental illness: A review. *European Archives of Psychiatry and Clinical Neuroscience*, *269*(1), 107–120. https://doi.org/10.1007/s00406-018-0970-7.
Lu, H.-C., & Mackie, K. (2016). An introduction to the endogenous cannabinoid system. *Biological Psychiatry*, *79*(7), 516–525. https://doi.org/10.1016/j.biopsych.2015.07.028.
Lucatch, A. M., Coles, A. S., Hill, K. P., & George, T. P. (2018). Cannabis and mood disorders. *Current Addiction Reports*, *5*(3), 336–345. https://doi.org/10.1007/s40429-018-0214-y.
Maccarrone, M., Guzmán, M., Mackie, K., Doherty, P., & Harkany, T. (2014). Programming of neural cells by (endo)cannabinoids: From physiological rules to emerging therapies. *Nature Reviews Neuroscience*, *15*(12), 786–801. https://doi.org/10.1038/nrn3846.

Mailleux, P., & Vanderhaeghen, J. J. (1992). Distribution of neuronal cannabinoid receptor in the adult rat brain: A comparative receptor binding radioautography and in situ hybridization histochemistry. *Neuroscience*, *48*(3), 655–668. https://doi.org/10.1016/0306-4522(92)90409-u.

Manitt, C., Mimee, A., Eng, C., Pokinko, M., Stroh, T., Cooper, H. M., et al. (2011). The netrin receptor DCC is required in the pubertal organization of mesocortical dopamine circuitry. *Journal of Neuroscience*, *31*(23), 8381–8394. https://doi.org/10.1523/JNEUROSCI.0606-11.2011.

Mateo, Y., Johnson, K. A., Covey, D. P., Atwood, B. K., Wang, H.-L., Zhang, S., et al. (2017). Endocannabinoid actions on cortical terminals orchestrate local modulation of dopamine release in the nucleus accumbens. *Neuron*, *96*(5), 1112–1126.e5. https://doi.org/10.1016/j.neuron.2017.11.012.

Mato, S., Del Olmo, E., & Pazos, A. (2003). Ontogenetic development of cannabinoid receptor expression and signal transduction functionality in the human brain. *The European Journal of Neuroscience*, *17*(9), 1747–1754. https://doi.org/10.1046/j.1460-9568.2003.02599.x.

McCutcheon, J. E., & Marinelli, M. (2009). Age matters. *The European Journal of Neuroscience*, *29*(5), 997–1014. https://doi.org/10.1111/j.1460-9568.2009.06648.x.

Mechoulam, R., Ben-Shabat, S., Hanus, L., Ligumsky, M., Kaminski, N. E., Schatz, A. R., et al. (1995). Identification of an endogenous 2-monoglyceride, present in canine gut, that binds to cannabinoid receptors. *Biochemical Pharmacology*, *50*(1), 83–90. https://doi.org/10.1016/0006-2952(95)00109-d.

Mechoulam, R., & Parker, L. A. (2013). The endocannabinoid system and the brain. *Annual Review of Psychology*, *64*(1), 21–47. https://doi.org/10.1146/annurev-psych-113011-143739.

Mechoulam, R., & Shvo, Y. (1963). Hashish. I. The structure of cannabidiol. *Tetrahedron*, *19*(12), 2073–2078. https://doi.org/10.1016/0040-4020(63)85022-x.

Miech, R. A., Johnston, L. D., O'Malley, P. M., Bachman, J. G., Schulenberg, J. E., & Patrick, M. E. (2018). Monitoring the future national survey results on drug use, 1975–2017. Volume I, secondary school students. In *Institute for Social Research* Institute for Social Research. https://eric.ed.gov/?id=ED589763.

Miech, R. A., Patrick, M. E., O'Malley, P. M., Johnston, L. D., & Bachman, J. G. (2020). Trends in reported marijuana vaping among US adolescents, 2017-2019. *JAMA*, *323*(5), 475–476. https://doi.org/10.1001/jama.2019.20185.

Munro, S., Thomas, K. L., & Abu-Shaar, M. (1993). Molecular characterization of a peripheral receptor for cannabinoids. *Nature*, *365*(6441), 61–65. https://doi.org/10.1038/365061a0.

Nakazawa, K., Zsiros, V., Jiang, Z., Nakao, K., Kolata, S., Zhang, S., et al. (2012). GABAergic interneuron origin of schizophrenia pathophysiology. *Neuropharmacology*, *62*(3), 1574–1583. https://doi.org/10.1016/j.neuropharm.2011.01.022.

Naneix, F., Marchand, A. R., Di Scala, G., Pape, J.-R., & Coutureau, E. (2012). Parallel maturation of goal-directed behavior and dopaminergic systems during adolescence. *Journal of Neuroscience*, *32*(46), 16223–16232. https://doi.org/10.1523/JNEUROSCI.3080-12.2012.

Nieh, E. H., Vander Weele, C. M., Matthews, G. A., Presbrey, K. N., Wichmann, R., Leppla, C. A., et al. (2016). Inhibitory input from the lateral hypothalamus to the ventral tegmental area disinhibits dopamine neurons and promotes behavioral activation. *Neuron*, *90*(6), 1286–1298. https://doi.org/10.1016/j.neuron.2016.04.035.

Nolan, S. O., Zachry, J. E., Johnson, A. R., Brady, L. J., Siciliano, C. A., & Calipari, E. S. (2020). Direct dopamine terminal regulation by local striatal microcircuitry. *Journal of Neurochemistry*, *155*(5), 475–493. https://doi.org/10.1111/jnc.15034.

O'Carroll, R. (2000). Cognitive impairment in schizophrenia. *Advances in Psychiatric Treatment*, *6*(3), 161–168. https://doi.org/10.1192/apt.6.3.161.

O'Donnell, P. (2011). Adolescent onset of cortical disinhibition in schizophrenia: Insights from animal models. *Schizophrenia Bulletin*, *37*(3), 484–492. https://doi.org/10.1093/schbul/sbr028.

Oleson, E. B., Beckert, M. V., Morra, J. T., Lansink, C. S., Cachope, R., Abdullah, R. A., et al. (2012). Endocannabinoids shape accumbal encoding of cue-motivated behavior via CB1 receptor activation in the ventral tegmentum. *Neuron*, *73*(2), 360–373. https://doi.org/10.1016/j.neuron.2011.11.018.

O'Shea, M., Singh, M. E., McGregor, I. S., & Mallet, P. E. (2004). Chronic cannabinoid exposure produces lasting memory impairment and increased anxiety in adolescent but not adult rats. *Journal of Psychopharmacology*, *18*(4), 502–508. https://doi.org/10.1177/0269881104047277.

Paladini, C. A., & Tepper, J. M. (2016). Chapter 17—Neurophysiology of substantia nigra dopamine neurons: Modulation by GABA and glutamate. In H. Steiner, & K. Y. Tseng (Eds.), *Vol. 24. Handbook of behavioral neuroscience* (pp. 335–360). Elsevier. https://doi.org/10.1016/B978-0-12-802206-1.00017-9.

Parker, L. A., Burton, P., Sorge, R. E., Yakiwchuk, C., & Mechoulam, R. (2004). Effect of low doses of delta9-tetrahydrocannabinol and cannabidiol on the extinction of cocaine-induced and amphetamine-induced conditioned place preference learning in rats. *Psychopharmacology*, *175*(3), 360–366. https://doi.org/10.1007/s00213-004-1825-7.

Paus, T., Keshavan, M., & Giedd, J. N. (2008). Why do many psychiatric disorders emerge during adolescence? *Nature Reviews Neuroscience*, *9*(12), 947–957. https://doi.org/10.1038/nrn2513.

Peters, K. Z., Cheer, J. F., & Tonini, R. (2021). Modulating the neuromodulators: Dopamine, serotonin, and the endocannabinoid system. *Trends in Neurosciences*, *44*(6), 464–477. https://doi.org/10.1016/j.tins.2021.02.001.

Peters, K. Z., Oleson, E. B., & Cheer, J. F. (2021). A brain on cannabinoids: The role of dopamine release in reward seeking and addiction. *Cold Spring Harbor Perspectives in Medicine*, *11*(1), a039305. https://doi.org/10.1101/cshperspect.a039305.

Pistis, M., Ferraro, L., Pira, L., Flore, G., Tanganelli, S., Gessa, G. L., et al. (2002). Delta(9)-tetrahydrocannabinol decreases extracellular GABA and increases extracellular glutamate and dopamine levels in the rat prefrontal cortex: An in vivo microdialysis study. *Brain Research*, *948*(1–2), 155–158. https://doi.org/10.1016/s0006-8993(02)03055-x.

Raber, J. C., Elzinga, S., & Kaplan, C. (2015). Understanding dabs: Contamination concerns of cannabis concentrates and cannabinoid transfer during the act of dabbing. *The Journal of Toxicological Sciences*, *40*(6), 797–803. https://doi.org/10.2131/jts.40.797.

Rao, S. G., Williams, G. V., & Goldman-Rakic, P. S. (2000). Destruction and creation of spatial tuning by disinhibition: GABA(A) blockade of prefrontal cortical neurons engaged by working memory. *Journal of Neuroscience*, *20*(1), 485–494.

Reichelt, A. C. (2016). Adolescent maturational transitions in the prefrontal cortex and dopamine signaling as a risk factor for the development of obesity and high fat/high sugar diet induced cognitive deficits. *Frontiers in Behavioral Neuroscience*, *10*, 189. https://doi.org/10.3389/fnbeh.2016.00189.

Renard, J., Krebs, M.-O., Jay, T. M., & Le Pen, G. (2013). Long-term cognitive impairments induced by chronic cannabinoid exposure during adolescence in rats: A strain comparison. *Psychopharmacology*, *225*(4), 781–790. https://doi.org/10.1007/s00213-012-2865-z.

Renard, J., Krebs, M.-O., Le Pen, G., & Jay, T. M. (2014). Long-term consequences of adolescent cannabinoid exposure in adult psychopathology. *Frontiers in Neuroscience*, *8*. https://doi.org/10.3389/fnins.2014.00361.

Renard, J., Rushlow, W. J., & Laviolette, S. R. (2018). Effects of adolescent THC exposure on the prefrontal GABAergic system: Implications for schizophrenia-related psychopathology. *Frontiers in Psychiatry*, *9*, 281. https://doi.org/10.3389/fpsyt.2018.00281.

Renard, J., Szkudlarek, H. J., Kramar, C. P., Jobson, C. E. L., Moura, K., Rushlow, W. J., et al. (2017). Adolescent THC exposure causes enduring prefrontal cortical disruption of GABAergic inhibition and dysregulation of sub-cortical dopamine function. *Scientific Reports*, 7(1), 1–14. https://doi.org/10.1038/s41598-017-11645-8.

Rodríguez de Fonseca, F., Ramos, J. A., Bonnin, A., & Fernández-Ruiz, J. J. (1993). Presence of cannabinoid binding sites in the brain from early postnatal ages. *Neuroreport*, *4*(2), 135–138. https://doi.org/10.1097/00001756-199302000-00005.

Rubino, T., Zamberletti, E., & Parolaro, D. (2012). Adolescent exposure to cannabis as a risk factor for psychiatric disorders. *Journal of Psychopharmacology (Oxford, England)*, *26*(1), 177–188. https://doi.org/10.1177/0269881111405362.

Russo, E., & Guy, G. W. (2006). A tale of two cannabinoids: The therapeutic rationale for combining tetrahydrocannabinol and cannabidiol. *Medical Hypotheses*, *66*(2), 234–246. https://doi.org/10.1016/j.mehy.2005.08.026.

Schneider, M., & Koch, M. (2003). Chronic pubertal, but not adult chronic cannabinoid treatment impairs sensorimotor gating, recognition memory, and the performance in a progressive ratio task in adult rats. *Neuropsychopharmacology: Official Publication of the American College of Neuropsychopharmacology*, *28*(10), 1760–1769. https://doi.org/10.1038/sj.npp.1300225.

Schneider, M., Schömig, E., & Leweke, F. M. (2008). Acute and chronic cannabinoid treatment differentially affects recognition memory and social behavior in pubertal and adult rats. *Addiction Biology*, *13*(3–4), 345–357. https://doi.org/10.1111/j.1369-1600.2008.00117.x.

Schoedel, K. A., Chen, N., Hilliard, A., White, L., Stott, C., Russo, E., et al. (2011). A randomized, double-blind, placebo-controlled, crossover study to evaluate the subjective abuse potential and cognitive effects of nabiximols oromucosal spray in subjects with a history of recreational cannabis use. *Human Psychopharmacology*, *26*(3), 224–236. https://doi.org/10.1002/hup.1196.

Schultz, W. (2016). Dopamine reward prediction-error signalling: A two-component response. *Nature Reviews Neuroscience*, *17*(3), 183–195. https://doi.org/10.1038/nrn.2015.26.

Solinas, M., Belujon, P., Fernagut, P. O., Jaber, M., & Thiriet, N. (2019). Dopamine and addiction: What have we learned from 40 years of research. *Journal of Neural Transmission*, *126*(4), 481–516. https://doi.org/10.1007/s00702-018-1957-2.

Spear, L. P. (2000). The adolescent brain and age-related behavioral manifestations. *Neuroscience and Biobehavioral Reviews*, *24*(4), 417–463. https://doi.org/10.1016/s0149-7634(00)00014-2.

Stamatakis, A. M., Jennings, J. H., Ung, R. L., Blair, G. A., Weinberg, R. J., Neve, R. L., et al. (2013). A unique population of ventral tegmental area neurons inhibits the lateral habenula to promote reward. *Neuron*, *80*(4), 1039–1053. https://doi.org/10.1016/j.neuron.2013.08.023.

Stefanis, N. C., Delespaul, P., Henquet, C., Bakoula, C., Stefanis, C. N., & Van Os, J. (2004). Early adolescent cannabis exposure and positive and negative dimensions of psychosis. *Addiction (Abingdon, England)*, *99*(10), 1333–1341. https://doi.org/10.1111/j.1360-0443.2004.00806.x.

Stogner, J. M., & Miller, B. L. (2015). Assessing the dangers of 'dabbing': Mere marijuana or harmful new trend? *Pediatrics*, *136*(1), 1–3. https://doi.org/10.1542/peds.2015-0454.

Sugiura, T., Kondo, S., Sukagawa, A., Nakane, S., Shinoda, A., Itoh, K., et al. (1995). 2-Arachidonoylglycerol: A possible endogenous cannabinoid receptor ligand in brain. *Biochemical and Biophysical Research Communications*, *215*(1), 89–97. https://doi.org/10.1006/bbrc.1995.2437.

Threlfell, S., & Cragg, S. J. (2011). Dopamine signaling in dorsal versus ventral striatum: The dynamic role of cholinergic interneurons. *Frontiers in Systems Neuroscience*, *5*, 11. https://doi.org/10.3389/fnsys.2011.00011.

Threlfell, S., Lalic, T., Platt, N. J., Jennings, K. A., Deisseroth, K., & Cragg, S. J. (2012). Striatal dopamine release is triggered by synchronized activity in cholinergic interneurons. *Neuron*, *75*(1), 58–64. https://doi.org/10.1016/j.neuron.2012.04.038.

Tseng, K. Y., Chambers, R. A., & Lipska, B. K. (2009). The neonatal ventral hippocampal lesion as a heuristic neurodevelopmental model of schizophrenia. *Behavioural Brain Research*, *204*(2), 295–305. https://doi.org/10.1016/j.bbr.2008.11.039.

Tseng, K.-Y., & O'Donnell, P. (2006). Dopamine modulation of prefrontal cortical interneurons changes during adolescence. *Cerebral Cortex*, *17*(5), 1235–1240. https://doi.org/10.1093/cercor/bhl034.

Tsou, K., Brown, S., Sañudo-Peña, M. C., Mackie, K., & Walker, J. M. (1998). Immunohistochemical distribution of cannabinoid CB1 receptors in the rat central nervous system. *Neuroscience*, *83*(2), 393–411. https://doi.org/10.1016/s0306-4522(97)00436-3.

Tunbridge, E. M., Weickert, C. S., Kleinman, J. E., Herman, M. M., Chen, J., Kolachana, B. S., et al. (2007). Catechol-o-methyltransferase enzyme activity and protein expression in human prefrontal cortex across the postnatal lifespan. *Cerebral Cortex (New York, N.Y.: 1991)*, *17*(5), 1206–1212. https://doi.org/10.1093/cercor/bhl032.

Twitchell, W., Brown, S., & Mackie, K. (1997). Cannabinoids inhibit N- and P/Q-type calcium channels in cultured rat hippocampal neurons. *Journal of Neurophysiology*, *78*(1), 43–50. https://doi.org/10.1152/jn.1997.78.1.43.

Uhlhaas, P. J., & Singer, W. (2011). The development of neural synchrony and large-scale cortical networks during adolescence: Relevance for the pathophysiology of schizophrenia and neurodevelopmental hypothesis. *Schizophrenia Bulletin*, *37*(3), 514–523. https://doi.org/10.1093/schbul/sbr034.

Van Waes, V., Beverley, J. A., Siman, H., Tseng, K. Y., & Steiner, H. (2012). CB1 cannabinoid receptor expression in the striatum: Association with corticostriatal circuits and developmental regulation. *Frontiers in Pharmacology*, *3*, 21. https://doi.org/10.3389/fphar.2012.00021.

Viveros, M., Llorente, R., Suarez, J., Llorente-Berzal, A., López-Gallardo, M., & Rodriguez de Fonseca, F. (2012). The endocannabinoid system in critical neurodevelopmental periods: Sex differences and neuropsychiatric implications. *Journal of Psychopharmacology*, *26*(1), 164–176. https://doi.org/10.1177/0269881111408956.

Volkow, N. D., Han, B., Einstein, E. B., & Compton, W. M. (2021). Prevalence of substance use disorders by time since first substance use among Young people in the US. *JAMA Pediatrics*, *175*(6), 640–643. https://doi.org/10.1001/jamapediatrics.2020.6981.

Volkow, N. D., Wise, R. A., & Baler, R. (2017). The dopamine motive system: Implications for drug and food addiction. *Nature Reviews Neuroscience*, *18*(12), 741–752. https://doi.org/10.1038/nrn.2017.130.

Voorn, P., Kalsbeek, A., Jorritsma-Byham, B., & Groenewegen, H. J. (1988). The pre- and postnatal development of the dopaminergic cell groups in the ventral mesencephalon and the dopaminergic innervation of the striatum of the rat. *Neuroscience*, *25*(3), 857–887. https://doi.org/10.1016/0306-4522(88)90041-3.

Vosberg, D. E., Leyton, M., & Flores, C. (2020). The Netrin-1/DCC guidance system: Dopamine pathway maturation and psychiatric disorders emerging in adolescence. *Molecular Psychiatry*, *25*(2), 297–307. https://doi.org/10.1038/s41380-019-0561-7.

Wang, H., Treadway, T., Covey, D. P., Cheer, J. F., & Lupica, C. R. (2015). Cocaine-induced endocannabinoid mobilization in the ventral tegmental area. *Cell Reports*, *12*(12), 1997–2008. https://doi.org/10.1016/j.celrep.2015.08.041.

Watabe-Uchida, M., Eshel, N., & Uchida, N. (2017). Neural circuitry of reward prediction error. *Annual Review of Neuroscience*, *40*(1), 373–394. https://doi.org/10.1146/annurev-neuro-072116-031109.

Wilson, R. I., & Nicoll, R. A. (2001). Endogenous cannabinoids mediate retrograde signalling at hippocampal synapses. *Nature*, *410*(6828), 588–592. https://doi.org/10.1038/35069076.

Wise, R. A., & Bozarth, M. A. (1985). Brain mechanisms of drug reward and euphoria. *Psychiatric Medicine*, *3*(4), 445–460.
Wise, R. A., & Robble, M. A. (2020). Dopamine and addiction. *Annual Review of Psychology*, *71*(1), 79–106. https://doi.org/10.1146/annurev-psych-010418-103337.
World Drug Report. (2019). *World drug report 2019*. United Nations Publication. Sales No. E.19.XI.8.
Yee, B. K., Keist, R., von Boehmer, L., Studer, R., Benke, D., Hagenbuch, N., et al. (2005). A schizophrenia-related sensorimotor deficit links α3-containing GABAA receptors to a dopamine hyperfunction. *Proceedings of the National Academy of Sciences of the United States of America*, *102*(47), 17154–17159. https://doi.org/10.1073/pnas.0508752102.
Zhang, H.-Y., Gao, M., Liu, Q.-R., Bi, G.-H., Li, X., Yang, H.-J., et al. (2014). Cannabinoid CB2 receptors modulate midbrain dopamine neuronal activity and dopamine-related behavior in mice. *Proceedings of the National Academy of Sciences of the United States of America*, *111*(46), E5007–E5015. https://doi.org/10.1073/pnas.1413210111.
Zlebnik, N. E., & Cheer, J. F. (2016). Drug-induced alterations of endocannabinoid-mediated plasticity in brain reward regions. *The Journal of Neuroscience*, *36*(40), 10230–10238. https://doi.org/10.1523/JNEUROSCI.1712-16.2016.

CHAPTER FOUR

The stoned age: Sex differences in the effects of adolescent cannabinoid exposure on prefrontal cortex structure and function in animal models

D.E. Ginder[a,†], H.R. Wright[b,†], and R.J. McLaughlin[a,b,*]
[a]Department of Psychology, Washington State University, Pullman, WA, United States
[b]Department of Integrative Physiology and Neuroscience, Washington State University, Pullman, WA, United States
*Corresponding author: e-mail address: ryan.mclaughlin@wsu.edu

Contents

1. Introduction 122
2. The adolescent brain is exquisitely sensitive to exogenous cannabinoids 124
3. Effects of adolescent cannabinoid exposure on prefrontal cortex (PFC)-dependent behavior 126
4. Effects of adolescent cannabinoid exposure on prefrontal cortex (PFC) structure and function 129
   4.1 Region-specific structural alterations following adolescent cannabinoid exposure 130
   4.2 Synaptic effects of adolescent cannabinoid exposure on the medial-PFC (mPFC) 131
   4.3 Sex-dependent molecular alterations following adolescent cannabinoid exposure 133
   4.4 Effects of adolescent cannabinoid exposure on non-neuronal cells 135
5. Conclusions 136
References 138

Abstract

Cannabis is the most used drug during adolescence, which is a period of enhanced cortical plasticity and synaptic remodeling that supports behavioral, cognitive, and emotional maturity. In this chapter, we review preclinical studies indicating that adolescent exposure to cannabinoids has lasting effects on the morphology and synaptic organization of the prefrontal cortex and associated circuitry, which may lead to

[†] These authors contributed equally to this work.

International Review of Neurobiology, Volume 161
ISSN 0074-7742
https://doi.org/10.1016/bs.irn.2021.07.005

cognitive dysfunction later in life. Additionally, we reviewed sex differences in the effects of adolescent cannabinoid exposure with a focus on brain systems that support cognitive functioning. The body of evidence indicates enduring sex-specific effects in behavior and organization of corticolimbic circuitry, which appears to be influenced by species, strain, drug, route of administration, and window/pattern of drug exposure. Caution should be exercised when extrapolating these results to humans. Adopting models that more closely resemble human cannabis use will provide more translationally relevant data concerning the long-term effects of cannabis use on the adolescent brain.

Abbreviations

2-AG	2-arachidonoylglycerol
11-OH-THC	11-hydroxy-delta-9-tetrahydrocannabinol
AEA	anandamide
BDNF	brain-derived neurotrophic factor
CB1	type-1 cannabinoid receptor
CB2	type-2 cannabinoid receptor
COX2	cyclooxygenase-2
CREB	cyclic adenosine monophosphate response element-binding protein
CSTADS	Canadian Student Tobacco, Alcohol and Drug Use Survey
DMRs	differentially methylated regions
eCB	endocannabinoid
FAAH	fatty acid amide hydrolase
GABA	gamma-Aminobutyric acid
iNOS	inducible nitric oxide synthase
i.p.	intraperitoneal
mPFC	medial prefrontal cortex
mTOR	mammalian target of rapamycin
NAc	nucleus accumbens
NMDA	N-Methyl-D-aspartate
pCREB	phosphorylated cyclic adenosine monophosphate response element-binding protein
PFC	prefrontal cortex
PND	postnatal day
THC	delta-9-tetrahydrocannabinol
TNF-α	tumor necrosis factor alpha
VAMP2	vesicle associated membrane protein 2
VTA	ventral tegmental area
WIN	55,212
Wnt	Wingless

1. Introduction

Cannabis is the most commonly used drug among adolescents in North America, with a lifetime prevalence of nearly double that of all other illicit drugs combined (NSDUH, 2020). As of 2020, 43.7% of 12th graders in the United States have reported using cannabis in their lifetime (NSDUH, 2020). Perhaps more concerning, the prevalence of *heavy* cannabis use during adolescence has seen a threefold increase in the past 25 years, with an estimated 6.9% of 12th graders reporting *daily* use in the US (NSDUH, 2020). These trends have been coupled with steady declines in the social stigma and perceived risks of cannabis use over the past decade (Okaneku, Vearrier, McKeever, LaSala, & Greenberg, 2015; Schuermeyer et al., 2014). Similar trends are also evident in Canada, where cannabis was legalized for recreational purposes at a national level in October 2018. Data from statistics for Canada's school-based Canadian Student Tobacco, Alcohol and Drug Use Survey (CSTADS) indicate that approximately 28% of teenagers in Grade 12, and 53% for young adults in the age range of 20–24, have used cannabis in the past year (CSTADS, 2018). Moreover, roughly 75% of those surveyed in CSTADS who had ever used cannabis reported using it in the past 3 months, with 32% of that group reporting daily use (CSTADS, 2018). These alarming statistics have reignited public concerns over the effects of adolescent cannabis use and have highlighted the pressing need to better understand its long-term repercussions and the neurobiological mechanisms by which these occur.

Despite the public consensus that cannabis use during adolescence has uniformly negative consequences that increase the risk for cognitive impairment later in life, review of the scientific literature presents a more complex picture. This is in part because the vast majority of human studies on this topic employ cross-sectional designs comparing cannabis users to non-users, and as such, a direct causal relationship has yet to be established (Scott et al., 2018, 2019; Weiland et al., 2015). Rodent models provide an excellent opportunity to answer complex questions in a high throughput manner under tightly controlled conditions in the absence of confounding variables. Depending on the outcome, rodent models offer differing levels of utility, but with respect to brain development, they offer a very valuable approach. The developmental organization of the rodent brain parallels the human brain in many ways, including the ontogeny of neurocircuitry connecting subcortical and cortical structures (Casey, Pattwell, Glatt, & Lee, 2013; Spear, 2000). With respect to adolescence, like humans, rodents exhibit a non-linear pattern of frontal cortical development that is highlighted by an overproduction of synapses, which are then pruned back in an experience-dependent manner (Spear, 2000). As such, rodent models represent a valuable approach for examining the lasting impacts of adolescent cannabinoid exposure on brain development and behavior. With that said, the primary objective of this mini-review is not to provide an exhaustive overview of all studies conducted on the effects of cannabis and cannabinoid administration during adolescence (see Stringfield & Torregrossa, 2021, for an excellent recent review of the human and rodent literature), but rather, to summarize the preclinical literature examining effects of adolescent cannabinoid exposure, with a focus on aspects of the cortical and striatal systems that support cognitive and motivational outcomes.

There is emerging evidence that while males are traditionally more at risk for using cannabis and developing cannabis-related problems (Hasin, Shmulewitz, & Sarvet, 2019), that male-female gap has narrowed in recent years, particularly among adolescent users (Johnson et al., 2015; Johnston et al., 2019). Moreover, like other drugs of abuse, women demonstrate a "telescoping effect" that results in more rapid progression from first use to cannabis use disorder (Ehlers et al., 2010; Hernandez-Avila, Rounsaville, & Kranzler, 2004; Khan et al., 2013; Schepis et al., 2011; Westermeyer & Boedicker, 2000). Accordingly, women enter treatment for cannabis-related problems after fewer years of use and less cumulative use than men (Hernandez-Avila et al., 2004). However, observed sex differences in the effects of adolescent cannabinoid use/exposure have not been systematically reviewed, which presents a major impediment to better understanding if and why these sex differences exist. Thus, a secondary objective of this review is to note evidence of sex differences in the effects of adolescent cannabinoid use/exposure.

2. The adolescent brain is exquisitely sensitive to exogenous cannabinoids

Adolescence is a critical neurodevelopmental period that is characterized by enhanced cortical plasticity and remodeling that supports behavioral, cognitive, and emotional maturity. Given the plastic nature of the brain during this developmental period, it can be exquisitely sensitive to perturbation by external stimuli. For example, it has been well established that phenomena such as head trauma (Toledo et al., 2012), substance use (Spear, 2018; Squeglia, Jacobus, & Tapert, 2009), and significant life stress (Andersen & Teicher, 2008) can impact the normative trajectory of brain development to exert indelible changes in brain structure. Structural alterations in cortical and subcortical brain regions, following these experiences during adolescence, are thought to be a primary mechanism for disrupting cognitive and behavioral function. With respect to cannabis use, a large body of evidence has indicated that adolescents that engage in heavy cannabis use often show alterations in brain development that are associated with impaired cognitive impairments (see Fischer, Tapert, Louie, Schatzberg, & Singh, 2020, for review). However, it remains unknown whether these alterations are caused by cannabis use or alternately reflect pre-existing conditions that increase vulnerability for problematic cannabis use (Jacobus & Tapert, 2014).

Importantly, adolescent brain development differs by sex. The female cortex reaches a point of macrostructural maturity faster than the male cortex (Lenroot et al., 2007; Mutlu et al., 2013), and males show an age-related increase in resting state functional connectivity in the prefrontal cortex (PFC), whereas females show an age-related decrease (Zuo et al., 2010). Additionally, subcortical regions involved in emotional regulation, such as the amygdala and hippocampus, also show rates of maturation that differ by sex (Frere et al., 2020; Goddings et al., 2014). The striatum also undergoes significant remodeling during adolescence and these processes also follow a sex-specific trajectory (Hammerslag & Gulley, 2016). As with the PFC, female development of the striatum occurs more rapidly than in males, with volume changes remaining steady after 12 years of age (Raznahan et al., 2014). Furthermore, males often continue to experience striatal volume and area changes into early adulthood, whereas females have already reached a relatively consistent state (Raznahan et al., 2014). Studies in rats also indicate sex-specific fluctuations in dopamine receptor density during adolescence. Male rats experience greater alterations in both D1 and D2 receptors in the ventral striatum during adolescent development, with large increases in D1/D2 receptor density followed by synaptic pruning (Andersen, Rutstein, Benzo, Hostetter, & Teicher, 1997). In contrast, females experience a slight increase in D1/D2 receptor density, which remains unchanged into adulthood (Andersen et al., 1997). This sex-dependent organization of cortical and striatal networks during adolescence could make males and females differentially susceptible to cannabis-induced perturbations, thereby producing unique cognitive and behavioral effects later in life. Along these lines, compelling clinical data indicate that sex may even moderate the relationship between adolescent cannabis use and PFC volume (Medina et al., 2009), which supports the idea that sex is a crucial variable to consider in studies examining effects of adolescent cannabinoid exposure.

The cannabis plant contains hundreds of molecules, but it is primarily delta-9-tetrahydrocannabinol (THC) that mediates the psychoactive properties of the plant (Mechoulam, Hanuš, Pertwee, & Howlett, 2014). THC exerts its biological effects through direct interaction with the endocannabinoid (eCB) system, a chemical signaling system in the brain and body that regulates a host of physiological processes (see Katona & Freund, 2012 for review). Generally speaking, THC activates type-1 cannabinoid (CB1) receptors, which influence synaptic transmission and information processing in the brain. From a neurodevelopmental perspective, CB1 receptors are widely expressed in the brain, and their expression peaks approximately at the onset of adolescence and declines into adulthood (Heng, Beverley, Steiner, & Tseng, 2011). In rodents, CB1 receptor expression is highest at the onset of adolescence (around postnatal day [PND] 30), particularly in the medial prefrontal cortex (mPFC) and striatum, followed by a decline in expression into adulthood (around PND 70) (Meyer, Lee, & Gee, 2018). The eCB system also displays sexual dimorphism, with fluctuations in eCB activity coinciding with estrous phases (González et al., 2000), which modulates the release of gonadal hormones in both humans and rodents (Meyer et al., 2018). Interestingly, CB1 receptor density begins increasing earlier in female rodents compared to their male counterparts (Rodríguez de Fonseca, Ramos, Bonnin, & Fernández-Ruiz, 1993). In addition to regulating synaptic transmission and gonadal hormone release, eCB signaling is also known to play a role in circuit formation by modulating long-range axonal pathfinding and patterning of synaptic contacts for the appropriate development of cortical circuits (Berghuis et al., 2007; Maccarrone, Guzmán, Mackie, Doherty, & Harkany, 2014). This positions the eCB system at the center of critical neurodevelopmental processes; and, as such, interfering with eCB-mediated processes via exogenous THC administration during adolescence could produce long-lasting alterations within corticostriatal and corticolimbic circuits, resulting in behavioral and cognitive impairments later in life.

3. Effects of adolescent cannabinoid exposure on prefrontal cortex (PFC)-dependent behavior

Many studies have been conducted using preclinical animal models to examine the effects of adolescent cannabinoid exposure on cognitive endpoints in adulthood. The majority of studies employing THC have used a consistent administration protocol, whereby rats receive escalating intraperitoneal doses of THC (2.5 mg/kg PND 35–37; 5 mg/kg PND 38–41; 10 mg/kg PND 42–45) or vehicle, followed by a washout period until testing around PND 70–75. This method is designed to model the pattern of use seen in heavy human adolescent cannabis users (Rubino et al., 2008). Studies examining sex-dependent effects of this escalating THC exposure regimen on PFC-dependent working memory tasks have revealed mixed results. In the radial arm maze task, deficits in spatial working memory have been documented in THC-exposed rats of both sexes (Rubino, Realini, Braida, Alberio, et al., 2009; Rubino, Realini, Braida, Guidi, et al., 2009). However, in the novel object recognition task, Quinn et al. (2008) have documented impaired discrimination in THC-exposed male Wistar rats (females were not tested), whereas two groups have independently shown effects specifically in female rats receiving this escalating THC dose regimen (Llorente-Berzal et al., 2013; Zamberletti et al., 2014). However, it should be noted that others have not observed any adolescent THC- (or cannabis smoke-) induced impairments in novel object recognition memory in male or female Long Evans rats that were tested in adulthood (Bruijnzeel et al., 2019). Thus, effects of adolescent THC administration on working memory performance have been reliably observed in rodent studies (but see Bruijnzeel et al., 2019, for conflicting evidence), but the results do not paint a clear picture of whether males or females are differentially impacted.

Effects of synthetic cannabinoid exposure during adolescence produce clearer evidence of a sex-dependent effect on memory. For instance, adolescent administration of the ultra-potent CB1 receptor agonist CP55,940 (0.4 mg/kg from PND 28–43) produced sex-specific alterations, with male (more so than female) rats showing an increase in working memory as measured by a water maze task (Higuera-Matas et al., 2009). Similarly, an escalating dose regimen of CP55,940 (0.15–3 mg/kg) from PND 29–50 also produced deficits selectively in male rats (Renard, Krebs, Jay, & Le Pen, 2013). These deficits were recapitulated in studies using male rats that were treated with another synthetic CB1 receptor agonist, WIN 55212–2 (WIN; 1.2 mg/kg/day), from PND 35–50 (Abush & Akirav, 2012; Kirschmann, McCalley, Edwards, & Torregrossa, 2017; Schneider & Koch, 2003). Moreover, Mateos et al. (2011) showed that male (but not female) rats treated with CP55,940 during adolescence exhibited impaired novel object recognition and increased hippocampal CB1 receptor activity when tested in adulthood, whereas only female rats exhibited impaired object location memory, with no changes in novel object recognition or CB1 receptor activity. Additionally, it may be that these effects are partially driven by the unintended stress produced by high doses of synthetic CB1 receptor agonists. When rats are given volitional control over intravenous WIN administration during adolescence, they achieve an average daily intake of only 0.216 mg/kg WIN and consequently, neither males nor females show deficits in novel object recognition memory in adulthood at this dose (Kirschmann, McCalley, et al., 2017). In fact, WIN self-administration actually led to *enhanced* working memory performance in an operant delayed match-to-sample task (Kirschmann, McCalley, et al., 2017), which is consistent with another study from the same group showing that low-dose treatment with WIN during adolescence enhances working memory performance in female rats tested in adulthood (Kirschmann, Pollock, et al., 2017). Finally, a follow-up study from this group also revealed that male rats that were trained to self-administer WIN during adolescence exhibited slightly enhanced working memory performance in the delayed match-to-sample task in adulthood; and, surprisingly, better performance in this task was positively correlated with total WIN self-administered (Kirschmann, McCalley, et al., 2017). These effects underscore the importance of considering dose, drug, and route of administration when attempting to draw comparisons regarding the effects of adolescent cannabinoid exposure across studies.

In addition to cognitive outcomes, cannabinoid-induced neuroadaptations could also be implicated in aberrant drug-seeking behavior as well. Chronic adolescent cannabinoid administration increases opiate self-administration in male rats (Ellgren, Spano, & Hurd, 2007) and increases cocaine self-administration in adult female rats, but the latter, surprisingly, had no effect in adult male rats (Higuera-Matas et al., 2008). Moreover, 3 mg/kg WIN exposure during early adolescence has been shown to increase alcohol preference in male mice (Frontera, Gonzalez Pini, Messore, & Brusco, 2018), while male rats that were exposed to THC during adolescence have been shown to self-administer more WIN in adulthood compared to THC-naïve control rats (Scherma et al., 2016). These changes in drug-seeking behavior may be due to alterations in ventral tegmental area (VTA) dopamine transmission, as in vivo recording studies have revealed that THC-treated adolescent rats exhibit a dose-dependent decrease in WIN-evoked VTA dopamine firing in adulthood, with a corresponding decrease in evoked dopamine release in the nucleus accumbens (NAc) shell (Scherma et al., 2016). One mechanism by which this may occur is through cannabinoid-induced alterations in eCB-mediated long-term depression (LTD) of inhibitory transmission in VTA GABA neurons, as reported in male CB1 knockout mice (Friend et al., 2017). However, there may be sex differences in these neuroadaptations. For example, female rats exposed to THC during adolescence were found to have reduced CB1 receptor density and/or CB-1/G-protein coupling in the VTA, NAc, and amygdala (Rubino et al., 2008). These changes in the VTA and NAc were exclusively seen in females, with males experiencing reductions in hippocampal CB1/G-protein coupling and amygdala CB1 receptor density (Rubino et al., 2008). Comparatively, female rats also experience reductions in CB1 receptor binding and anandamide (AEA) (but not 2-arachidonoylglycerol [2-AG]) content in the PFC following escalating doses of THC (Rubino et al., 2015). Studies have further confirmed an upregulation of CB1 receptors on GABAergic terminals and a downregulation of CB1 receptors on glutamatergic terminals following THC exposure selectively in adolescent female mice that express a genetic polymorphism in the fatty acid amide hydrolase gene ($FAAH^{C/A}$) that results in compromised AEA degradation (Burgdorf et al., 2020). These data highlight how cannabinoid-mediated disinhibition of VTA dopamine projections to the NAc may contribute to sex-specific brain alterations that drive aberrant and persistent drug-use and -seeking behavior in adulthood.

4. Effects of adolescent cannabinoid exposure on prefrontal cortex (PFC) structure and function

A large body of literature has demonstrated a central involvement of cortical and striatal dopamine signaling in working memory and flexible decision making (Floresco, 2013; see Floresco, Zhang, & Enomoto, 2009; Klanker, Feenstra, & Denys, 2013 for detailed reviews). The ventral striatum serves as a functional interface between limbic and motor systems (Floresco, 2013), and the mPFC provides top-down control via glutamatergic projections to striatal regions to guide decision making via indirect eCB-mediated modulation of dopamine release (Mateo et al., 2017). Thus, interference with eCB-mediated control of the corticostriatal pathway via exogenous cannabinoid administration could alter the structure and function of these synapses, thereby setting the stage for cognitive and motivational alterations. In the following section, we will briefly review evidence demonstrating morphological, synaptic, and molecular alterations following adolescent cannabinoid exposure and note sex-specific effects where possible. We conclude by describing evidence indicating effects of adolescent cannabinoid exposure on non-neuronal cell populations (i.e., glial cells and inflammatory markers), which is important to consider in future studies.

4.1 Region-specific structural alterations following adolescent cannabinoid exposure

Chronic cannabinoid exposure during adolescence induces long-lasting changes to the structure of mPFC neurons and their synaptic connections in rats, including reduced basal dendritic arborization of pyramidal neurons in layer II/III (Rubino et al., 2015), and impaired hippocampal-input driven synaptic plasticity in the male rat mPFC (Rubino, Realini, Braida, Guidi, et al., 2009). Additionally, dendritic length and spine formation in hippocampal granule cells is reduced in proximal and distal dendrites from male rats exposed to THC during adolescence (Rubino, Realini, Braida, Guidi, et al., 2009). In contrast, females had decreased spine density at distal basal dendrites (as opposed to proximal basal or apical) specifically in layer II/III mPFC pyramidal neurons (Rubino et al., 2015) that project to other cortical areas and are thought to be a key node in the working memory network (Goldman-Rakic, 1995). However, the effects of cannabinoid administration in rats during adolescence on dendritic morphology have been somewhat inconsistent, as others have shown that cannabinoids can *increase* dendritic length, spine density and the number of dendritic branches in mPFC pyramidal neurons (Carvalho, Reyes, Ramalhosa, Sousa, & Van Bockstaele, 2016). Notably, these changes are in direct opposition to the ventral striatum, which experiences widespread reductions in spine density in both adolescence and adulthood following adolescent cannabinoid administration (Carvalho et al., 2016). Thus, mPFC and NAc neurons undergo major structural changes following developmental cannabinoid exposure that may be sex-dependent and could have negative ramifications for corticostriatal-dependent behavior.

RNA-sequencing of prelimbic mPFC layer III cells from adult male rats exposed to THC in adolescence has revealed robust changes in the developmental transcription trajectory (i.p. 1.5 mg/kg, every third day PND 28–49) (Miller et al., 2019). For instance, vehicle-treated rats showed expected changes for developmental age in genes related to cell morphology, morphogenesis, and signal transduction that were surprisingly absent in THC-treated rats. In fact, pyramidal cell developmental transcriptomes exhibited very little overlap between the treatment groups. In addition to changes associated with actin cytoskeleton and dendritic development, THC exposure was associated with impaired epigenetic mechanisms like chromatin modification and histone methylation across development (Miller et al., 2019). Enrichment analysis of the differentially expressed genes revealed a functional association with Kmt2a, a chromatin methyl transferase, and the histone it interacts with, H3k4me3, which is implicated in cellular processes linked to neurodevelopment that are coincidentally associated with the development of psychiatric conditions such as major depression (Cruceanu et al., 2013). THC-treated rats also exhibited different dendritic arborization, spine density, and premature spine pruning compared to vehicle-treated rats (Miller et al., 2019). Another study found that adolescent THC exposure in female rats (i.p. 2.5–10 mg/kg from PND 35–45) modified mPFC H3k9me3, a histone that regulates genes related to synaptic plasticity, and impaired performance in a novel object recognition task (Prini et al., 2018). Importantly, pharmacological blockade of H3k9me3 during development prevented the negative effects of THC on these cognitive endpoints (Prini et al., 2018). These results suggest that sex differences in adolescent THC exposure may result from differential effects on epigenetic regulation of gene expression, and that these long lasting epigenetic changes may persist across generations (Szutorisz et al., 2014; Watson et al., 2015), at least in the striatum. The male offspring of parents exposed to THC during adolescence (i.p. 1.5 mg/kg, every third day PND 28–49) exhibit 1027 differentially methylated regions (DMRs) in the NAc, both core and shell, when compared to the offspring of rats treated with vehicle as adolescents (Watson et al., 2015). Watson et al. (2015) also reported that functional networks impacted by these DMRs are involved in the regulation of glutamatergic synapses such as glutamatergic receptors, G proteins, ion channels, and scaffolding proteins. These studies indicated that adolescent THC exposure induced heritable, sex-specific alterations in the epigenetic regulation of genes that affected the structure and activity of corticostriatal synapses.

4.2 Synaptic effects of adolescent cannabinoid exposure on the medial-PFC (mPFC)

As mentioned above, the eCB system fundamentally contributes to adolescent neurodevelopment, particularly in the PFC (Meyer et al., 2018). Excitatory and inhibitory transmission in the PFC is under the control of eCBs (see McLaughlin, Hill, & Gorzalka, 2014, for review), and both glutamatergic and GABAergic synapses undergo considerable plasticity during adolescence, which can be perturbed following adolescent cannabinoid exposure. For instance, eCB-mediated LTD of excitatory transmission is reduced in the mPFC in adult female rats and mice following adolescent cannabinoid exposure, suggesting decreased eCB-mediated control of glutamatergic transmission (Lovelace et al., 2015; Rubino et al., 2015).

Moreover, escalating doses of THC during adolescence increased the expression of GluA1 subunits in α-amino-3-hydroxy-5-methyl-4-isoxazolepropionic acid (AMPA) receptors as well as GluN2B subunits in N-Methyl-D-aspartate (NMDA) receptors in adulthood (Rubino et al., 2015), which may help explain reduced LTD at glutamatergic synapses in the mPFC previously documented (Lovelace et al., 2015; Rubino et al., 2015). Remarkably, daily administration of the FAAH inhibitor URB597 in adult rats (PND 75–105) rescued deficits in (1) mPFC CB1 receptor density, (2) eCB-mediated LTD, (3) density and dendritic arborization of newborn hippocampal neurons, and (4) synaptic plasticity within the dentate gyrus in a CB1 receptor-dependent manner (Cuccurazzu et al., 2018). Thus, re-establishing a normative baseline of AEA signaling could reverse some of the neurobiological and behavioral deficits that have been observed in female rodents following adolescent cannabinoid exposure.

At the synaptic level, the vesicle-associated protein synaptophysin and postsynaptic anchoring postsynaptic density protein 95 (PSD95) have been shown to be reduced in the mPFC of adult female rodents exposed to THC during adolescence (Rubino, Realini, Braida, Alberio, et al., 2009). Notably, this reduction was unique to the mPFC, as it was not seen in the hippocampus. In contrast, male rats exposed to THC in adolescence showed altered expression of proteins involved in synaptic transmission in the hippocampus rather than mPFC (Rubino, Realini, Braida, Guidi, et al., 2009). Decreased hippocampal PSD95 expression paralleled decreased NMDA receptor levels throughout the hippocampus. Since PSD95 mediates the anchoring of NMDA receptors in the synapse, these effects may be closely linked. Vesicle associated protein VAMP2, a critical mediator of neurotransmitter release and vesicle trafficking that interacts with synaptophysin (Pennuto, Bonanomi, Benfenati, & Valtorta, 2003), was similarly decreased.

Early adolescent exposure to WIN not only enhances excitatory transmission, but also disrupts inhibitory networks of the mPFC, at least in male rats, as evidenced by a disruption in gamma oscillations that are comprised of aggregated synchronous activity of fast-spiking interneuron populations (Cass et al., 2014; Renard, Szkudlarek, et al., 2017). Accordingly, 2 mg/kg WIN exposure from PND 35–45 (but not PND 50–55 or PND 75–80) decreased GABA transmission onto layer V pyramidal neurons in the mPFC of male rodents (Cass et al., 2014), which suggests a specific window of vulnerability for cannabinoid-induced alterations at GABAergic synapses. Accordingly, male mice exhibited reductions in cortical oscillations at gamma, beta, alpha, and theta bandwidths (Raver, Haughwout, & Keller, 2013). It has been speculated that this compromised mPFC disinhibition could be an underlying mechanism behind working memory deficits seen in male, but not female, rats following synthetic CB1 receptor agonist treatment during adolescence (Poulia et al., 2020; Raver et al., 2013). However, it should also be noted that basal GABA levels, as well as GAD67 expression in parvalbumin- and cholecystokinin-positive neurons, are also reduced in the mPFC of female rats following escalating THC injections during adolescence (Zamberletti et al., 2014). Thus, systematic interrogation of sex differences in mPFC GABA transmission following adolescent cannabinoid exposure will be required before we can fully understand whether these effects are indeed sexually dimorphic.

Overall, it appears that adolescent cannabinoid administration produces region-, and sex-dependent effects on cortical cell morphology and synaptic transmission. This causes exaggerated mPFC activation, perhaps as a result of diminished GABAergic input to mPFC pyramidal cells. This could, in turn, give rise to the cognitive deficits described above. It should be noted, however, that glutamatergic inputs from the mPFC to the dorsal striatum unexpectedly show reduced activity following escalating doses of THC in female rodents, which questions the hypothesis of uniformly overactive cortical-striatal neurotransmission (Zamberletti et al., 2014). It remains unknown whether this effect is unique to female rodents, as similar endpoints have not yet been evaluated in male rodents. Interestingly, chronic administration of increasing doses of THC during adolescence sensitizes the locomotor response to acute administration of the NMDA receptor antagonist phencyclidine in female (Zamberletti et al., 2014) but not male rats (Zamberletti et al., 2016) when tested in adulthood, which suggests that cannabinoid-induced neuroadaptations in motor pathways may be more specific to adolescent female rodents.

4.3 Sex-dependent molecular alterations following adolescent cannabinoid exposure

Studies designed to characterize the molecular underpinnings of cannabinoid-induced behavioral alterations have pointed to a role for phosphorylated cyclic adenosine monophosphate response element-binding protein (pCREB), which regulates the transcription of proteins that are fundamentally involved in neurodevelopment and synaptic plasticity. THC-treated female rodents showed a reduction in pCREB levels in the mPFC and hippocampus; where, notably, many antidepressants have been shown to increase CREB activity (Blendy, 2006). THC-treated female rats also exhibited decreased neurogenesis in the hippocampus, another hallmark of depression in humans (Realini et al., 2011). Additionally, pCREB levels were increased in the

NAc of THC-treated female rodents, the opposite of which has been shown to produce antidepressant-like effects (Newton et al., 2002). In contrast to female rodents, THC-treated male rodents did not differ in pCREB levels from vehicle treated males. These data suggest that females may be more susceptible to the deleterious effects of adolescent THC exposure that result in a pronounced depressive-like phenotype in adulthood.

In addition to pCREB, brain-derived neurotrophic factor (BDNF), a known mediator of cellular survival and synapse maintenance, has also been implicated in altered neurodevelopment of frontal brain regions (Poulia et al., 2020; Reichardt, 2006). Adolescent male and female rats treated with the same THC administration protocol, as above, exhibited differences in BDNF protein levels in the PFC when tested during adulthood. Western blot analysis determined that male rodents have decreased PFC BDNF expression whereas female rodents showed an increase. Since BDNF transcription is mediated by CREB family transcription factor activity (Esvald et al., 2020), altered pCREB levels may disrupt the tightly controlled autoregulation of BDNF transcripts that can account for the differential developmental responses to cannabinoids between males and females, at least in rodents.

It is tempting to posit that altered CREB and BDNF signaling may be due to neurodevelopmental disruption, which results in reduced synaptic activity. Proteomic analysis of THC-treated female rat PFC synaptosomes revealed a reduction in the mitochondrial protein cytochrome *b*-c1 complex subunits 1 and 2, and adenosine triphosphate synthase alpha and beta subunits, which suggests reduced synaptosomal mitochondria and less metabolically active synaptosomes (Rubino, Realini, Braida, Alberio, et al., 2009). Further evidence supporting this conclusion includes reduced glycolysis enzymes pyruvate kinase isozymes M1/M2, fructose bisphosphate aldolase isoform A, and glyceraldehyde-3-phosphate dehydrogenase, as well as reduced molecular chaperone heat shock cognate 71 kDa protein, a regulator of proteostasis during periods of oxidative stress (Rubino, Realini, Braida, Alberio, et al., 2009). These data support a general pattern of disruption for various metabolic markers in the mPFC of female rodents after adolescent THC exposure, whereas male rodents showed more prominent alterations in the hippocampus. However, one study indicated that adolescent THC exposure disrupts mammalian target of rapamycin (mTOR) and wingless (Wnt) signaling in the PFC of male rats, which are critical for synaptic plasticity (Renard, Rosen, et al., 2017). Decreased phosphorylated Akt-thr308, glycogen synthase kinase-3, β-catenin, ribosomal protein S6 kinase, and mTOR were also reported (Renard, Rosen, et al., 2017). Similar to findings in post-mortem PFC tissue from patients with schizophrenia (Beasley et al., 2001; Kozlovsky, Belmaker, & Agam, 2001) and major depression (Jernigan et al., 2011), alterations in Wnt signaling may underlie some of the observed cognitive and behavioral abnormalities.

4.4 Effects of adolescent cannabinoid exposure on non-neuronal cells

Astrocytes are known to be key mediators of synaptic transmission and metabolic activity in neighboring neurons, particularly in the hippocampus where stimulation of astrocytic CB1 receptors results in the release of excitatory gliotransmitters at tripartite synapses (Navarrete & Araque, 2008). It is possible this is another mechanism by which adolescent THC exposure disrupts circuit development and synapse maturity. There is additional evidence for a contribution of non-neuronal cells in the effects of cannabinoids on cortical neurodevelopment. For instance, escalating doses of THC exposure during adolescence induce a persistent neuroinflammatory state in the mPFC of female rats (male rats were not tested), that is characterized by increased expression of the microglia marker Iba1, upregulation of CB2 receptors on microglia, a reduction in the anti-inflammatory cytokine IL-10, and an increase in the pro-inflammatory markers TNF-α, iNOS, and COX-2 (Zamberletti, Gabaglio, Prini, Rubino, & Parolaro, 2015). Notably, these effects are specific to the mPFC, with no significant group differences observed in the hippocampus or amygdala. In contrast, male rats exhibited increased glial fibrillary acidic protein (GFAP) expression and a neuroinflammatory state similar to female rats, but in the hippocampus rather than the mPFC (Zamberletti et al., 2016). Chronic neuroinflammation that is mediated by glia are synapto- and cyto-toxic; and, as such, these changes are receiving more attention as key contributors to neurodegenerative and neuropsychiatric diseases. Strikingly, treating female rodents with ibudilast to block glial activation during adolescent THC exposure prevented mPFC neuroinflammation and rescued performance in the novel object recognition task in adulthood, but not altered passive behavior in the forced swim test (Zamberletti et al., 2015). These results suggest a novel mechanism in female rodents whereby adolescent THC exposure provokes mPFC neuroinflammation, presumably mediated by microglia and astrocytes, damaging neurons and their synaptic connections, with a concomitant memory impairment that is independent of increased passive coping behavior. Unfortunately, the ibudilast experiment has yet to be conducted in pretreated male rodents to test the contributions of hippocampal microglia, nor were the effects of ibudilast tested on sucrose preference or the morphological characteristics of mPFC neurons.

5. Conclusions

As the rates of daily, heavy cannabis use continue to rise across North America, there is an increased urgency to better understand the long-term effects of adolescent cannabinoid exposure on the brain, behavior, and cognitive functioning. The PFC is rich in CB1 receptors and undergoes substantial remodeling during adolescence, thus making it is exquisitely sensitive to cannabinoid-induced perturbations that may also differ by sex. Preclinical studies have generally supported the notion that adolescent cannabinoid administration causes enduring effects on working memory and novel object recognition memory, particularly in males that are administered high doses of synthetic cannabinoids, though these effects may be influenced by the species, strain, drug employed, route of administration, and developmental window and pattern of drug exposure. Both male and female rodents that receive cannabinoids during adolescence show alterations in drug-seeking behavior later in life, but emerging evidence indicates that females may experience more reward-related effects following adolescent THC administration than males. This appears to interact with genetic polymorphisms in components of the eCB system that synergize to augment preference for THC (and possibly other drugs) later in adulthood (Burgdorf et al., 2020). These findings are in line with the possibility that cannabinoid exposure during adolescence accelerates the "telescoping effect" of THC exposure that has been routinely demonstrated in female rodents and humans.

At the neurobiological level, cannabinoid-induced alterations in the structure and function of neurons in the mPFC results in long-term adaptations in both glutamatergic and GABAergic synapses. This culminates in increased disinhibition and hyperactivity of mPFC pyramidal output neurons. However, fully powered comparisons by sex are lacking and as such, it remains unknown whether these effects are more pronounced in either sex or whether they give rise to unique, sex-specific behavioral alterations. The effects of adolescent cannabinoid exposure at the synaptic level are supported by studies demonstrating alterations in transcription factors (pCREB) and neurotrophic factors (BDNF) in the mPFC of females, which differs from results in males that tend to be more restricted to the hippocampus.

Emerging evidence further indicates that non-neuronal cells in the mPFC are also impacted by adolescent cannabinoid exposure in females, whereas these cells are again more negatively impacted in the hippocampus of males. Remarkably, restoring eCB signaling via the inhibition of FAAH has emerged as a promising strategy for mitigating the long-term neurobiological and behavioral effects of adolescent cannabinoid exposure in rodent studies.

As with all preclinical studies, we must proceed with caution when attempting to translate findings from animal models to human populations. Except for a few notable exceptions using cannabis smoke (Bruijnzeel et al., 2019) or self-administration (Kirschmann, McCalley, et al., 2017; Kirschmann, Pollock, et al., 2017), virtually all of the studies described in this review used forced injections of isolated cannabis constituents or synthetic CB1 receptor agonists, which exhibit a fundamentally different pharmacological profile compared to volitional inhalation of whole-plant cannabis preparations, the most common form of cannabis use in humans (McLaughlin, 2018). In the studies described herein, the doses administered are also exceptionally high, which further limits the translational value of these studies. Differences in the metabolism of THC between adolescents and adults could also be an important consideration. For instance, studies have shown 50–70% higher brain concentrations of THC metabolites in adolescent mice relative to adult mice (Torrens et al., 2020). This is an important consideration when assessing sex differences as well, since studies now indicate that adolescent female rodents exhibit enhanced metabolism of THC compared to males, resulting in elevated production of the psychoactive and highly potent THC metabolite 11-OH-THC (Ruiz et al., 2021; Wiley & Burston, 2014). This may lead to greater unintended CB1 receptor activation in adolescent females under circumstances where metabolites are likely to accumulate, like that seen after a large bolus injection of THC. Thus, to draw meaningful conclusions regarding the effects of adolescent cannabis use and potential sex differences in these effects, it is absolutely critical that we adopt more translationally relevant models that recapitulate the drug and route of administration that is most common among human users (see Freels et al., 2020, for validation of a novel model of cannabis vapor self-administration in male rats and Glodosky et al., 2020, for a demonstration of its feasibility in females). Adopting models that more closely resemble human adolescent cannabis use, as well as the behavioral and cognitive endpoints of interest, will undoubtedly provide us with a more trustworthy indication of the long-term effects of cannabis on the developing brain. This can be hopefully leveraged to inform prospective and current cannabis users on the actual risk of using cannabis during this sensitive neurodevelopmental period.

References

Abush, H., & Akirav, I. (2012). Short- and long-term cognitive effects of chronic cannabinoids administration in late-adolescence rats. *PLoS One*, 7, 1–12. https://doi.org/10.1371/journal.pone.0031731.

Andersen, S. L., Rutstein, M., Benzo, J. M., Hostetter, J. C., & Teicher, M. H. (1997). Sex differences in dopamine receptor overproduction and elimination. *Neuroreport*, *8*(6). https://doi.org/10.1097/00001756-199704140-00034.

Andersen, S. L., & Teicher, M. H. (2008). Stress, sensitive periods and maturational events in adolescent depression. *Trends in Neurosciences*, *31*(4), 183–191. https://doi.org/10.1016/j.tins.2008.01.004.

Beasley, C., Cotter, D., Khan, N., Pollard, C., Sheppard, P., Varndell, I., et al. (2001). Glycogen synthase kinase-3beta immunoreactivity is reduced in the prefrontal cortex in schizophrenia. *Neuroscience Letters*, *302*, 117–120. https://doi.org/10.1016/s0304-3940(01)01688-3.

Berghuis, P., Rajnicek, A. M., Morozov, Y. M., Ross, R. A., Mulder, J., Urbán, G. M., et al. (2007). Hardwiring the brain: Endocannabinoids shape neuronal connectivity. *Science*, *316*(5828), 1212–1216. https://doi.org/10.1126/science.1137406.

Blendy, J. A. (2006). The role of CREB in depression and antidepressant treatment. *Biological Psychiatry*, *59*(12), 1144–1150. https://doi.org/10.1016/j.biopsych.2005.11.003.

Bruijnzeel, A. W., Knight, P., Panunzio, S., Xue, S., Bruner, M. M., Wall, S. C., et al. (2019). Effects in rats of adolescent exposure to cannabis smoke or THC on emotional behavior and cognitive function in adulthood. *Psychopharmacology*, *236*(9), 2773–2784. https://doi.org/10.1007/s00213-019-05255-7.

Burgdorf, C. E., Jing, D., Yang, R., Huang, C., Hill, M. N., Mackie, K., et al. (2020). Endocannabinoid genetic variation enhances vulnerability to THC reward in adolescent female mice. *Science Advances*, *6*(7), eaay1502. https://doi.org/10.1126/sciadv.aay1502.

Canadian Tobacco, Alcohol and Drug Use Survey. (2018). https://www.canada.ca/en/health-canada/services/canadian-tobacco-alcohol-drugs-survey/2017-summary.html.

Carvalho, A. F., Reyes, B. A. S., Ramalhosa, F., Sousa, N., & Van Bockstaele, E. J. (2016). Repeated administration of a synthetic cannabinoid receptor agonist differentially affects cortical and accumbal neuronal morphology in adolescent and adult rats. *Brain Structure and Function*, *221*(1), 407–419. https://doi.org/10.1007/s00429-014-0914-6.

Casey, B. J., Pattwell, S. S., Glatt, C. E., & Lee, F. S. (2013). Treating the developing brain: Implications from human imaging and mouse genetics. *Annual Review of Medicine*, *64*, 427–439. https://doi.org/10.1146/annurev-med-052611-130408.

Cass, D. K., Flores-Barrera, E., Thomases, D. R., Vital, W. F., Caballero, A., & Tseng, K. Y. (2014). CB1 cannabinoid receptor stimulation during adolescence impairs the maturation of GABA function in the adult rat prefrontal cortex. *Molecular Psychiatry*, *19*(5), 536–543. https://doi.org/10.1038/mp.2014.14.

Cruceanu, C., Alda, M., Nagy, C., Freemantle, E., Rouleau, G. A., & Turecki, G. (2013). H3K4 tri-methylation in synapsin genes leads to different expression patterns in bipolar disorder and major depression. *International Journal of Neuropsychopharmacology*, *16*(2), 289–299. https://doi.org/10.1017/S1461145712000363.

Cuccurazzu, B., Zamberletti, E., Nazzaro, C., Prini, P., Trusel, M., Grilli, M., et al. (2018). Adult cellular neuroadaptations induced by adolescent THC exposure in female rats are rescued by enhancing anandamide signaling. *International Journal of Neuropsychopharmacology*, *21*(11), 1014–1024. https://doi.org/10.1093/ijnp/pyy057.

Ehlers, C. L., Gizer, I. R., Vieten, C., Gilder, D. A., Stouffer, G. M., Lau, P., et al. (2010). Cannabis dependence in the San Francisco family study: Age of onset of use, DSM-IV symptoms, withdrawal, and heritability. *Addictive Behaviors*, *35*(2), 102–110. https://doi.org/10.1016/j.addbeh.2009.09.009.

Ellgren, M., Spano, S. M., & Hurd, Y. L. (2007). Adolescent cannabis exposure alters opiate intake and opioid limbic neuronal populations in adult rats. *Neuropsychopharmacology*, *32*(3), 607–615. https://doi.org/10.1038/sj.npp.1301127.

Esvald, E. E., Tuvikene, J., Sirp, A., Patil, S., Bramham, C. R., & Timmusk, T. (2020). CREB family transcription factors are major mediators of BDNF transcriptional autoregulation in cortical neurons. *Journal of Neuroscience*, *40*(7), 1405–1426. https://doi.org/10.1523/JNEUROSCI.0367-19.2019.

Fischer, A. S., Tapert, S. F., Louie, D. L., Schatzberg, A. F., & Singh, M. K. (2020). Cannabis and the developing adolescent brain. *Current Treatment Options in Psychiatry*, 7(2), 144–161. https://doi.org/10.1007/s40501-020-00202-2.

Floresco, S. B. (2013). Prefrontal dopamine and behavioral flexibility: Shifting from an "inverted-U" toward a family of functions. *Frontiers in Neuroscience*, 7, 62. https://doi.org/10.3389/fnins.2013.00062.

Floresco, S. B., Zhang, Y., & Enomoto, T. (2009). Neural circuits subserving behavioral flexibility and their relevance to schizophrenia. *Behavioural Brain Research*, *204*(2), 396–409. https://doi.org/10.1016/j.bbr.2008.12.001.

Freels, T. G., Baxter-Potter, L. N., Lugo, J. M., Glodosky, N. C., Wright, H. R., Baglot, S. L., et al. (2020). Vaporized cannabis extracts have reinforcing properties and support conditioned drug-seeking behavior in rats. *Journal of Neuroscience*, *40*(9), 1897–1908. https://doi.org/10.1523/JNEUROSCI.2416-19.2020.

Frere, P.B., Vetter, N.C., Artiges, E., Filippi, I., Miranda, R., Vulser, H., Paillère-Martinot, M.L., Ziesch, V., Conrod, P., Cattrell, A., Walter, H., Gallinat, J., Bromberg, U., Jurk, S., Menningen, E., Frouin, V., Papadopoulos Orfanos, D., Stringaris, A., Penttilä, J., van Noort, B., Grimmer, Y., Schumann, G., Smolka, M.N., Martinot, J., Lemaître, H., & Imagen consortium (2020). Sex effects on structural maturation of the limbic system and outcomes on emotional regulation during adolescence. NeuroImage, 210, 116441. https://doi.org/10.1016/j.neuroimage.2019.116441.

Friend, L., Weed, J., Sandoval, P., Nufer, T., Ostlund, I., & Edwards, J. G. (2017). CB1-dependent long-term depression in ventral tegmental area GABA neurons: A novel target for marijuana. *Journal of Neuroscience*, *37*(45), 10943–10954. https://doi.org/10.1523/JNEUROSCI.0190-17.2017.

Frontera, J. L., Gonzalez Pini, V. M., Messore, F. L., & Brusco, A. (2018). Exposure to cannabinoid agonist WIN 55,212-2 during early adolescence increases alcohol preference and anxiety in CD1 mice. *Neuropharmacology*, *137*, 268–274. https://doi.org/10.1016/j.neuropharm.2018.05.018.

Glodosky, N. C., Cuttler, C., Freels, T. G., Wright, H. R., Rojas, M. J., Baglot, S. L., et al. (2020). Cannabis vapor self-administration elicits sex- and dose-specific alterations in stress reactivity in rats. *Neurobiology of Stress*, *13*, 100260. https://doi.org/10.1016/j.ynstr.

Goddings, A. L., Mills, K. L., Clasen, L. S., Giedd, J. N., Viner, R. M., & Blakemore, S. J. (2014). The influence of puberty on subcortical brain development. *NeuroImage*, *88*, 242–251. https://doi.org/10.1016/j.neuroimage.2013.09.073.

Goldman-Rakic, P. S. (1995). Cellular basis of working memory. *Neuron*, *14*(3), 477–485. https://doi.org/10.1016/0896-6273(95)90304-6.

González, S., Bisogno, T., Wenger, T., Manzanares, J., Milone, A., Berrendero, F., et al. (2000). Sex steroid influence on cannabinoid CB1 receptor mRNA and endocannabinoid levels in the anterior pituitary gland. *Biochemical and Biophysical Research Communications*, *270*(1), 260–266. https://doi.org/10.1006/bbrc.2000.2406.

Hammerslag, L. R., & Gulley, J. M. (2016). Sex differences in behavior and neural development and their role in adolescent vulnerability to substance use. *Behavioural Brain Research*, *298*, 15–26. https://doi.org/10.1016/j.bbr.2015.04.008.

Hasin, D. S., Shmulewitz, D., & Sarvet, A. L. (2019). Time trends in US cannabis use and cannabis use disorders overall and by sociodemographic subgroups: A narrative review and new findings. *The American Journal of Drug and Alcohol Abuse*, *45*(6), 623–643. https://doi.org/10.1080/00952990.2019.1569668.

Heng, L., Beverley, J. A., Steiner, H., & Tseng, K. Y. (2011). Differential developmental trajectories for CB1 cannabinoid receptor expression in limbic/associative and sensorimotor cortical areas. *Synapse*, *65*(4), 278–286. https://doi.org/10.1002/syn.20844.

Hernandez-Avila, C. A., Rounsaville, B. J., & Kranzler, H. R. (2004). Opioid-, cannabis- and alcohol-dependent women show more rapid progression to substance abuse treatment. *Drug and Alcohol Dependence*, *74*(3), 265–272. https://doi.org/10.1016/j.drugalcdep.2004.02.001.

Higuera-Matas, A., Botreau, F., Miguéns, M., Del Olmo, N., Borcel, E., Pérez-Alvarez, L., et al. (2009). Chronic periadolescent cannabinoid treatment enhances adult hippocampal PSA-NCAM expression in male Wistar rats but only has marginal effects on anxiety, learning and memory. *Pharmacology, Biochemistry, and Behavior*, *93*(4), 482–490. https://doi.org/10.1016/j.pbb.2009.06.013.

Higuera-Matas, A., Soto-Montenegro, M. L., del Olmo, N., Miguéns, M., Torres, I., Vaquero, J. J., et al. (2008). Augmented acquisition of cocaine self-administration and altered brain glucose metabolism in adult female but not male rats exposed to a cannabinoid agonist during adolescence. *Neuropsychopharmacology*, *33*(4), 806–813. https://doi.org/10.1038/sj.npp.1301467.

Jacobus, J., & Tapert, S. F. (2014). Effects of cannabis on the adolescent brain. *Current Pharmaceutical Design*, *20*(13), 2186–2193. https://doi.org/10.2174/13816128113199990426.

Jernigan, C. S., Goswami, D. B., Austin, M. C., Iyo, A. H., Chandran, A., Stockmeier, C. A., et al. (2011). The mTOR signaling pathway in the prefrontal cortex is compromised in major depressive disorder. *Progress in Neuropsychopharmacology and Biological Psychiatry*, *35*, 1774–1779. https://doi.org/10.1016/j.pnpbp.2011.05.010.

Johnson, R. M., Fairman, B., Gilreath, T., Xuan, Z., Rothman, E. F., Parnham, T., et al. (2015). Past 15-year trends in adolescent marijuana use: Differences by race/ethnicity and sex. *Drug and Alcohol Dependence*, *155*, 8–15. https://doi.org/10.1016/j.drugalcdep.2015.08.025.

Johnston, L. D., Miech, R. A., O'Malley, P. M., Bachman, J. G., Schulenberg, J. E., & Patrick, M. E. (2019). *Monitoring the future national survey results on drug use 1975–2018: Overview, key findings on adolescent drug use.* Ann Arbor: Institute for Social Research, University of Michigan.

Katona, I., & Freund, T. F. (2012). Multiple functions of endocannabinoid signaling in the brain. *Annual Review of Neuroscience*, *35*, 529–558. https://doi.org/10.1146/annurev-neuro-062111-150420.

Khan, S. S., Secades-Villa, R., Okuda, M., Wang, S., Pérez-Fuentes, G., Kerridge, B. T., et al. (2013). Gender differences in cannabis use disorders: Results from the National Epidemiologic Survey of alcohol and related conditions. *Drug and Alcohol Dependence*, *130*(1–3), 101–108. https://doi.org/10.1016/j.drugalcdep.2012.10.015.

Kirschmann, E. K., McCalley, D. M., Edwards, C. M., & Torregrossa, M. M. (2017). Consequences of adolescent exposure to the cannabinoid receptor agonist WIN55,212-2 on working memory in female rats. *Frontiers in Behavioral Neuroscience*, *11*, 137. https://doi.org/10.3389/fnbeh.2017.00137.

Kirschmann, E. K., Pollock, M. W., Nagarajan, V., & Torregrossa, M. M. (2017). Effects of adolescent cannabinoid self-administration in rats on addiction-related behaviors and working memory. *Neuropsychopharmacology*, *42*, 989–1000. https://doi.org/10.1038/npp.2016.178.

Klanker, M., Feenstra, M., & Denys, D. (2013). Dopaminergic control of cognitive flexibility in humans and animals. *Frontiers in Neuroscience*, 7, 201. https://doi.org/10.3389/fnins.2013.00201.

Kozlovsky, N., Belmaker, R. H., & Agam, G. (2001). Low GSK-3 activity in frontal cortex of schizophrenic patients. *Schizophrenia Research*, *52*, 101–105. https://doi.org/10.1016/s0920-9964(00)00174-2.

Lenroot, R. K., Gogtay, N., Greenstein, D. K., Wells, E. M., Wallace, G. L., Clasen, L. S., et al. (2007). Sexual dimorphism of brain developmental trajectories during childhood and adolescence. *NeuroImage*, *36*(4), 1065–1073. https://doi.org/10.1016/j.neuroimage.2007.03.053.

Llorente-Berzal, A., Puighermanal, E., Burokas, A., Ozaita, A., Maldonado, R., Marco, E. M., et al. (2013). Sex-dependent psychoneuroendocrine effects of THC and MDMA in an animal model of adolescent drug consumption. *PLoS One*, *8*(11), e78386. https://doi.org/10.1371/journal.pone.0078386.

Lovelace, J. W., Corches, A., Vieira, P. A., Hiroto, A. S., Mackie, K., & Korzus, E. (2015). An animal model of female adolescent cannabinoid exposure elicits a long-lasting deficit in presynaptic long-term plasticity. *Neuropharmacology*, *99*(5), 242–255. https://doi.org/10.1016/j.neuropharm.2015.04.034.

Maccarrone, M., Guzmán, M., Mackie, K., Doherty, P., & Harkany, T. (2014). Programming of neural cells by (endo) cannabinoids: From physiological rules to emerging therapies. *Nature Reviews Neuroscience*, *15*(12), 786–801. https://doi.org/10.1038/nrn3846.

Mateo, Y., Johnson, K. A., Covey, D. P., Atwood, B. K., Wang, H. L., Zhang, S., et al. (2017). Endocannabinoid actions on cortical terminals orchestrate local modulation of dopamine release in the nucleus accumbens. *Neuron*, *96*(5), 1112–1126.e5. https://doi.org/10.1016/j.neuron.2017.11.012.

Mateos, B., Borcel, E., Loriga, R., Luesu, W., Bini, V., Llorente, R., et al. (2011). Adolescent exposure to nicotine and/or the cannabinoid agonist CP 55,940 induces gender-dependent long-lasting memory impairments and changes in brain nicotinic and CB(1) cannabinoid receptors. *Journal of Psychopharmacology*, *25*(12). https://doi.org/10.1177/0269881110370503. 1676–1590.

McLaughlin, R. J. (2018). Toward a translationally relevant preclinical model of cannabis use. *Neuropsychopharmacology*, *43*(1), 213. https://doi.org/10.1038/npp.2017.191.

McLaughlin, R. J., Hill, M. N., & Gorzalka, B. B. (2014). A critical role for prefrontocortical endocannabinoid signaling in the regulation of stress and emotional behavior. *Neuroscience & Biobehavioral Reviews*, *42*, 116–131. https://doi.org/10.1016/j.neubiorev.2014.02.006.

Mechoulam, R., Hanuš, L. O., Pertwee, R., & Howlett, A. C. (2014). Early phytocannabinoid chemistry to endocannabinoids and beyond. *Nature Reviews Neuroscience*, *15*(11), 757–764. https://doi.org/10.1038/nrn3811.

Medina, K. L., McQueeny, T., Nagel, B. J., Hanson, K. L., Yang, T. T., & Tapert, S. F. (2009). Prefrontal cortex morphometry in abstinent adolescent marijuana users: Subtle gender effects. *Addiction Biology*, *14*(4), 457–468. https://doi.org/10.1111/j.1369-1600.2009.00166.x.

Meyer, H. C., Lee, F. S., & Gee, D. G. (2018). The role of the endocannabinoid system and genetic variation in adolescent brain development. *Neuropsychopharmacology Reviews*, *43*, 21–33. https://doi.org/10.1038/npp.2017.143.

Miller, M. L., Chadwick, B., Dickstein, D. L., Purushothaman, I., Egervari, G., Rahman, T., et al. (2019). Adolescent exposure to Δ9-tetrahydrocannabinol alters the transcriptional trajectory and dendritic architecture of prefrontal pyramidal neurons. *Molecular Psychiatry*, *24*(4), 588–600. https://doi.org/10.1038/s41380-018-0243-x.

Mutlu, A. K., Schneider, M., Debbané, M., Badoud, D., Eliez, S., & Schaer, M. (2013). Sex differences in thickness, and folding developments throughout the cortex. *NeuroImage*, *82*, 200–207. https://doi.org/10.1016/j.neuroimage.2013.05.076.

Navarrete, M., & Araque, A. (2008). Endocannabinoids mediate neuron-astrocyte communication. *Neuron*, *57*(6), 883–893. https://doi.org/10.1016/j.neuron.2008.01.029.

Newton, S. S., Thome, J., Wallace, T. L., Shirayama, Y., Schlesinger, L., Sakai, N., et al. (2002). Inhibition of cAMP response element-binding protein or dynorphin in the nucleus accumbens produces an antidepressant-like effect. *Journal of Neuroscience*, *22*(24), 10883–10890. https://doi.org/10.1523/JNEUROSCI.22-24-10883.2002.

Okaneku, J., Vearrier, D., McKeever, R. G., LaSala, G. S., & Greenberg, M. I. (2015). Change in perceived risk associated with marijuana use in the United States from 2002 to 2012. *Clinical Toxicology*, *53*(3), 151–155. https://doi.org/10.3109/15563650.2015.1004581.

Pennuto, M., Bonanomi, D., Benfenati, F., & Valtorta, F. (2003). Synaptophysin I controls the targeting of VAMP2/synaptobrevin II to synaptic vesicles. *Molecular Biology of the Cell*, *14*(12), 4909–4919. https://doi.org/10.1091/mbc.e03-06-0380.

Poulia, N., Delis, F., Brakatselos, C., Lekkas, P., Kokras, N., Dalla, C., et al. (2020). Escalating low-dose Δ9-tetrahydrocannabinol exposure during adolescence induces differential behavioral and neurochemical effects in male and female adult rats. *European Journal of Neuroscience*, *52*(1), 2681–2693. https://doi.org/10.1111/ejn.14598.

Prini, P., Rusconi, F., Zamberletti, E., Gabaglio, M., Penna, F., Fasano, M., et al. (2018). Adolescent THC exposure in female rats leads to cognitive deficits through a mechanism involving chromatin modifications in the prefrontal cortex. *Journal of Psychiatry & Neuroscience*, *43*(2), 87–101. https://doi.org/10.1503/jpn.170082.

Quinn, H. R., Matsumoto, I., Callaghan, P. D., Long, L. E., Arnold, J. C., Gunasekaran, N., et al. (2008). Adolescent rats find repeated delta(9)-THC less aversive than adult rats but display greater residual cognitive deficits and changes in hippocampal protein expression following exposure. *Neuropsychopharmacology*, *33*(5), 1113–1126. https://doi.org/10.1038/sj.npp.1301475.

Raver, S, Haughwout, S, & Keller, A. (2013). Adolescent cannabinoid exposure permanently suppresses cortical oscillations in adult mice. *Neuropsychopharmacology*, *38*(12), 2338–2347. https://doi.org/10.1038/npp.2013.164.

Raznahan, A., Shaw, P. W., Lerch, J. P., Clasen, L. S., Greenstein, D., Berman, R., et al. (2014). Longitudinal four-dimensional mapping of subcortical anatomy in human development. *Proceedings of the National Academy of Sciences U S A*, *111*(4), 1592–1597. https://doi.org/10.1073/pnas.1316911111.

Realini, N., Vigano, D., Guidali, C., Zamberletti, E., Rubino, T., & Parolaro, D. (2011). Chronic URB597 treatment at adulthood reverted most depressive-like symptoms induced by adolescent exposure to THC in female rats. *Neuropharmacology*, *60*(2–3), 235–243. https://doi.org/10.1016/j.neuropharm.2010.09.003.

Reichardt, L. F. (2006). Neurotrophin-regulated signaling pathways. *Philosophical Transactions of the Royal Society B: Biological Sciences*, *361*(1473), 1545–1564. https://doi.org/10.1098/rstb.2006.1894.

Renard, J., Krebs, M. O., Jay, T. M., & Le Pen, G. (2013). Long-term cognitive impairments induced by chronic cannabinoid exposure during adolescence in rats: A strain comparison. *Psychopharmacology*, *225*, 781–790. https://doi.org/10.1007/s00213-012-2865-z.

Renard, J., Rosen, L. G., Loureiro, M., De Oliveira, C., Schmid, S., Rushlow, W. J., et al. (2017). Adolescent cannabinoid exposure induces a persistent sub-cortical hyper-dopaminergic state and associated molecular adaptations in the prefrontal cortex. *Cerebral Cortex*, *27*(2), 1297–1310. https://doi.org/10.1093/cercor/bhv335.

Renard, J., Szkudlarek, H. J., Kramar, C. P., Jobson, C., Moura, K., Rushlow, W. J., et al. (2017). Adolescent THC exposure causes enduring prefrontal cortical disruption of GABAergic inhibition and dysregulation of sub-cortical dopamine function. *Scientific Reports*, 7(1), 11420. https://doi.org/10.1038/s41598-017-11645-8.

Rodríguez de Fonseca, F., Ramos, J. A., Bonnin, A., & Fernández-Ruiz, J. J. (1993). Presence of cannabinoid binding sites in the brain from early postnatal ages. *Neuroreport*, *4*(2), 135–138. https://doi.org/10.1097/00001756-199302000-00005.

Rubino, T., Prini, P., Piscitelli, F., Zamberletti, E., Trusel, M., Melis, M., et al. (2015). Adolescent exposure to THC in female rats disrupts developmental changes in the prefrontal cortex. *Neurobiology of Disease*, *73*, 60–69. https://doi.org/10.1016/j.nbd.2014.09.015.

Rubino, T., Realini, N., Braida, D., Alberio, T., Capurro, V., Viganò, D., et al. (2009). The depressive phenotype induced in adult female rats by adolescent exposure to THC is associated with cognitive impairment and altered neuroplasticity in the prefrontal cortex. *Neurotoxicity Research*, *15*(4), 291–302. https://doi.org/10.1016/j.nbd.2014.09.015.

Rubino, T., Realini, N., Braida, D., Guidi, S., Capurro, V., Viganò, D., et al. (2009). Changes in hippocampal morphology and neuroplasticity induced by adolescent THC treatment are associated with cognitive impairment in adulthood. *Hippocampus*, *19*(8), 763–772. https://doi.org/10.1002/hipo.20554.

Rubino, T., Realini, N., Guidali, C., Braida, D., Capurro, V., Castiglioni, C., et al. (2008). Chronic Δ 9-tetrahydrocannabinol during adolescence provokes sex-dependent changes in the emotional profile in adult rats: Behavioral and biochemical correlates. *Neuropsychopharmacology*, *33*(11), 2760–2771. https://doi.org/10.1038/sj.npp.1301664.

Ruiz, C. M., Torrens, A., Castillo, E., Perrone, C. R., Cevallos, J., Inshishian, V. C., et al. (2021). Pharmacokinetic, behavioral, and brain activity effects of Δ^9-tetrahydrocannabinol in adolescent male and female rats. *Neuropsychopharmacology*, *46*(5), 959–969. https://doi.org/10.1038/s41386-020-00839-w.

Schepis, T. S., Desai, R. A., Cavallo, D. A., Smith, A. E., McFetridge, A., Liss, T. B., et al. (2011). Gender differences in adolescent marijuana use and associated psychosocial characteristics. *Journal of Addiction Medicine*, *5*(1), 65–73. https://doi.org/10.1097/ADM.0b013e3181d8dc62.

Scherma, M., Dessì, C., Muntoni, A. L., Lecca, S., Satta, V., Luchicchi, A., et al. (2016). Adolescent Δ 9-tetrahydrocannabinol exposure alters WIN55,212-2 self-administration in adult rats. *Neuropsychopharmacology*, *41*(5), 1416–1426. https://doi.org/10.1038/npp.2015.295.

Schneider, M., & Koch, M. (2003). Chronic pubertal, but not adult chronic cannabinoid treatment impairs sensorimotor gating, recognition memory, and the performance in a progressive ratio task in adult rats. *Neuropsychopharmacology*, *28*, 1760–1769. https://doi.org/10.1038/sj.npp.1300225.

Schuermeyer, J., Salomonsen-Sautel, S., Price, R. K., Balan, S., Thurstone, C., Min, S. J., et al. (2014). Temporal trends in marijuana attitudes, availability and use in Colorado compared to non-medical marijuana states: 2003–11. *Drug and Alcohol Dependence*, *140*, 145–155. https://doi.org/10.1016/j.drugalcdep.2014.04.016.

Scott, J. C., Rosen, A., Moore, T. M., Roalf, D. R., Satterthwaite, T. D., Calkins, M. E., et al. (2019). Cannabis use in youth is associated with limited alterations in brain structure. *Neuropsychopharmacology*, *44*(8), 1362–1369. https://doi.org/10.1038/s41386-019-0347-2.

Scott, J. C., Slomiak, S. T., Jones, J. D., Rosen, A., Moore, T. M., & Gur, R. C. (2018). Association of cannabis with cognitive functioning in adolescents and young adults: A systematic review and meta-analysis. *JAMA Psychiatry*, *75*(6), 585–595. https://doi.org/10.1001/jamapsychiatry.2018.0335.

Spear, L. P. (2000). The adolescent brain and age-related behavioral manifestations. *Neuroscience and Biobehavioral Reviews*, *24*(4), 417–463. https://doi.org/10.1016/s0149-7634(00)00014-2.

Spear, L. P. (2018). Effects of adolescent alcohol consumption on the brain and behaviour. *Nature Reviews Neuroscience*, *19*(4), 197. https://doi.org/10.1016/s0149-7634(00)00014-2.

Squeglia, L. M., Jacobus, J., & Tapert, S. F. (2009). The influence of substance use on adolescent brain development. *Clinical EEG and Neuroscience*, *40*(1), 31–38. https://doi.org/10.1177/155005940904000110.

Stringfield, S. J., & Torregrossa, M. M. (2021). Disentangling the lasting effects of adolescent cannabinoid exposure. *Progress in Neuropsychopharmacology & Biological Psychiatry*, *104*, 110067. https://doi.org/10.1016/j.pnpbp.2020.110067.

Substance Abuse and Mental Health Services Administration. (2020). *Key substance use and mental health indicators in the United States: Results from the 2019 National Survey on drug use and health (HHS publication no. PEP20-07-01-001, NSDUH series H-55)*. Rockville, MD: Center for Behavioral Health Statistics and Quality, Substance Abuse and Mental Health Services Administration. Retrieved from https://www.samhsa.gov/data/.

Szutorisz, H., DiNieri, J., Sweet, E., Egervari, G., Michaelides, M., Carter, J., et al. (2014). Parental THC exposure leads to compulsive heroin-seeking and altered striatal synaptic plasticity in the subsequent generation. *Neuropsychopharmacology*, *39*(6), 1315–1323. https://doi.org/10.1038/npp.2013.352.

Toledo, E., Lebel, A., Becerra, L., Minster, A., Linnman, C., Maleki, N., et al. (2012). The young brain and concussion: Imaging as a biomarker for diagnosis and prognosis. *Neuroscience & Biobehavioral Reviews*, *36*(6), 1510–1531. https://doi.org/10.1016/j.neubiorev.2012.03.007.

Torrens, A., Vozella, V., Huff, H., McNeil, B., Ahmed, F., Ghidini, A., et al. (2020). Comparative pharmacokinetics of Δ^9-tetrahydrocannabinol in adolescent and adult male mice. *The Journal of Pharmacology and Experimental Therapeutics*, *374*(1), 151–160. https://doi.org/10.1124/jpet.120.265892.

Watson, C. T., Szutorisz, H., Garg, P., Martin, Q., Landry, J. A., Sharp, A. J., et al. (2015). Genome-wide DNA methylation profiling reveals epigenetic changes in the rat nucleus accumbens associated with cross-generational effects of adolescent THC exposure. *Neuropsychopharmacology*, *40*(13), 2993–3005. https://doi.org/10.1038/npp.2015.155.

Weiland, B. J., Thayer, R. E., Depue, B. E., Sabbineni, A., Bryan, A. D., & Hutchison, K. E. (2015). Daily marijuana use is not associated with brain morphometric measures in adolescents or adults. *Journal of Neuroscience*, *35*(4), 1505–1512. https://doi.org/10.1523/JNEUROSCI.2946-14.2015.

Westermeyer, J., & Boedicker, A. E. (2000). Course, severity, and treatment of substance abuse among women versus men. *The American Journal of Drug and Alcohol Abuse*, *26*(4), 523–535. https://doi.org/10.1081/ada-100101893.

Wiley, J. L., & Burston, J. J. (2014). Sex differences in Δ9-tetrahydrocannabinol metabolism and in vivo pharmacology following acute and repeated dosing in adolescent rats. *Neuroscience Letters*, *576*, 51–55. https://doi.org/10.1016/j.neulet.2014.05.057.

Zamberletti, E., Beggiato, S., Steardo, L., Prini, P., Antonelli, T., Ferraro, L., et al. (2014). Alterations of prefrontal cortex GABAergic transmission in the complex psychotic-like phenotype induced by adolescent delta-9-tetrahydrocannabinol exposure in rats. *Neurobiology of Disease*, *63*, 35–47. https://doi.org/10.1016/j.nbd.2013.10.028.

Zamberletti, E., Gabaglio, M., Grilli, M., Prini, P., Catanese, A., Pittaluga, A., et al. (2016). Long-term hippocampal glutamate synapse and astrocyte dysfunctions underlying the altered phenotype induced by adolescent THC treatment in male rats. *Pharmacological Research*, *111*, 459–470. https://doi.org/10.1016/j.phrs.2016.07.008.

Zamberletti, E., Gabaglio, M., Prini, P., Rubino, T., & Parolaro, D. (2015). Cortical neuroinflammation contributes to long-term cognitive dysfunctions following adolescent delta-9-tetrahydrocannabinol treatment in female rats. *European Neuropsychopharmacology*, *25*(12), 2404–2415. https://doi.org/10.1016/j.euroneuro.2015.09.021.

Zuo, X. N., Kelly, C., Di Martino, A., Mennes, M., Margulies, D. S., Bangaru, S., et al. (2010). Growing together and growing apart: Regional and sex differences in the lifespan developmental trajectories of functional homotopy. *Journal of Neuroscience*, *30*(45), 15034–15043. https://doi.org/10.1523/JNEUROSCI.2612-10.2010.

CHAPTER FIVE

Adolescent opioid abuse: Role of glial and neuroimmune mechanisms

S. Rahman[a,*], Z.I. Rahman[b], P.J. Ronan[c,d], K. Lutfy[e], and R.L. Bell[f]
[a]Department of Pharmaceutical Sciences, South Dakota State University, Brookings, SD, United States
[b]University of Minnesota Medical School, Minneapolis, MN, United States
[c]Department of Psychiatry and Basic Biomedical Sciences, University of South Dakota Sanford School of Medicine, Sioux Falls, SD, United States
[d]Research Service, Sioux Falls VA Healthcare System, Sioux Falls, SD, United States
[e]Department of Pharmaceutical Sciences, College of Pharmacy, Western University of Health Sciences, Pomona, CA, United States
[f]Department of Psychiatry, Indiana University School of Medicine, Indianapolis, IN, United States
*Corresponding author: e-mail address: shafiqur.rahman@sdstate.edu

Contents

1. Introduction 148
2. Glial cells and brain development 150
3. Neuroimmune response and adolescent brain development 151
4. Opioid exposure and glial activation 152
5. Opioid exposure and glial neuroimmune signaling 154
6. Glial targets, opioid abuse, and therapeutic significance 157
7. Conclusions 158
References 158

Abstract

Opioids are widely prescribed for pain management, and prescription opioid misuse in adolescents has become a major epidemic in the United States and worldwide. Emerging data indicate that adolescence represents a critical period of brain development, and exposure to opioids during adolescence may increase the risk of addiction in adulthood. There is growing evidence that disruptions in brain glial function may be implicated in numerous chronic neuropathologies. Evidence suggests that glial mechanisms have an important role in the development and maintenance of opioid abuse and the risk for addiction. This review will describe glial and neuroimmune mechanisms involved in opioid use disorders during adolescence, which may increase substance use disorder liability later in life. Moreover, this review will identify some important neuro-glial targets, involved in opioid abuse and addiction, to develop future preventions and treatment strategies.

International Review of Neurobiology, Volume 161
ISSN 0074-7742
https://doi.org/10.1016/bs.irn.2021.07.003

Abbreviations

CNS central nervous system
CPP conditioned place preference
GFAP glial fibrillary acidic protein
GPCRs G-protein coupled receptors
Iba1 ionized calcium binding adaptor molecule 1
MAPKs mitogen-activated protein kinases
MORs μ opioid receptors
MyD88 myeloid differentiation primary response 88
NAc nucleus accumbens
NF-κB nuclear factor-κB
NMUPO nonmedical use of prescription opioids
TLRs toll-like receptors
TLR4 toll-like receptor 4

1. Introduction

Prescription opioid misuse and opioid use disorders (OUDs) represent a public health concern in the United States and worldwide (e.g., McCabe et al., 2019; Peacock et al., 2018). Evidence suggests an association between medical use of prescription opioids and subsequent nonmedical use of prescription opioids (NMUPO) during adolescence and young adulthood (McCabe, Schulenberg, McCabe, & Veliz, 2020; McCabe et al., 2019; McCabe, Veliz, & Schulenberg, 2016; Miech, Johnston, O'Malley, Keyes, & Heard, 2015). The prevalence rates of substance use disorders (SUDs) are significantly high among adolescents who report medical use of prescription opioids and history of NMUPO based on epidemiological data (McCabe et al., 2016, 2020). However, the neurobiological mechanisms of opioid abuse during adolescence remain unclear.

Adolescence in humans is a critical stage of brain maturation and behavioral development (e.g., Klinker, Cross, & Leslie, 2019; Lacagnina, Rivera, & Bilbo, 2017; Salmanzadeh et al., 2020; Schepis, Adinoff, & Rao, 2008). Emerging evidence indicates that this unique developmental phase is highly sensitive to numerous environmental influences, including opioid abuse (Bava, Jacobus, Thayer, & Tapert, 2013; Bell, Franklin, Hauser, & Engleman, 2013; Schwarz & Bilbo, 2013; Schwarz, Hutchinson, & Bilbo, 2011). A growing body of evidence suggests that adolescent opioid exposure contributes to long-lasting changes in brain development, which subsequently compromises adult behavior and increases the vulnerability to addiction and cognitive dysfunction (Hanson, Cummins, Tapert, & Brown, 2011; Kuhn, Wilson, & Swartzwelder, 2013; Moss, Chen, & Yi, 2014; Rutherford, Mayes, & Potenza, 2010; Vestal-Laborde, Eschenroeder, Bigbee, Robinson, & Sato-Bigbee, 2014). Thus, the trajectory and maturation of the central nervous system (CNS) development during adolescence increases its vulnerability to the effects of opioids. Simply put, exposure to licit and illicit drugs during adolescence disrupts CNS maturation, which leads to long-term behavioral dysfunction (Guerri & Pascual, 2019; Winters, Fahnhorst, Botzet, Lee, & Lalone, 2012). Adolescent opioid use also induces behavioral despair after sustained abstinence in animal models, suggesting potential comorbidity between depression and opiate addiction (Lutz, Reiss, Ouagazzal, & Kieffer, 2013), although it is difficult to disentangle anhedonia due to opioid withdrawal from the former. Previous data have also shown that adolescent morphine treatment significantly facilitates the development of tolerance to its analgesic effects and increases withdrawal signs in rodent models (Salmanzadeh, Azizi, & Semnanian, 2017). This suggests that neuroadaptation in the CNS may continue into adulthood. However, morphine treatment during adolescence had parameter-specific effects in adulthood, such that it did not affect morphine-induced onset or peak response of lateral paragigantocellular neurons (Salmanzadeh, Azizi, Ahmadi Soleimani, Pachenari, & Semnanian, 2018). Studies also evaluated the effect of opioid drug administration on impulsive-like behavior. Results from behavioral tests, related to reaction-time, 24 h and 25 days after administration of morphine to adolescent rats found a profound effect on motor impulsive behavior later in adulthood (Moazen, Azizi, Salmanzadeh, & Semnanian, 2018). Regarding pharmacological effects of chronic morphine administration during adolescence, this treatment increased pain responsiveness, in the formalin test, in adult rats compared to controls (Ghasemi, Pachenari, Semnanian, & Azizi, 2018). Adolescent oxycodone exposure resulted in increased sensitivity to the rewarding effects of opioids in adult mice (Sanchez, Carpenter, Yohn, & Blendy, 2016). Taken together, the neurobehavioral evidence suggests that opioid exposure during adolescence can induce a variety of effects different from those observed following exposure in adulthood (Spear, 2000, 2015). Therefore, adolescent opioid exposure can adversely influence neural, behavioral, and cognitive functioning long after this early-life exposure. While the role of neuronal mechanisms is somewhat known, growing evidence suggests that disruptions in brain glial function are implicated in chronic neuroadaptations. This review will describe the glial and neuroimmune mechanisms involved in opioid use and/or abuse during adolescence that have been shown to increase drug abuse liability later in life. Moreover, this review will explore important neuro-glial targets involved in opioid abuse and addiction-related disorders.

2. Glial cells and brain development

Glial cells are widely distributed throughout the CNS and are broadly classified based on morphological and functional characteristics (Bachtell, Jones, Heinzerling, Beardsley, & Comer, 2017; Klinker et al., 2019; Lacagnina et al., 2017; Rahman & Alzarea, 2019). The primary type of glial cells in the CNS include astrocytes, microglia, oligodendrocytes, and ependymal cells. Astrocytes are known to provide numerous structural and metabolic support functions for neurons and regulate the extracellular environment by clearing neurotransmitters, buffering ion concentrations, and releasing signaling molecules (Sofroniew & Vinters, 2010). In addition, astrocytes also play an essential role in immune cell trafficking and activation in the CNS (c.f., Rahman & Alzarea, 2019). Microglia are part of the innate immune system known as resident macrophages that maintain immunological functions in the CNS (c.f., Rahman & Alzarea, 2019). Functionally, microglia are immunocompetent cells that detect pathogens and engage in phagocytosis, lysosomal degradation, and secretion of various pro- and anti-inflammatory substances (c.f., Graeber, 2010). Oligodendrocytes are glial cells that wrap around neuronal axons, forming a myelin sheath that insulates the nerve fiber from the extracellular fluid and allows for more efficient propagation of axonal signals (Nave, 2010). Ependymal cells line the spinal cord and brain ventricles, including the blood-brain barrier (BBB), where these cells produce, secrete, and circulate cerebrospinal fluid. Collectively, glial cells are important and are underappreciated targets, or regulators, of CNS function (Bachtell et al., 2017; Klinker et al., 2019; Lacagnina et al., 2017; Menassa & Gomez-Nicola, 2018; Rahman & Alzarea, 2019; Reemst, Noctor, Lucassen, & Hol, 2016).

The past few decades have witnessed a greater interest in the importance of glia in CNS development, plasticity and disease, including SUDs (Klinker et al., 2019; Lacagnina et al., 2017; Zuchero & Barres, 2015). For instance, although the study of brain development has largely focused on neuronal development in the past, neuronal development coincides with the development of neuroglia, mediating their role in CNS development (Menassa & Gomez-Nicola, 2018; Reemst et al., 2016). Quantification methods indicate that at least 50% of all cells in the human brain are glia, with considerable variations and differences between studies and different brain regions, as well as differences in neuro-glial ratios across development (Azevedo et al., 2009; Lyck et al., 2009). For example, the glia to neuron ratio of the cerebral cortex is approximately 3.76, whereas it is only 0.23 in the cerebellum (Azevedo et al., 2009). Additionally, the finding that the human brain is made up of at least 50% glial cells has also been observed in primates and rodents, indicating a critical role for glia in numerous brain functions across multiple species (c.f., Reemst et al., 2016). For example, radial gill cells direct microglial migration, after which microglia control radial glial differentiation. On the other hand, astrocytes affect microglia maturation, which can affect radial glial cells and neuronal functionality. Thus, as a highly interactive/integrated network, the developing CNS is strongly dependent on appropriate and coordinated signaling between neurons and glia, which affects adolescent brain function (c.f., Menassa & Gomez-Nicola, 2018; Reemst et al., 2016). Although microglia are the resident macrophages of the brain, recent research demonstrates that they have functions in addition to biological defense (Wu, Dissing-Olesen, MacVicar, & Stevens, 2015). Similarly, astrocytes have functions beyond their supportive role to also influence neurocircuit development (Kim, Nabekura, & Koizumi, 2017). Both microglia and astrocytes exhibit significant changes in morphology, gene expression, and function following drug exposure, as seen by increasing evidence that microglia and astrocytes can perpetuate the detrimental effects of drugs of abuse on the brain (Klinker et al., 2019; Lacagnina et al., 2017; Melbourne, Thompson, Peng, & Nixon, 2019). Thus, drug exposure during the critical period of adolescence may alter the interactive functions of glial cells with neurons well into adulthood.

3. Neuroimmune response and adolescent brain development

The innate immune responses, including toll-like receptors (TLRs), of neuroglia mediate many actions of drugs-of-abuse on the brain, which contributes to drug addiction-related behaviors (c.f., Guerri & Pascual, 2019). For example, activated microglia produce and release numerous pro-inflammatory cytokines and other mediators, such as prostanoids, as well as free radicals associated with neuroinflammation (Kettenmann, Hanisch, Noda, & Verkhratsky, 2011). Furthermore, both microglia and astrocytes express innate immune receptors, such as TLRs and cytoplasmic NOD-like immune receptors (NLRs) (Alfonso-Loeches, Ureña-Peralta, Morillo-Bargues, et al., 2016; Blanco, Perez-Arago, Fernandez-Lizarbe, & Guerri, 2008). Stimulation of TLRs triggers signaling pathways, such as the mammalian target of rapamycin (mTOR) and nuclear transcription factor (NF-κB) pathways, which causes the production of cytokines, interleukins, and other inflammatory mediators (Montesinos, Pascual, Rodríguez-Arias, Miñarro, & Guerri, 2016). These receptors participate in neuroinflammation and the long-term behavioral and neuropathological dysfunction induced by drug use and/or abuse during adolescence (Hutchinson et al., 2012; Zhu et al., 2018). Although astrocytes and microglia are the main immune cells that express several TLRs (Okun, Griffioen, & Mattson, 2011), neurons can also express TLR4. However, the signaling pathways associated with a TLR4 response differ between cell types (Leow-Dyke, Allen, Denes, et al., 2012; Okun et al., 2011). Under normal conditions, the release of cytokines by glial cells modulates neuronal function by regulating synaptic and neural plasticity. Nevertheless, pro-inflammatory cytokines can impair synaptic plasticity and may underlie cognitive disturbances, memory dysfunction, as well as mood disorders (Khairova, Machado-Vieira, Du, & Manji, 2009). Here we discuss evidence indicating the participation of the neuroimmune response of glial cells, the influence of the innate immune TLRs in neuroinflammation, and long-term behavioral dysfunction associated with opioid use and/or abuse in adolescence.

4. Opioid exposure and glial activation

Accumulating evidence indicates that opioids stimulate glial cells via different opioid receptors (Fig. 1) expressed on the cell membrane (c.f., Zhang, Largent-Milnes, & Vanderah, 2020). The opioid receptors include μ, δ, and κ subtypes and these receptor subtypes are known heterotrimeric Gi/o protein-coupled receptors (GPCRs; Corbett, Henderson, McKnight, & Paterson, 2006; Waldhoer, Bartlett, & Whistler, 2004). Furthermore, μ opioid receptors (MORs) are the primary receptor subtype mediating the rewarding and reinforcing effects of opioids. Previous studies have shown that these opioid receptors are expressed on both microglia (Chao et al., 1996; Dobrenis, Makman, & Stefano, 1995; Mika et al., 2014) and astrocytes (Gurwell et al., 1996; Meyer et al., 2017; Nam et al., 2018; Woo et al., 2018). Interestingly, activated glial opioid receptors produce mixed results (c.f., Zhang et al., 2020). For example, several studies have

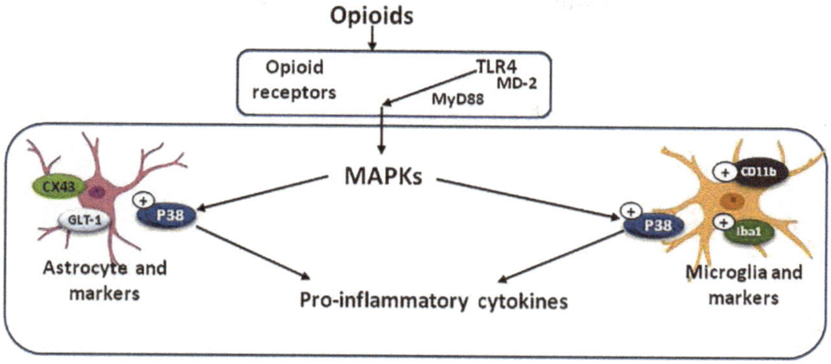

Fig. 1 Opioids and putative glial mechanisms. Opioids activate microglia and astrocytes via opioid receptors, and the toll-like receptor (TLR4). Opioids are direct agonists at TLR4 and bind to a unique binding site in the myeloid differentiation factor 2 (MD-2) domain of the TLR4 receptor. TLR4 activation causes the translocation of the transcription factor nuclear factor kappa-light-chain-enhancer of activated B cells (NF-κβ) to the nucleus via the myeloid differentiation primary response 88 (MyD88) pathway and MyD88-independent pathways. The stimulation of TLR4 by opioids causes the activity of intracellular mitogen-activated protein kinases (MAPKs) that control the formation of microglial or astrocytic markers and pro-inflammatory cytokines. Opioid-induced glial activation also involves other receptors and signaling mechanisms. Protein P38 is associated with MAPK signaling. See text for detail. Astrocyte activation markers connexin 43 (CX43) and glutamate transporter-1 (GLT-1); Microglial activation markers such as cluster of differentiation molecule 11b (CD11b), Ionized calcium binding adaptor molecule 1 (Iba1).

reported that stimulation of glial opioid receptors can reduce functional cellular activity (Chao et al., 1997; Hu, Chao, Hegg, Thayer, & Peterson, 2000). Furthermore, previous studies have shown that activation of opioid receptors can stimulate immune system (Muscoli et al., 2010). A recent study also found that activating astrocytic MORs induced glutamate release from astrocytes in the hippocampus (Nam et al., 2019). This resulted in a conditioned place preference (CPP), demonstrating a stimulatory role of Gi/o protein-mediated signaling by MORs and other GPCRs (Nam et al., 2019). It is possible that unique downstream cell signaling pathways are coupled with opioid receptors under distinct pathophysiological conditions, which may contribute to this inconsistency (Al-Hasani & Bruchas, 2011). Whether opioid glial activation and subsequent internal cell signaling pathways across development are similar, remain to be investigated.

Additional studies have demonstrated that administration of opioid drugs increases the expression of glial fibrillary acidic protein (GFAP), an activation marker for astrocytes, and ionized calcium binding adaptor molecule 1 (Iba1), an activation marker for microglia, in animal models (Beitner-Johnson, Guitart, & Nestler, 1993; García-Pérez, Luisa Laorden, Núñez, & Victoria Milanés, 2014). Furthermore, morphine administration altered glial signaling facilitating the release of various cytokines and chemokines (Schwarz et al., 2011; Zhang et al., 2017). These findings suggest a strong association between opioids and neuroinflammatory signaling.

There is increasing evidence that TLR4 is a potential target where opioids (Fig. 1) can cause an innate immune response (Zhang et al., 2020). For example, opioid administration stimulates mitogen-activated protein kinase (MAPK) signaling downstream of TLR4, including p38, JNK, and ERK (Wang et al., 2012; Zhang et al., 2012). Interestingly, genetic deletion of TLR4 blocks this effect, suggesting a critical role for the TLR4 in opioid effects. Additionally, opioid-induced TLR4 stimulation promotes the production of pro-inflammatory signals from glial cells, which can be prevented by glial activation inhibitors or selective TLR4 antagonists (Eidson & Murphy, 2013; Liang, Chu, Jiang, & Yuan, 2016). It is noteworthy to mention that morphine metabolites with no opioid receptor activity can bind and activate TLR4 (Due et al., 2012). This finding suggests that opioid-induced central immune signaling does not necessarily require activity at classical opioid receptors and their downstream signaling. Overall, these studies indicate that adolescence is a sensitive period for opioid-induced changes in the TLR4 signaling pathway as well as microglial function of microglia in MCL reward circuitry (c.f., Guerri & Pascual, 2019). This alteration in microglial function enhances the risk for adverse consequences later in life, since opioid exposure induces long-term changes in the neurocircuitry mediating OUDs (Schwarz & Bilbo, 2013). Given the role of the innate immune system in neuropsychiatric disorders, it is possible that adolescent opioid abuse may influence these disorders later in life through a complex mechanism, which will require further investigation (c.f., Guerri & Pascual, 2019).

5. Opioid exposure and glial neuroimmune signaling

As previously described, microglia and astrocytes are the primary cells in the CNS mediating an immune response; which, at least in part, protects the CNS from the effects of stress and pathogens by upregulating inflammatory signaling (c.f., Guerri & Pascual, 2019). Several studies indicate that the brain integrates neuronal and immune communication/signaling and that peripheral immune dysregulation and inflammation can alter brain function as well (c.f., Guerri & Pascual, 2019; Pavlov, Chavan, & Tracey, 2018). Pharmacological manipulation of TLR4 inhibits opioid-induced dopamine release in the nucleus accumbens (NAc) as well as its CPP and self-administration (Chen et al., 2017; Hutchinson et al., 2012). The latter authors reported that the rewarding effects of opioids are mediated, at least in part, via a TLR4-triggered intracellular pathway. Similarly, intra-NAc injection of a p38 inhibitor was found to prevent the acquisition and maintenance of a morphine-induced CPP (Zhang et al., 2012). Furthermore, deletion of TLR4 prevented the phosphorylation of p38 and JNK, downstream signaling of TLR4-associated MyD88, a molecular inflammatory pathway (Hutchinson et al., 2012). A more recent study reported that selectively blocking microglial MyD88 activity did not change the acquisition of opioid reward but did prolong the extinction of this reward memory in preclinical models (Rivera et al., 2019). These results suggest that microglial TLR4-MyD88 activity (Fig. 1) plays an important role in opioid-induced rewarding effects and reinforcement behavior (Jacobsen, Watkins, & Hutchinson, 2014).

Further studies have indicated that TLR4's effects are partially mediated by proinflammatory cytokine tumor necrosis factor alpha (TNFα) activity (Eidson, Inoue, Young, Tansey, & Murphy, 2017). These authors also indicated that TNFα, an inflammatory pathway end-product, and the sequestration of soluble TNFα prevent TLR4-mediated opioid-induced neuroinflammation. Thus, TNFα may modulate opioid reward processing since it contributes to altered opioid sensitivity in humans (Reyes-Gibby et al., 2008). Additionally, Niwa et al. (2007) reported that TNFα disrupted a morphine-induced CPP. Interleukin-1β (IL-1β) is another important proinflammatory cytokine thought to mediate TLR4-induced alterations in neuronal plasticity (Guerri & Pascual, 2019). Like TNFα, IL-1β is also a product of TLR4-signaling since its production involves the activation of TLR4 with a concomitant increase in pro-IL-1β expression (Hutchinson et al., 2011). IL-1β has been found to play a role in morphine-induced reward-related behavior (Zhang et al., 2020). For example, single nucleotide polymorphism in the *IL1B* gene was found to be associated with an increased risk of opioid, and alcohol, dependence in humans (Liu, Hutchinson, White, Somogyi, & Coller, 2009). Previous studies revealed that opioid addiction shares similar characteristics with chronic pain, indicating possible common mechanisms and neuroadaptations between these two neuropathologies (Elman & Borsook, 2016). In addition, two polymorphisms of brain-derived neurotrophic factor (BDNF) were found to be associated with the age of onset in heroin-dependent patients, suggesting an important modulatory role for BDNF in opioid abuse and addiction (Jia et al., 2011; Meng et al., 2012). Taylor et al. (2016) found that chronic opioid use activates microglia and its release of BDNF in the ventral tegmental area (VTA). A previous study showed a single dose of BDNF into the VTA of naive animals recapitulated opioid-associated withdrawal behavior (Vargas-Perez et al., 2009). This supports a role for microglial-derived BDNF in the regulation of opioid-induced rewarding effects and related behavior.

Besides the inflammatory activity discussed above, previous reports have indicated that opioid-induced neuroimmune activation via TLR4 contributes to deficits in glutamate turnover by inhibiting glutamate uptake (Eidson et al., 2017). In addition, chronic opioid exposure reduces the expression of both astrocytic glutamate transporters (GLT-1) as well as glutamate aspartate transporters, likely diminishing synaptic clearance of glutamate (Ozawa, Nakagawa, Shige, Minami, & Satoh, 2001; Shen, Scofield, Boger, Hensley, & Kalivas, 2014). Thus, restoring deficits in GLT-1 levels could prevent many chronic opioid-induced behavioral effects (c.f., Bachtell et al., 2017). Further studies indicate that administration of the GLT-1 activator, MS-153, attenuated opioid-induced CPP; whereas overexpression of GLT-1 in the NAc prevented an opioid-induced CPP (Fujio et al., 2005). Interestingly, several studies suggest that injection of the beta-lactam antibiotic, ceftriaxone, upregulates endogenous GLT-1 levels and may effectively reduce opioid-induced hyperthermia, tolerance to this effect, and reinstatement of heroin-seeking behaviors (Rawls et al., 2010; Shen et al., 2014). Further evidence has indicated that reducing opioid-induced increases in glutamate with chronic administration of the cysteine pro-drug, *N*-acetylcysteine, effectively reduces heroin-seeking behaviors (Zhou & Kalivas, 2008). Importantly, both ceftriaxone and *N*-acetylcysteine display anti-inflammatory properties making it difficult to separate their ability to suppress immune activation from their effects on restoring glutamate homeostasis (c.f., Bachtell et al., 2017). Taken together, these data indicate that opioid-induced neuroimmune signaling and disrupted glutamate regulation may be important mechanisms that alter the behavioral effects of acute and/or chronic opioid administration. Therefore, given the critical role of glial neuroimmune mechanisms in opioid effects, it is clear that adolescent opioid use likely alters neuroimmune signaling controlling neural development (e.g., Brenhouse & Schwarz, 2016).

6. Glial targets, opioid abuse, and therapeutic significance

Given the above, it is clear that glial activity significantly influences the abuse potential of opioids. Therefore, glial cells are potential targets to develop drugs that prevent or treat opioid abuse. As of now, the following agents have been investigated, which target glial pathways to reverse glial activation and associated neuroinflammatory and/or other neuroimmune mechanisms.

For example, peroxisome proliferator-activated receptor-γ (PPARγ) is a nuclear receptor expressed by both microglia and astrocytes (Warden et al., 2016; Zhang et al., 2020). Several preclinical studies indicate that pioglitazone, a PPARγ agonist, effectively suppresses opioid reward-related behavior in animal models (de Guglielmo et al., 2017, 2015). PPARγ may represent a novel therapeutic strategy for treating opioid addiction, as the activation of this nuclear receptor significantly suppresses the production of pro-inflammatory cytokines from microglia and astrocytes (Kielian & Drew, 2003). Consistent with this, a recent study reported that pioglitazone has pharmacological efficacy in altering the addiction liability of oxycodone in opioid users (Jones et al., 2016). A small molecule Ibudilast, a glial modulator and non-selective phosphodiesterase (PDE) inhibitor, reduces neurobiological markers associated with opioid-induced pro-inflammatory responses, while it attenuates both antagonist-precipitated and abstinence-induced morphine withdrawal in animal models (Hutchinson et al., 2009). More recently, ibudilast was found to reduce opioid-related withdrawal symptoms (Cooper, Johnson, Pavlicova, et al., 2016). These findings show the potential clinical utility of glial modulators for treating opioid withdrawal in humans. Minocycline, a tetracycline antibiotic, inhibits the activation of microglia. In preclinical studies, minocycline prevented the development of tolerance to opioid analgesia and opioid-induced hyperalgesia (Arout, Waters, MacLean, Compton, & Sofuoglu, 2019). This exploratory investigation sought to determine if minocycline changes pain threshold and tolerance in individuals with OUDs maintained on an opioid agonist treatment (Arout et al., 2019). These studies indicate minocycline is a potential therapeutic for OUDs and other SUDs. Further preclinical and clinical investigations will provide additional insights into its potential therapeutic utility.

7. Conclusions

In this review, we have explored evidence that opioid abuse during adolescence can modify brain functions and produce drug-specific neurobehavioral consequences. The preclinical and clinical data discussed above indicate that exposure to opioids during adolescence can cause long-lasting neurobiological, cognitive, affective, and behavioral dysfunctions. The net result is an impaired brain function that lasts into adulthood. Moreover, the development of multiple brain regions is particularly sensitive to adolescent opioid exposure. In addition, we have reviewed that neuroimmune signaling pathways mediate opioid-induced glial activation and associated cellular and molecular mechanisms. Furthermore, recent studies have revealed an important role for neuroimmune mechanisms, including glial activation, in many actions of opioids in the brain. These actions include neural damage, cognitive dysfunction, and alterations to reward neurocircuitry. Therefore, targeting neuroimmune signaling, including glial activation is a viable therapeutic strategy for future prevention and treatment of OUDs.

References

Alfonso-Loeches, S., Ureña-Peralta, J., Morillo-Bargues, M. J., et al. (2016). Ethanol-induced TLR4/NLRP3 neuroinflammatory response in microglial cells promotes leukocyte infiltration across the BBB. *Neurochemical Research*, *41*, 193–209. https://doi.org/10.1007/s11064-015-1760-5.

Al-Hasani, R., & Bruchas, M. R. (2011). Molecular mechanisms of opioid receptor-dependent signaling and behavior. *Anesthesiology*, *115*, 1363–1381. https://doi.org/10.1097/ALN.0b013e318238bba6.

Arout, C. A., Waters, A. J., MacLean, R. R., Compton, P., & Sofuoglu, M. (2019). Minocycline does not affect experimental pain or addiction-related outcomes in opioid maintained patients. *Psychopharmacology*, *236*, 2857–2866. https://doi.org/10.1007/s00213-018-5146-7.

Azevedo, F. A. C., Carvalho, L. R. B., Grinberg, L. T., Farfel, J. M., Ferretti, R. E. L., Leite, R. E. P., et al. (2009). Equal numbers of neuronal and nonneuronal cells make the human brain an isometrically scaled-up primate brain. *Journal of Comparative Neurology*, *513*, 532–541. https://doi.org/10.1002/cne.21974.

Bachtell, R. K., Jones, J. D., Heinzerling, K. G., Beardsley, P. M., & Comer, S. D. (2017). Glial and neuroinflammatory targets for treating substance use disorders. *Drug and Alcohol Dependence*, *180*, 156–170. https://doi.org/10.1016/j.drugalcdep.2017.08.003.

Bava, S., Jacobus, J., Thayer, R. E., & Tapert, S. F. (2013). Longitudinal changes in white matter integrity among adolescent substance users. *Alcoholism: Clinical and Experimental Research*, *37, Suppl 1*(Suppl. 1), E181–E189. https://doi.org/10.1111/j.1530-0277.2012.01920.x.

Beitner-Johnson, D., Guitart, X., & Nestler, E. J. (1993). Glial fibrillary acidic protein and the mesolimbic dopamine system: Regulation by chronic morphine and Lewis-Fischer strain differences in the rat ventral tegmental area. *Journal of Neurochemistry*, *61*, 1766–1773. https://doi.org/10.1111/j.1471-4159.1993.tb09814.x.

Bell, R. L., Franklin, K. M., Hauser, S., & Engleman, E. A. (2013). Next stop dependence-binge drinking on the road to alcoholism: Preclinical findings on its neurobiology from rat animal models. In S. B. Harris (Ed.), *Binge eating and binge drinking* (pp. 1–60). New York: Nova Science Publishers Inc.

Blanco, A. M., Perez-Arago, A., Fernandez-Lizarbe, S., & Guerri, C. (2008). Ethanol mimics ligand-mediated activation and endocytosis of IL-1RI/TLR4 receptors via lipid rafts caveolae in astroglial cells. *Journal of Neurochemistry*, *106*, 625–639. https://doi.org/10.1111/j.1471-4159.2008.05425.x.

Brenhouse, H. C., & Schwarz, J. M. (2016). Immunoadolescence: Neuroimmune development and adolescent behavior. *Neuroscience and Biobehavioral Reviews*, *70*, 288–299. https://doi.org/10.1016/j.neubiorev.2016.05.035.

Chao, C. C., Gekker, G., Hu, S., Sheng, W. S., Shark, K. B., Bu, D. F., et al. (1996). Kappa opioid receptors in human microglia downregulate human immunodeficiency virus 1 expression. *Proceedings of the National Academy of Sciences of the United States of America*, *93*, 8051–8056. https://doi.org/10.1073/pnas.93.15.8051.

Chao, C. C., Hu, S., Shark, K. B., Sheng, W. S., Gekker, G., & Peterson, P. K. (1997). Activation of mu opioid receptors inhibits microglial cell chemotaxis. *The Journal of Pharmacology and Experimental Therapeutics*, *281*, 998–1004. https://pubmed.ncbi.nlm.nih.gov/9152411/.

Chen, J.-X., Huang, K.-M., Liu, M., Jiang, J.-X., Liu, J.-P., Zhang, Y.-X., et al. (2017). Activation of TLR4/STAT3 signaling in VTA contributes to the acquisition and maintenance of morphine-induced conditioned place preference. *Behavioural Brain Research*, *335*, 151–157. https://doi.org/10.1016/j.bbr.2017.08.022.

Cooper, Z. D., Johnson, K. W., Pavlicova, M., et al. (2016). The effects of ibudilast, a glial activation inhibitor, on opioid withdrawal symptoms in opioid-dependent volunteers. *Addiction Biology*, *21*, 895–903. https://doi.org/10.1111/adb.12261.

Corbett, A. D., Henderson, G., McKnight, A. T., & Paterson, S. J. (2006). 75 years of opioid research: The exciting but vain quest for the holy grail. *British Journal of Pharmacology*, *147*(Suppl. 1), S153–S162. https://doi.org/10.1038/sj.bjp.0706435.

de Guglielmo, G., Kallupi, M., Scuppa, G., Demopulos, G., Gaitanaris, G., & Ciccocioppo, R. (2017). Pioglitazone attenuates the opioid withdrawal and vulnerability to relapse to heroin seeking in rodents. *Psychopharmacology*, *234*, 223–234. https://doi.org/10.1007/s00213-016-4452-1.

de Guglielmo, G., Melis, M., De Luca, M. A., Kallupi, M., Li, H. W., Niswender, K., et al. (2015). PPARγ activation attenuates opioid consumption and modulates mesolimbic dopamine transmission. *Neuropsychopharmacology*, *40*, 927–937. https://doi.org/10.1038/npp.2014.268.

Dobrenis, K., Makman, M. H., & Stefano, G. B. (1995). Occurrence of the opiate alkaloid-selective μ3 receptor in mammalian microglia, astrocytes and Kupffer cells. *Brain Research*, *686*, 239–248. https://doi.org/10.1016/0006-8993(95)00452-v.

Due, M. R., Piekarz, A. D., Wilson, N., Feldman, P., Ripsch, M. S., Chavez, S., et al. (2012). Neuroexcitatory effects of morphine-3-glucuronide are dependent on Toll-like receptor 4 signaling. *Journal of Neuroinflammation*, *9*, 200. https://doi.org/10.1186/1742-2094-9-200.

Eidson, L. N., Inoue, K., Young, L. J., Tansey, M. G., & Murphy, A. Z. (2017). Toll-like receptor 4 mediates morphine-induced neuroinflammation and tolerance via soluble tumor necrosis factor signaling. *Neuropsychopharmacology*, *42*, 661–670. https://doi.org/10.1038/npp.2016.131.

Eidson, L. N., & Murphy, A. Z. (2013). Blockade of Toll-like receptor 4 attenuates morphine tolerance and facilitates the pain relieving properties of morphine. *The Journal of Neuroscience*, *33*, 15952–15963. https://doi.org/10.1523/JNEUROSCI.1609-13.2013.

Elman, I., & Borsook, D. (2016). Common brain mechanisms of chronic pain and addiction. *Neuron*, *89*, 11–36. https://doi.org/10.1016/j.neuron.2015.11.027.

Fujio, M., Nakagawa, T., Sekiya, Y., Ozawa, T., Suzuki, Y., Minami, M., et al. (2005). Gene transfer of GLT-1, a glutamate transporter, into the nucleus accumbens shell attenuates methamphetamine- and morphine-induced conditioned place preference in rats. *The European Journal of Neuroscience*, *22*, 2744–2754. https://doi.org/10.1111/j.1460-9568.2005.04467.x.

García-Pérez, D., Luisa Laorden, M., Núñez, C., & Victoria Milanés, M. (2014). Glial activation and midkine and pleiotrophin transcription in the ventral tegmental area are modulated by morphine administration. *Journal of Neuroimmunology*, *274*, 244–248. https://doi.org/10.1016/j.jneuroim.2014.07.017.

Ghasemi, E., Pachenari, N., Semnanian, S., & Azizi, H. (2018). Adolescent morphine exposure increases nociceptive behaviors in rat model of formalin test. *Developmental Psychobiology*, *61*, 254–260. https://doi.org/10.1002/dev.21790.

Graeber, M. B. (2010). Changing face of microglia. *Science*, *330*, 783–788. https://doi.org/10.1126/science.1190929.

Guerri, C., & Pascual, M. (2019). Impact of neuroimmune activation induced by alcohol or drug abuse on adolescent brain development. *International Journal of Developmental Neuroscience*, 77, 89–98. https://doi.org/10.1016/j.ijdevneu.2018.11.006.

Gurwell, J. A., Duncan, M. J., Maderspach, K., Stiene-Martin, A., Elde, R. P., & Hauser, K. F. (1996). Kappa-opioid receptor expression defines a phenotypically distinct subpopulation of astroglia: Relationship to Ca2+ mobilization, development, and the antiproliferative effect of opioids. *Brain Research*, *737*, 175–187. https://doi.org/10.1016/0006-8993(96)00728-7.

Hanson, K. L., Cummins, K., Tapert, S. F., & Brown, S. A. (2011). Changes in neuropsychological functioning over 10 years following adolescent substance abuse treatment. *Psychology of Addictive Behaviors: Journal of the Society of Psychologists in Addictive Behaviors*, *25*, 127–142. https://doi.org/10.1037/a0022350.

Hu, S., Chao, C. C., Hegg, C. C., Thayer, S., & Peterson, P. K. (2000). Morphine inhibits human microglial cell production of, and migration towards, RANTES. *Journal of Psychopharmacology*, *14*, 238–243. https://doi.org/10.1177/026988110001400307.

Hutchinson, M. R., Lewis, S. S., Coats, B. D., Skyba, D. A., Crysdale, N. Y., Berkelhammer, D. L., et al. (2009). Reduction of opioid withdrawal and potentiation of acute opioid analgesia by systemic AV411 (ibudilast). *Brain, Behavior, and Immunity*, *23*, 240–250. https://doi.org/10.1016/j.bbi.2008.09.012.

Hutchinson, M. R., Northcutt, A. L., Hiranita, T., Wang, X., Lewis, S. S., Thomas, J., et al. (2012). Opioid activation of toll-like receptor 4 contributes to drug reinforcement. *The Journal of Neuroscience*, *32*, 11187–111200. https://doi.org/10.1523/JNEUROSCI.0684-12.2012.

Hutchinson, M. R., Shavit, Y., Grace, P. M., Rice, K. C., Maier, S. F., & Watkins, L. R. (2011). Exploring the neuroimmunopharmacology of opioids: An integrative review of mechanisms of central immune signaling and their implications for opioid analgesia. *Pharmacological Reviews*, *63*, 772–810. https://doi.org/10.1124/pr.110.004135.

Jacobsen, J. H. W., Watkins, L. R., & Hutchinson, M. R. (2014). Discovery of a novel site of opioid action at the innate immune pattern-recognition receptor TLR4 and its role in addiction. *International Review of Neurobiology*, *118*, 129–163. https://doi.org/10.1016/B978-0-12-801284-0.00006-3.

Jia, W., Shi, J. G., Wu, B., Ao, L., Zhang, R., & Zhu, Y. S. (2011). Polymorphisms of brain-derived neurotrophic factor associated with heroin dependence. *Neuroscience Letters*, *495*, 221–224.

Jones, J. D., Sullivan, M. A., Manubay, J. M., Mogali, S., Metz, V. E., Ciccocioppo, R., et al. (2016). The effects of pioglitazone, a PPARγ receptor agonist, on the abuse liability of oxycodone among nondependent opioid users. *Physiology & Behavior*, *159*, 33–39. https://doi.org/10.1016/j.physbeh.2015.10.006.

Kettenmann, H., Hanisch, U. K., Noda, M., & Verkhratsky, A. (2011). Physiology of microglia. *Physiological Reviews*, *91*, 461–553. https://doi.org/10.1152/physrev.00011.2010.

Khairova, R. A., Machado-Vieira, R., Du, J., & Manji, H. K. (2009). A potential role for pro-inflammatory cytokines in regulating synaptic plasticity in major depressive disorder. *The International Journal of Neuropsychopharmacology*, *12*, 561–578. https://doi.org/10.1017/S1461145709009924.

Kielian, T., & Drew, P. D. (2003). Effects of peroxisome proliferator-activated receptor-? Agonists on central nervous system inflammation. *Journal of Neuroscience Research*, *71*, 315–325. https://doi.org/10.1002/jnr.10501.

Kim, S. K., Nabekura, J., & Koizumi, S. (2017). Astrocyte-mediated synapse remodeling in the pathological brain. *Glia*, *65*, 1719–1727. https://doi.org/10.1002/glia.23169.

Klinker, K. E., Cross, S. J., & Leslie, F. M. (2019). Glial mechanisms underlying substance use disorders. *The European Journal of Neuroscience*, *50*, 2574–2589. https://doi.org/10.1111/ejn.14163.

Kuhn, C. M., Wilson, W., & Swartzwelder, S. (2013). Enduring effects of adolescent drug exposure. *Biological Psychiatry*, *74*, 480–481. https://doi.org/10.1016/j.biopsych.2013.07.030.

Lacagnina, M. J., Rivera, P. D., & Bilbo, S. D. (2017). Glial and neuroimmune mechanisms as critical modulators of drug use and abuse. *Neuropsychopharmacology*, *42*, 156–177. https://doi.org/10.1038/npp.2016.121.

Leow-Dyke, S., Allen, C., Denes, A., et al. (2012). Neuronal toll-like receptor 4 signaling induces brain endothelial activation and neutrophil transmigration *in vitro*. *Journal of Neuroinflammation*, *9*, 230. https://doi.org/10.1186/1742-2094-9-230.

Liang, Y., Chu, H., Jiang, Y., & Yuan, L. (2016). Morphine enhances IL-1β release through toll-like receptor 4-mediated endocytic pathway in microglia. *Purinergic Signal*, *12*, 637–645. https://doi.org/10.1007/s11302-016-9525-4.

Liu, L., Hutchinson, M. R., White, J. M., Somogyi, A. A., & Coller, J. K. (2009). Association of IL-1B genetic polymorphisms with an increased risk of opioid and alcohol dependence. *Pharmacogenetics and Genomics*, *19*, 869–876. https://doi.org/10.1097/FPC.0b013e328331e68f.

Lutz, P. E., Reiss, D., Ouagazzal, A. M., & Kieffer, B. L. (2013). A history of chronic morphine exposure during adolescence increases despair-like behaviour and strain-dependently promotes sociability in abstinent adult mice. *Behavioural Brain Research*, *243*, 44–52. https://doi.org/10.1016/j.bbr.2012.12.049.

Lyck, L., Santamaria, I. D., Pakkenberg, B., Chemnitz, J., Schrøder, H. D., Finsen, B., et al. (2009). An empirical analysis of the precision of estimating the numbers of neurons and glia in human neocortex using a fractionator-design with sub-sampling. *Journal of Neuroscience Methods*, *182*, 143–156. https://doi.org/10.1016/j.jneumeth.2009.06.003.

McCabe, S. E., Schulenberg, J., McCabe, V. V., & Veliz, P. T. (2020). Medical use and misuse of prescription opioids in US 12th-grade youth: School-level correlates. *Pediatrics*, *146*, e20200387. https://doi.org/10.1542/peds.2020-0387.

McCabe, S. E., Veliz, P. T., Boyd, C. J., Schepis, T. S., McCabe, V. V., & Schulenberg, J. E. (2019). A prospective study of nonmedical use of prescription opioids during adolescence and subsequent substance use disorder symptoms in early midlife. *Drug and Alcohol Dependence*, *194*, 377–385. https://doi.org/10.1016/j.drugalcdep.2018.10.027.

McCabe, S. E., Veliz, P., & Schulenberg, J. E. (2016). Adolescent context of exposure to prescription opioids and substance use disorder symptoms at age 35: A national longitudinal study. *Pain*, *157*, 2173–2178. https://doi.org/10.1097/j.pain.0000000000000624.

Melbourne, J. K., Thompson, K. R., Peng, H., & Nixon, K. (2019). Its complicated: The relationship between alcohol and microglia in the search for novel pharmacotherapeutic targets for alcohol use disorders. *Progress in Molecular Biology and Translational Science*, *167*, 179–221. https://doi.org/10.1016/bs.pmbts.2019.06.011.

Menassa, D. A., & Gomez-Nicola, D. (2018). Microglial dynamics during human brain development. *Frontiers in Immunology*, *9*, 1014. https://doi.org/10.3389/fimmu.2018.01014.

Meng, C., Lan, J., Wang, Y., Song, M., Gao, X., Ran, L., et al. (2012). Influence of brain-derived neurotrophic factor genetic polymorphisms on the ages of onset for heroin dependence in a Chinese population. *Genetic Testing and Molecular Biomarkers*, *16*, 1044–1050. https://doi.org/10.1089/gtmb.2012.0016.

Meyer, L. C., Paisley, C. E., Mohamed, E., Bigbee, J. W., Kordula, T., Richard, H., et al. (2017). Novel role of the nociceptin system as a regulator of glutamate transporter expression in developing astrocytes. *Glia*, *65*, 2003–2023. https://doi.org/10.1002/glia.23210.

Miech, R., Johnston, L., O'Malley, P. M., Keyes, K. M., & Heard, K. (2015). Prescription opioids in adolescence and future opioid misuse. *Pediatrics*, *136*, e1169–e1177. https://doi.org/10.1542/peds.2015-1364.

Mika, J., Popiolek-Barczyk, K., Rojewska, E., Makuch, W., Starowicz, K., & Przewlocka, B. (2014). Delta-opioid receptor analgesia is independent of microglial activation in a rat model of neuropathic pain. *PLoS One*, *9*, e104420. 10 https://doi.org/10.1371/journal.pone.0104420.

Moazen, P., Azizi, H., Salmanzadeh, H., & Semnanian, S. (2018). Adolescent morphine exposure induces immediate and long-term increases in impulsive behavior. *Psychopharmacology*, *235*, 3423–3434. https://doi.org/10.1007/s00213-018-5051-0.

Montesinos, J., Pascual, M., Rodríguez-Arias, M., Miñarro, J., & Guerri, C. (2016). Involvement of TLR4 in the long-term epigenetic changes, rewarding and anxiety effects induced by intermittent ethanol treatment in adolescence. *Brain, Behavior, and Immunity*, *53*, 159–171. https://doi.org/10.1016/j.bbi.2015.12.006.

Moss, H. B., Chen, C. M., & Yi, H.-Y. (2014). Early adolescent patterns of alcohol, cigarettes, and marijuana polysubstance use and young adult substance use outcomes in a nationally representative sample. *Drug and Alcohol Dependence*, *136*, 51–62. https://doi.org/10.1016/j.drugalcdep.2013.12.011.

Muscoli, C., Doyle, T., Dagostino, C., Bryant, L., Chen, Z., Watkins, L. R., et al. (2010). Counter-regulation of opioid analgesia by glial-derived bioactive sphingolipids. *The Journal of Neuroscience*, *30*, 15400–15408. https://doi.org/10.1523/JNEUROSCI.2391-10.2010.

Nam, M. H., Han, K. S., Lee, J., Won, W., Koh, W., Bae, J. Y., et al. (2019). Activation of astrocytic μ-opioid receptor causes conditioned place preference. *Cell Reports*, *28*, 1154–1166.e5. https://doi.org/10.1016/j.celrep.2019.06.071.

Nam, M. H., Han, K. S., Lee, J., Bae, J. Y., An, H., Park, S., et al. (2018). Expression of μ-opioid receptor in CA1 hippocampal astrocytes. *Experimental Neurobiology*, *27*, 120–128. https://doi.org/10.5607/en.2018.27.2.120.

Nave, K. A. (2010). Myelination and support of axonal integrity by glia. *Nature*, *468*, 244–252. https://doi.org/10.1038/nature09614.

Niwa, M., Nitta, A., Yamada, Y., Nakajima, A., Saito, K., Seishima, M., et al. (2007). Tumor necrosis factor-α and its inducer inhibit morphine-induced rewarding effects and sensitization. *Biological Psychiatry*, *62*, 658–668. https://doi.org/10.1016/j.biopsych.2006.10.009.

Okun, E., Griffioen, K. J., & Mattson, M. P. (2011). Toll-like receptor signaling in neural plasticity and disease. *Trends in Neurosciences*, *34*, 269–281. https://doi.org/10.1016/j.tins.2011.02.005.

Ozawa, T., Nakagawa, T., Shige, K., Minami, M., & Satoh, M. (2001). Changes in the expression of glial glutamate transporters in the rat brain accompanied with morphine dependence and naloxone-precipitated withdrawal. *Brain Research*, *905*, 254–258. https://doi.org/10.1016/s0006-8993(01)02536-7.

Pavlov, V. A., Chavan, S. S., & Tracey, K. J. (2018). Molecular and functional neuroscience in immunity. *Annual Review of Immunology, 36*, 783–812. https://doi.org/10.1146/annurev-immunol-042617-053158.

Peacock, A., Leung, J., Larney, S., Colledge, S., Hickman, M., Rehm, J., et al. (2018). Global statistics on alcohol, tobacco and illicit drug use: 2017 status report. *Addiction, 113*, 1905–1926. https://doi.org/10.1111/add.14234.

Rahman, S., & Alzarea, S. (2019). Glial mechanisms underlying major depressive disorder: Potential therapeutic opportunities. *Progress in Molecular Biology and Translational Science, 167*, 159–178. https://doi.org/10.1016/bs.pmbts.2019.06.010.

Rawls, S. M., Zielinski, M., Patel, H., Sacavage, S., Baron, D. A., & Patel, D. (2010). Beta-lactam antibiotic reduces morphine analgesic tolerance in rats through GLT-1 transporter activation. *Drug and Alcohol Dependence, 107*, 261–263. https://doi.org/10.1016/j.drugalcdep.2009.10.010.

Reemst, K., Noctor, S. C., Lucassen, P. J., & Hol, E. M. (2016). The indispensable roles of microglia and astrocytes during brain development. *Frontiers in Human Neuroscience, 10*, 566. https://doi.org/10.3389/fnhum.2016.00566.

Reyes-Gibby, C. C., El Osta, B., Spitz, M. R., Parsons, H., Kurzrock, R., Wu, X., et al. (2008). The influence of tumor necrosis factor-alpha −308 G/A and IL-6-174 G/C on pain and analgesia response in lung cancer patients receiving supportive care. *Cancer Epidemiology, Biomarkers & Prevention, 17*, 3262–3267. https://doi.org/10.1158/1055-9965.EPI-08-0125.

Rivera, P. D., Hanamsagar, R., Kan, M. J., Tran, P. K., Stewart, D., Jo, Y. C., et al. (2019). Removal of microglial-specific MyD88 signaling alters dentate gyrus doublecortin and enhances opioid addiction-like behaviors. *Brain, Behavior, and Immunity, 76*, 104–115. https://doi.org/10.1016/j.bbi.2018.11.010.

Rutherford, H. J. V., Mayes, L. C., & Potenza, M. N. (2010). Neurobiology of adolescent substance use disorders: Implications for prevention and treatment. *Child and Adolescent Psychiatric Clinics of North America, 19*, 479–492. https://doi.org/10.1016/j.chc.2010.03.003.

Salmanzadeh, H., Ahmadi-Soleimani, S. M., Pachenari, N., Azadi, M., Halliwell, R. F., Rubino, T., et al. (2020). Adolescent drug exposure: A review of evidence for the development of persistent changes in brain function. *Brain Research Bulletin, 156*, 105–117. https://doi.org/10.1016/j.brainresbull.2020.01.007.

Salmanzadeh, H., Azizi, H., Ahmadi Soleimani, S. M., Pachenari, N., & Semnanian, S. (2018). Chronic adolescent morphine exposure alters the responses of lateral paragigantocellular neurons to acute morphine administration in adulthood. *Brain Research Bulletin, 137*, 178–186. https://doi.org/10.1016/j.brainresbull.2017.12.007.

Salmanzadeh, H., Azizi, H., & Semnanian, S. (2017). Adolescent chronic escalating morphine administration induces long lasting changes in tolerance and dependence to morphine in rats. *Physiology & Behavior, 174*, 191–196. https://doi.org/10.1016/j.physbeh.2017.03.014.

Sanchez, V., Carpenter, M. D., Yohn, N. L., & Blendy, J. A. (2016). Long-lasting effects of adolescent oxycodone exposure on reward-related behavior and gene expression in mice. *Psychopharmacology, 233*, 3991–4002. https://doi.org/10.1007/s00213-016-4425-4.

Schepis, T. S., Adinoff, B., & Rao, U. (2008). Neurobiological processes in adolescent addictive disorders. *The American Journal on Addictions, 17*, 6–23. https://doi.org/10.1080/10550490701756146.

Schwarz, J. M., & Bilbo, S. D. (2013). Adolescent morphine exposure affects long-term microglial function and later-life relapse liability in a model of addiction. *The Journal of Neuroscience, 33*, 961–971. https://doi.org/10.1523/JNEUROSCI.2516-12.2013.

Schwarz, J. M., Hutchinson, M. R., & Bilbo, S. D. (2011). Early-life experience decreases drug-induced reinstatement of morphine CPP in adulthood via microglial-specific epigenetic programming of anti-inflammatory IL-10 expression. *The Journal of Neuroscience, 31*, 17835–17847. https://doi.org/10.1523/JNEUROSCI.3297-11.2011.

Shen, H., Scofield, M. D., Boger, H., Hensley, M., & Kalivas, P. W. (2014). Synaptic glutamate spillover due to impaired glutamate uptake mediates heroin relapse. *The Journal of Neuroscience*, *34*, 5649–5657. https://doi.org/10.1523/JNEUROSCI.4564-13.2014.

Sofroniew, M. V., & Vinters, H. V. (2010). Astrocytes: Biology and pathology. *Acta Neuropathologica*, *119*, 7–35. https://doi.org/10.1007/s00401-009-0619-8.

Spear, L. P. (2000). The adolescent brain and age-related behavioral manifestations. *Neuroscience and Biobehavioral Reviews*, *24*, 417–463. https://doi.org/10.1016/s0149-7634(00)00014-2.

Spear, L. P. (2015). Adolescent alcohol exposure: Are there separable vulnerable periods within adolescence? *Physiology & Behavior*, *148*, 122–130. https://doi.org/10.1016/j.physbeh.2015.01.027.

Taylor, A. M. W., Castonguay, A., Ghogha, A., Vayssiere, P., Pradhan, A. A. A., Xue, L., et al. (2016). Neuroimmune regulation of GABAergic neurons within the ventral tegmental area during withdrawal from chronic morphine. *Neuropsychopharmacology*, *41*, 949–959. https://doi.org/10.1038/npp.2015.221.

Vargas-Perez, H., Ting-A Kee, R., Walton, C. H., Hansen, D. M., Razavi, R., Clarke, L., et al. (2009). Ventral tegmental area BDNF induces an opiate-dependent-like reward state in naive rats. *Science*, *324*, 1732–1734. https://doi.org/10.1126/science.1168501.

Vestal-Laborde, A. A., Eschenroeder, A. C., Bigbee, J. W., Robinson, S. E., & Sato-Bigbee, C. (2014). The opioid system and brain development: Effects of methadone on the oligodendrocyte lineage and the early stages of myelination. *Developmental Neuroscience*, *36*, 409–421. https://doi.org/10.1159/000365074.

Waldhoer, M., Bartlett, S. E., & Whistler, J. L. (2004). Opioid receptors. *Annual Review of Biochemistry*, *73*, 953–990. https://doi.org/10.1146/annurev.biochem.73.011303.073940.

Wang, X., Loram, L. C., Ramos, K., de Jesus, A. J., Thomas, J., Cheng, K., et al. (2012). Morphine activates neuroinflammation in a manner parallel to endotoxin. *Proceedings of the National Academy of Sciences of the United States of America*, *109*, 6325–6330. https://doi.org/10.1073/pnas.1200130109.

Warden, A., Truitt, J., Merriman, M., Ponomareva, O., Jameson, K., Ferguson, L. B., et al. (2016). Localization of PPAR isotypes in the adult mouse and human brain. *Scientific Reports*, *6*, 27618. https://doi.org/10.1038/srep27618.

Winters, K. C., Fahnhorst, T., Botzet, A., Lee, S., & Lalone, B. (2012). Brief intervention for drug-abusing adolescents in a school setting: Outcomes and mediating factors. *Journal of Substance Abuse Treatment*, *42*, 279–288. https://doi.org/10.1016/j.jsat.2011.08.005.

Woo, D. H., Bae, J. Y., Nam, M.-H., An, H., Ju, Y. H., Won, J., et al. (2018). Activation of astrocytic μ-opioid receptor elicits fast glutamate release through TREK-1-containing K2P channel in hippocampal astrocytes. *Frontiers in Cellular Neuroscience*, *12*, 319. https://doi.org/10.3389/fncel.2018.00319.

Wu, Y., Dissing-Olesen, L., MacVicar, B. A., & Stevens, B. (2015). Microglia: Dynamic mediators of synapse development and plasticity. *Trends in Immunology*, *36*, 605–613. https://doi.org/10.1016/j.it.2015.08.008.

Zhang, X.-Q., Cui, Y., Cui, Y., Chen, Y., Na, X.-D., Chen, F.-Y., et al. (2012). Activation of p38 signaling in the microglia in the nucleus accumbens contributes to the acquisition and maintenance of morphine-induced conditioned place preference. *Brain Behavior and Immunology*, *26*, 318–325. https://doi.org/10.1016/j.bbi.2011.09.017.

Zhang, H., Largent-Milnes, T. M., & Vanderah, T. W. (2020). Glial neuroimmune signaling in opioid reward. *Brain Research Bulletin*, *155*, 102–111. https://doi.org/10.1016/j.brainresbull.2019.11.012.

Zhang, Y., Liang, Y., Levran, O., Randesi, M., Yuferov, V., Zhao, C., et al. (2017). Alterations of expression of inflammation/immune-related genes in the dorsal and ventral striatum of adult C57BL/6J mice following chronic oxycodone self-administration: A RNA sequencing study. *Psychopharmacology*, *234*, 2259–2275. https://doi.org/10.1007/s00213-017-4657-y.

Zhou, W., & Kalivas, P. W. (2008). N-acetylcysteine reduces extinction responding and induces enduring reductions in cue- and heroin-induced drug-seeking. *Biological Psychiatry*, *63*, 338–340. https://doi.org/10.1016/j.biopsych.2007.06.008.

Zhu, R., Bu, Q., Fu, D., Shao, X., Jiang, L., Guo, W., et al. (2018). Toll-like receptor 3 modulates the behavioral effects of cocaine in mice. *Journal of Neuroinflammation*, *15*, 93. https://doi.org/10.1186/s12974-018-1130-8.

Zuchero, J. D., & Barres, B. A. (2015). Glia in mammalian development and disease. *Development*, *142*, 3805–3809. https://doi.org/10.1242/dev.129304.

CHAPTER SIX

Adolescent neuroimmune function and its interaction with alcohol

T.L. Doremus-Fitzwater[a,b,*] and T. Deak[b,c]
[a]Department of Psychology, Ithaca College, Ithaca, NY, United States
[b]Developmental Exposure Alcohol Research Center (DEARC), Binghamton, NY, United States
[c]Binghamton University-SUNY, Binghamton, NY, United States
[*]Corresponding author: e-mail address: tfitzwater@ithaca.edu

Contents

1. Introduction 168
2. Neuroimmune development and functioning during adolescence 172
3. Early adolescent sensitivity to ethanol-induced alterations in neuroimmune processes 178
   3.1 Acute ethanol effects 180
   3.2 Chronic ethanol effects in adolescence 183
4. Impact of stress and immune stimuli on adolescent neuroimmune functioning 187
5. Conclusions 191
Acknowledgments 192
Conflicts of interest 192
References 192

Abstract

Adolescence is an evolutionarily conserved developmental period associated with behavioral change, including increased risk-taking and alcohol use. Experimentation with alcohol typically begins in adolescence and transitions to binge-like patterns of consumption. Alcohol exposure during adolescence can alter normative changes in brain structure and function. Understanding mechanisms by which ethanol impacts neurodevelopmental processes is important for preventing and ameliorating the deleterious consequences of adolescent alcohol abuse. This review focuses on the neuroimmune system as a key contributor to ethanol-induced changes in adolescent brain and behavior. After brief review of neuroimmune system development, acute and chronic effects of ethanol on adolescent neuroimmune functioning are addressed. Comparisons between stress/immunological challenges and ethanol on adolescent neuroimmunity are reviewed, as cross-sensitization is relevant during adolescence. The mechanisms by which ethanol alters neuroimmune functioning are then discussed, as they may portend development of neuropathological consequences and thus increase vulnerability to subsequent challenges and potentiate addictive behaviors.

International Review of Neurobiology, Volume 161
ISSN 0074-7742
https://doi.org/10.1016/bs.irn.2021.08.006

Abbreviations

A1	A1-state astrocytes
A2	A2-state astrocytes
AIE	adolescent intermittent ethanol
AUD	alcohol use disorder
BBB	blood brain barrier
BrdU	Bromodeoxyuridine
CCL	chemokine (C–C motif) ligand
CCR	chemokine receptor
CD14	cluster of differentiation 14
CNS	central nervous system
DAMPs	damage associated molecular patterns
GD	gestational day
GFAP	glial fibrillary acidic protein
HMBG1	high-mobility group box 1
i.c.v.	intracerebroventricular
i.g.	intragastric
i.p.	intraperitoneal
IA2BC	intermittent access, 2-bottle choice
Iba-1	ionized calcium-binding adaptor molecule 1
IKK-β	inhibitor of nuclear factor kappa-B kinase subunit beta
IL-1Ra	interleukin-1 receptor antagonist
IL-1β	interleukin-1 beta
IL-6	interleukin-6
JAK/STAT	Janus kinase/signal transducer and activator of transcription
KO	knock-out
LPS	lipopolysaccharide
M1	M1-state microglia
M2	M2-state microglia
NF-κB	nuclear factor kappa-light-chain-enhancer of activated B cells
NIAAA	National Institute of Alcohol Abuse and Alcoholism
P	postnatal day
PAMPs	pathogen-associated molecular patterns
PFC	prefrontal cortex
PRRs	pathogen recognition receptors
RAGE	receptor for advanced glycation end products
RANGE	rapid alterations in neuroimmune gene expression
TLR	toll-like receptor
TNFα	tumor necrosis factor-alpha

1. Introduction

Adolescence is a critical period of development in which organisms across a variety of species transition from immaturity and dependence on caregivers to maturity and greater independence (*c.f.*, Casey, Duhoux, & Malter Cohen, 2010; Spear, 2000, 2011). Beginning at puberty, but lasting past the teenage years and into the early to mid-20's, this transitional period is characterized by dramatic alterations in physiology, hormones, neurodevelopment, and behavior (Blakemore, 2012; Spear, 2000, 2013). Pubertal processes lead to production of sex hormones, maturation of the gonads, and outward changes in physical appearance due to the development of secondary sexual characteristics (Marshall, 1975; McCormick, Green, & Simone, 2017; Sisk & Foster, 2004). Additionally, somatic growth accelerates as the adolescent growth spurt takes place (Marshall, 1975; McCormick et al., 2017). Concomitant with these pubertal alterations, adolescent neurodevelopmental processes occur as well. From increased myelination, to synaptic pruning, and alterations in specific neurotransmitter systems including but not limited to dopamine, serotonin, GABA, glutamate and endocannabinoids, the adolescent brain is undergoing marked transformation (*c.f.*, Blakemore, 2012; Casey et al., 2010; Giedd et al., 2006; Spear, 2000, 2013; Vijayakumar, Op de Macks, Shirtcliff, & Pfeifer, 2018). Not surprisingly, research suggests that maturational changes in physiological and neurological systems are likely contributors to the expression of adolescent-typical behaviors. From a behavioral standpoint, adolescence is associated with dramatic increases in social behavior directed toward peers, impulsivity, risk-taking, sensation-seeking, and novelty-directed behaviors (*c.f.*, Spear, 2018), as well as changes in emotional processes, executive functioning, and behavioral control (Casey, Heller, Gee, & Cohen, 2019; Constantinidis & Luna, 2019).

Importantly, the hormonal, neural, and behavioral alterations of adolescence appear to be evolutionarily conserved (Spear, 2000). In particular, increased risk-taking and novelty-seeking behaviors during adolescence are pervasive across many species, including humans, non-human primates, and rodents (Doremus-Fitzwater & Spear, 2016; Spear, 2000). It has been hypothesized that these seemingly dangerous and risky behaviors may actually be adaptive, as they promote emigration away from the family unit and help prevent genetic inbreeding (Spear, 2000). Alterations in neurotransmitter and hormone systems that promote the rewarding value of social stimuli and novelty, however, seems to also impart vulnerability toward drug and alcohol use during this ontogenetic period. Indeed, adolescence is a developmental epoch in which teenagers begin experimentation with alcohol (Windle et al., 2008). When surveying adolescents (15–19 years of age) worldwide, it was recently estimated that there are approximately 155 million teenagers who report being current drinkers (World Health Organization, 2018). Alcohol use commonly starts in early adolescence (Faden, 2006; Masten, Faden, Zucker, & Spear, 2009), with a self-reported 50–70% of 15 year-olds already reporting alcohol use (World Health Organization, 2018). Earlier onset of alcohol drinking behavior may be particularly threatening, as younger initiation of alcohol use is associated with a significantly greater risk for development of problematic drinking and alcohol use disorders (AUDs) later (Davis, Slutske, Piasecki, Martin, & Lynskey, 2020; DeWit, Adlaf, Offord, & Ogborne, 2000). In particular, it seems that adolescents who begin drinking alcohol between the ages of 11–14 are at a substantially elevated risk, as it was estimated that they are four times more likely to develop AUDs compared to peers that postponed drinking until after the age of 20 (Dawson, Goldstein, Chou, Ruan, & Grant, 2008; Ehlers, Slutske, Gilder, Lau, & Wilhelmsen, 2006).

Whereas some amount of alcohol consumption may be normative during the adolescent period, research is suggesting that heavier drinking is particularly problematic for adolescents, both in terms of its prevalence among this age group and its negative consequences on adolescent health, development, and behavior (Silveri, Dager, Cohen-Gilbert, & Sneider, 2016; Spear, 2018; Squeglia, Jacobus, & Tapert, 2014). Heavy episodic drinking is termed "binge drinking," and refers to ingestion of larger amounts of alcohol within a short time period, which typically results in a more rapid time course to intoxication. Binge drinking has been formally defined by the National Institute on Alcohol Abuse and Alcoholism (NIAAA) (2021) as 5 or more drinks within in a 2-h period for males, or 4 or more drinks for females (NIAAA). According to this NIAAA definition, such binge drinking behavior is said to have occurred when blood alcohol levels, at the end of 2 h, meet or exceed 80 mg/dL. Binge drinking is especially common from late adolescence to young adulthood (18–24 years of age) (Kanny, Naimi, Liu, Lu, & Brewer, 2018), but also occurs at younger ages. In a recent national survey, 3.8%, 8.5%, and 14.4% of eighth, tenth, and twelfth graders, respectively, reported at least one episode of binge drinking within the past 2 weeks (Johnston et al., 2020). The consequences of this type of drinking behavior are particularly damaging for adolescents, as excessive levels of alcohol use during this ontogenetic period seem to have an even greater effect on the likelihood of developing AUDs (Bonomo, 2005; Bonomo, Bowes, Coffey, Carlin, & Patton, 2004; Hingson, Heeren, & Winter, 2006), as well as negatively impacting adolescent brain development.

When examining patterns of use, it has become apparent that some individuals—both teens and adults—are drinking at levels that greatly exceed the definitions of binge drinking. Specifically, some adolescents are reporting alcohol consumption that is two to three times higher than binge levels, with 10% of 18–20 year-olds reporting 10 or more drinks per occasion within the past 2 weeks, and some of these individuals reporting levels as high as 15 or more drinks per occasion (Patrick et al., 2013; Patrick & Terry-McElrath, 2017). This pattern of high-intensity drinking appears to be promoted by the college environment, and to peak around the age of 25 (Patrick & Terry-McElrath, 2017). Although less is known about high intensity drinking, early initiation of alcohol consumption and high-intensity binge drinking behavior is especially hazardous for the developing adolescent brain given the multitude of maturational changes that are occurring during this ontogenetic period (Bava & Tapert, 2010; Casey, Getz, & Galvan, 2008; Olsson et al., 2016; Silveri, 2012; Spear, 2018).

Decades of research have demonstrated that alcohol exposure is an environmental stimulus that significantly alters numerous physiological, psychological, and peripheral and central nervous system processes. From cardiac and liver function, to its effects on the gut microbiome and endocrine pathways, alcohol has widespread effects across the whole body and on the brain. Within the brain, a major focus has been on alcohol's influence on the structure and function of nearly every neurotransmitter system, such as GABA, glutamate, dopamine, serotonin, acetylcholine, and the endocannabinoids (Bell et al., 2016). Additionally, alcohol has now been shown to impact a variety of hormones, neuropeptides, growth factors, enzymes, intracellular signaling molecules, and transcription factors (Abrahao, Salinas, & Lovinger, 2017). Moreover, excessive alcohol use can have deleterious neurological consequences on brain structure including neuronal cell death, gliosis, and white matter degeneration (Zahr & Pfefferbaum, 2017). Given these vast effects of ethanol on the nervous system, it is not surprising that ethanol-induced changes in peripheral and central immune system functions have emerged as yet another mechanism by which ethanol alters normal physiological and neural processes that might contribute to AUDs and ethanol-related tissue damage (Erickson, Grantham, Warden, & Harris, 2019). Ethanol's effects on immune activity were first identified in the peripheral immune system and investigated in the context of alcohol-related liver damage among chronic alcoholics (Romeo, Tio, Rahman, Chiappelli, & Taylor, 2001; Romeo, Tio, & Taylor, 2003; Szabo, 1999). Since then, research has shown that ethanol also potently activates central immune factors, both indirectly *via* (i) peripheral cytokine-induced activation of the vagus as well as, potentially, additional cranial nerves such as the glossopharyngeal nerve and others (Romeo et al., 2001, 2003); (ii) direct actions on circulating monocytes and leukocytes (Gao, Ahmad, Nagy, & Tsukamoto, 2019; Shi, DeLucia, Bao, & Zhang, 2019); or (iii) directly by altering activity of neurons and glial cells and their production of neuroimmune factors locally within the central nervous system (CNS) (Crews & Vetreno, 2016; Crews, Zou, & Qin, 2011; Mayfield, Ferguson, & Harris, 2013). Neuroinflammatory processes have now been linked to altered sensitivity to ethanol-induced behavioral changes such as ethanol-induced sedation (*e.g.*, Blednov, Benavidez, Black, Mayfield, & Harris, 2015; Wu et al., 2011) and memory impairment (*e.g.*, Pascual, Balino, Alfonso-Loeches, Aragon, & Guerri, 2011), as well as ethanol drinking patterns (*c.f.*, Erickson et al., 2019) and even ethanol-induced increases in neuronal cell death and diminished neurogenesis (Broadwater, Liu, Crews, & Spear, 2014; Crews, Mdzinarishvili, Kim, He, & Nixon, 2006; Nixon, 2006; Nixon & Crews, 2002; Obernier, Bouldin, & Crews, 2002).

Until more recently, investigations of ethanol-neuroimmune interactions focused on mature, young adult subjects. The recognition that adolescence represents a unique developmental period of ethanol sensitivity and vulnerability has led to increasing interest in identifying potential age-related differences in the neuroimmune effects of ethanol. Thus, the focus of this review will be to examine evidence of altered sensitivity of adolescents to ethanol-induced changes in neuroimmune processes. We begin by briefly discussing maturation of the neuroimmune system into adolescence, including the possible role that resident immune cells, such as microglia, might have on adolescent-typical changes in brain structure (*e.g.*, synaptic pruning). Next, we examine evidence from developmental studies which indicate that central inflammatory cells and signaling molecules may also be uniquely sensitive to the effects of both acute and chronic binge-level exposure to ethanol during adolescence as compared to adulthood. We then address limited data which suggests stress- and immune-related challenges may also differentially affect neuroimmunity during adolescence, and we consider the possibility of cross-sensitization with ethanol exposure. Throughout, we present potential mechanisms by which ethanol may influence neuroimmune processes and signaling within the adolescent brain as it relates to distinct cognitive and behavioral responses of adolescents to ethanol administration.

2. Neuroimmune development and functioning during adolescence

The first step toward understanding neuroimmune involvement in early ethanol-induced effects requires some consideration of what is meant by the term "neuroimmune" or "neuroinflammation," as these constructs have quite disparate meanings across investigators and studies. For decades, the CNS and immune system were thought to operate mostly independent of each other, in large part because neurons are fixed in place, whereas immune cells show broad migratory patterns as part of their routine surveillance activity (Kettenmann, Hanisch, Noda, & Verkhratsky, 2011; Lively & Schlichter, 2013). We now know that there is extensive communication between the CNS and immune system that occurs despite a certain amount of "immunological privilege" that the blood brain barrier (BBB) bestows on the CNS (Banks, 2016; Engelhardt, Vajkoczy, & Weller, 2017; Maier & Watkins, 1998). Despite the apparent compartmentalization that occurs due to the BBB, the CNS maintains a host of non-neuronal cell populations, including microglia, astrocytes, fibroblasts, dendritic cells and other immunologically-active cells that work in concert to support a healthy CNS, buffer against excitotoxicity, and rescue the brain parenchyma from damage (Dantzer, 2018). Many of these cells, especially microglia and dendritic cells, are embryonically-derived from the myeloid cell lineage, conferring a full complement of scavenging, phagocytosis, and antigen presentation characteristic of these cell types (Arcuri, Mecca, Bianchi, Giambanco, & Donato, 2017; Savina & Amigorena, 2007). Yolk-sac derived myeloid precursor cells populate the CNS during prenatal and early postnatal development, seeding a massive population of microglia that will eventually comprise about 10% of the total cells in the brain (Tay, Savage, Hui, Bisht, & Tremblay, 2017). Once seeded, microglia continue to replicate across the lifespan, with new microglia passing into the mature CNS at considerably lower rates than during embryonic development (Nikodemova et al., 2015; Tay et al., 2017), at least under normal (unchallenged) conditions. In contrast, other immunologically-derived cells such as T-cells and B-cells are largely absent from the CNS under normal conditions. Indeed, the presence of lymphoid-derived cells in the CNS is most often associated with emerging or established neuropathological conditions such as autoimmune disorders (Kang & McGavern, 2008; Matullo, O'Regan, Curtis, & Rall, 2011) or severe pathogenic infection of the CNS (Vincenti & Merkler, 2021). Together, these non-neuronal cell types perform vital maintenance and restorative functions in the CNS, and their primary means of communication is through soluble signaling agents originally identified as inter-cellular communication pathways that orchestrate peripheral inflammation and cell-mediated immunity, such as cytokines, chemokines and prostanoids (Aloisi, 2001; Antel, Becher, Ludwin, Prat, & Quintana, 2020; Colombo & Farina, 2016). In this way, glia and other non-neuronal cells—as well as the profile of inflammation-related signaling molecules they use to communicate—form the fundamental basis of the neuroimmune system.

Given their broad prevalence and rapid response rate in the CNS, microglia can be described as "first responders" during a variety of threats/challenges to the CNS (Aloisi, 2001; Antel et al., 2020; Arcuri et al., 2017; Kettenmann et al., 2011). Through microglial expression and release of multiple Damage Associated Molecular Patterns (DAMPs) and Pathogen Recognition Receptors (PRRs), chemokines and pro-inflammatory cytokines are rapidly induced (Aloisi, 2001; Kettenmann et al., 2011). As a class of molecules, chemokines are among the first to respond to damage, and serve to elicit ultrastructural changes in the morphology of immune-derived cells to promote motility and migratory activity of immune cells (*e.g.*, Kitabgi, Melik-Parsadaniantz, & Rostene, 2006). As such, a common feature of neuroinflammation includes assessment of physical characteristics of microglia, which display broad, extensively branched processes under resting conditions (Kettenmann et al., 2011; Lively & Schlichter, 2013). The activated state of microglia, in contrast, is evidenced by a more ameboid state in which processes are retracted to allow for greater cell motility. Microglia can therefore be visually categorized based on their physical appearance. More recent advances allow for automated assessments of branch points, process length, cell body diameter, and other biophysical characteristics of cells (*c.f.*, Perkins, Varlinskaya, & Deak, 2019). Morphology alone, however, does not say much about the activity of those cells, which can be (i) activated and forward-propagating of inflammation through the expression of pro-inflammatory cytokines; or (ii) activated, but expressing factors that squelch further inflammation (Franco & Fernandez-Suarez, 2015; Kettenmann et al., 2011; Lively & Schlichter, 2013; Orihuela, McPherson, & Harry, 2016). This pro- *versus* anti-inflammatory distinction has been referred to the M1 and M2 activational states of microglia, respectively, so named from a similar response profile initially observed among circulating monocytes and macrophages (Orecchioni, Ghosheh, Pramod, & Ley, 2019). The M1/M2 designation offers a useful and simplified framework for understanding microglial activity in the CNS, with these distinct cellular states corresponding to a pattern of expression of inflammation-related signaling molecules that are associated with pro-inflammatory (M1) and anti-inflammatory (M2) outcomes (Franco & Fernandez-Suarez, 2015; Orihuela et al., 2016). It should be noted, however, that this nomenclature has been challenged as overly simplistic (Ransohoff, 2016).

More recent studies using single cell gene expression profiling have suggested that microglia may exist in at least 9 distinct activational states relating to age, inflammatory challenge, and disease-state based upon the genes they express (Hammond et al., 2019). Thus, microglia are a prevalent cell type in the CNS that is often examined as an index of neuroimmune activation and/or neuroinflammation.

Once activated, microglia communicate through the secretion of soluble factors to other cells in the local vicinity, including astrocytes. Astrocytes are anatomically positioned to serve as key regulatory cells that form a communication bridge between the BBB, neurons, and other cell types (Jensen, Massie, & De Keyser, 2013; Molofsky & Deneen, 2015). Interestingly, astrocytic activational profiles follow a similar nomenclature as microglia, with A1 and A2 activational designations corresponding to neurotoxic (A1) and neuroprotective (A2) phenotypes, respectively (Ding et al., 2021). Like microglia, astrocytes also produce large quantities of cytokines and other inflammation-related molecules, though this is often in response to signals originating from microglia. For example, in a classic series of studies, Raghavendra, Tanga, and DeLeo (2004) demonstrated that initial inflammatory signals in a pain model occurred rapidly by microglia, but that they were quickly resolved within approximately 24h. The pro-inflammatory signal induced by astrocytes, however, was slower to develop compared to that in microglia, and the astrocytic activity was sustained for a more protracted period as well (Raghavendra et al., 2004). Studies like these aptly point to microglia as first responders, and position astrocytes later in the process of sustained neuroinflammation. In terms of timing, it is also relevant to note that morphological changes in microglia and other neuroimmune cells tend to lag substantially behind the expression of inflammation-related genes, often by 24–48h (Norden, Trojanowski, Villanueva, Navarro, & Godbout, 2016). Thus, neuroimmune activation is a dynamic process that begins quickly, but unfolds across not just hours, but days to weeks, based on the challenge that is encountered.

Since multiple cell types produce inflammation-related signaling molecules in the CNS, depending on the conditions and timing relative to the insult, many studies utilize tissue punching followed by homogenization procedures that aggregate gene and/or protein expression across many cell types. Like any technique, these approaches have both strengths and limitations. The primary advantage is the ability to rapidly assess inflammatory signaling within a localized tissue microenvironment, irrespective of the cellular source. Although conflicting data certainly exists, neurons express little, if any, of the primary pro-inflammatory cytokines (Licinio, 1997; Licinio & Frost, 2000), suggesting that cytokine signatures in mixed cell populations are largely attributable to microglia, astrocytes, and other non-neuronal cells contained in the tissue punch/extract, especially if fragments of the BBB are present in the tissue sample. Despite this, neurons express receptors for many pro-inflammatory cytokines as evidenced by rapid alterations in neurophysiological response profiles when exposed to cytokines such as IL-1β, TNFα, and IL-6 (Conti et al., 2007). Of course, the limitation of using mixed cell populations is that relatively small effects in a limited population of cells might be "diluted" by the lack of a significant effect in much more prevalent cells. Indeed, Harris and colleagues have shown that certain neuroimmune effects of ethanol were detectable in purified populations of microglia, but not in whole tissue homogenates (McCarthy, Farris, Blednov, Harris, & Mayfield, 2018). Thus, a variety of technical considerations must be considered carefully in evaluating studies and approaches that measure neuroimmune signaling effects.

Most germane to the present review are the effects of neuroimmune challenges before the brain has reached full maturity comparable to healthy adults. From a cellular standpoint, the CNS is thought to be populated by microglia progenitors embryonically between gestational day (GD) 8–9 that differentiate and mature into microglia, a process which is largely completed by the first week or two, postnatally, in rodents (Alliot, Godin, & Pessac, 1999; Ginhoux et al., 2010). Note that the first week of postnatal life in rats and mice is considered equivalent to the human third trimester based on morphology and function of the CNS (Clancy, Finlay, Darlington, & Anand, 2007; Quinn, 2005). From birth, microglia play an indispensable role in sculpting sexually dimorphic brain sites, particularly hypothalamic areas critical to social behavior (*c.f.*, Lenz & McCarthy, 2015). During adolescence, microglia are involved with extensive pruning of dendrites and paring of synapses, a process that continues to unfold into young adulthood (Brenhouse & Schwarz, 2016; Giedd et al., 1999; Keshavan, Giedd, Lau, Lewis, & Paus, 2014). Pruning and paring activities of microglia occurs widely throughout the CNS, and is particularly evident throughout cortical regions and other telencephalic structures. This involves complement-mediated "tagging" of synapses for elimination by phagocytosis (Wu, Dissing-Olesen, MacVicar, & Stevens, 2015). Thus, the natural role of microglia in sculpting neural architecture positions drugs like ethanol, and other damage-related events, in a primary position to disrupt ongoing cortical maturation.

In terms of neuroimmune reactivity, adolescents seem to display a blunted or muted induction of neuroimmune genes, regardless of the challenge used to evoke the inflammation. For instance, work from our lab has shown reduced induction of IL-6 and other neuroimmune genes after exposure to lipopolysaccharide (LPS)—a bacterial mimetic (Doremus-Fitzwater, Gano, Paniccia, & Deak, 2015). Early adolescent rats injected with LPS also failed to release odor cues that cause avoidance by other conspecifics, suggesting their response to LPS was not fully mature (Arakawa, Arakawa, & Deak, 2009). Interestingly, studies also showed that LPS, when injected intraperitoneally (i.p.), failed to transit into the bloodstream of adolescents, suggesting a fundamental difference in the absorption and/or distribution, as well as pharmacokinetics, of LPS across development (Doremus-Fitzwater et al., 2015). While the mechanism(s) underlying these differences remain elusive, it can be noted that vagal tone has been hypothesized as a key individual characteristic determining an LPS response (Romanovsky et al., 2000), with vagal tone also changing notably across early development (Kalia, 1992). Nevertheless, these issues raise the possibility that adolescent-adult differences in neuroimmune reactivity to LPS could be due to pharmacokinetic or other physiological variables, rather than a reflection of functional immaturity of the neuroimmune system in adolescents. To further test this issue, we recently examined neuroimmune reactivity to a brief footshock challenge which is known to increase IL-1β through microglial-dependent pathways (Blandino et al., 2009; Blandino, Barnum, & Deak, 2006). Once again, adolescent rats displayed a severely muted induction of neuroimmune genes in response to footshock, extending beyond IL-1 to other members of the IL-1 signaling family. In that series of experiments, it was demonstrated that these age differences could not be attributed to general physiological differences in body composition, drug pharmacokinetics, or disparate timing of neuroimmune gene induction between adolescents and adults, since multiple time-points were examined (Marsland, Parrella, Orlofsky, et al., 2021; Marsland, Parrella, Vore, et al., 2021). Moreover, we tested adolescents and adults of both sexes, arriving at the same conclusion—induction of neuroimmune genes was functionally impaired among adolescents, suggestive of immature neuroimmune function.

In the studies described above, it is critical to note that our comparisons largely centered on early adolescent rats ranging in age from postnatal day (P) 28-P34. This age range typically includes the onset of puberty, although sex differences in timing of this developmental event are evident (Spear, 2015). Reproductive maturation is driven by the pubertal surge in gonadal steroids and is usually complete among most female rats by P36, whereas most male rats require an additional 4–8 days to achieve genital maturation, suggesting full reproductive maturation in males between P40–44 (Spear, 2015). Though prior studies have shown that gonadal steroids have a relatively modest effect on alcohol intake, alcohol preference, and novelty-directed behaviors (Varlinskaya, Vetter-O'Hagen, & Spear, 2013), androgens and estrogens both have potent immunomodulatory properties (Cutolo et al., 2004; Homo-Delarche et al., 1991). It will be crucial for future studies to determine what role, if any, the pubertal surge in hormones might play in maturation, at least phenotypically, of neuroimmune function. Moreover, assessment of downstream, functional differences in sensitivity to cytokines would help determine the functional integrity and maturation of the adolescent neuroimmune system.

3. Early adolescent sensitivity to ethanol-induced alterations in neuroimmune processes

From an ontogenetic standpoint, even in animal models, adolescence is characterized by elevated ethanol consumption (Doremus, Brunell, Rajendran, & Spear, 2005; Garcia-Burgos, Gonzalez, Manrique, & Gallo, 2009; Maldonado, Finkbeiner, Alipour, & Kirstein, 2008; Vetter, Doremus-Fitzwater, & Spear, 2007). Both males and females exhibit this adolescent-related elevation in ethanol intake (Doremus et al., 2005), with consumption highest during the pre-adolescent/early adolescent period (Garcia-Burgos et al., 2009; Vetter et al., 2007) and then gradually declining into adulthood. Besides increased consumption patterns among adolescents, rats in this age range also show altered sensitivity to the behavioral sequelae following ethanol administration. For instance, adolescents exhibited diminished reactivity to many consequences of ethanol exposure, including the sedative/hypnotic (Silveri & Spear, 1998), motor-impairing (White et al., 2002), acute withdrawal (Doremus, Brunell, Varlinskaya, & Spear, 2003), aversive (Anderson, Varlinskaya, & Spear, 2010) and social-inhibitory (Varlinskaya & Spear, 2002) effects. These blunted ethanol responses do not represent a global insensitivity to alcohol during adolescence, however, as there are ethanol effects to which adolescents are contrarily more sensitive, such as ethanol-induced memory impairment (Markwiese, Acheson, Levin, Wilson, & Swartzwelder, 1998), ethanol-related deficits in hippocampal LTP (Pyapali, Turner, Wilson, & Swartzwelder, 1999), and the social stimulatory effects of acute ethanol exposure (Varlinskaya & Spear, 2002). Given the unique behavioral responsiveness observed when adolescents are exposed to ethanol, researchers have sought to explore whether adolescence might also represent a critical period of altered sensitivity to ethanol-induced changes in neuroimmune activation and signaling.

Research with adult rodents has demonstrated a complex interaction between ethanol and neuroimmune pathways, with this research in mature subjects having been previously reviewed extensively (Crews et al., 2006; Crews, Lawrimore, Walter, & Coleman, 2017; Crews & Vetreno, 2016; Cui, Grandison, & Noronha, 2011, 2013; Cui, Shurtleff, & Harris, 2014; Erickson et al., 2019; Mayfield et al., 2013). Briefly, it is now understood that manipulation of neuroimmune signaling *via* genetic knock-out (KO) models or pharmacological methods can alter (increase or decrease) voluntary ethanol drinking. For instance, in genetic KO studies, deletions of a wide variety of different neuroimmune-related genes (*e.g.*, toll-like receptor [TLR]-2, CD14, IKK-β, IL-6, IL-1Ra, CCR, CCL3) have been shown to induce significant reductions in ethanol intake, preference, and/or ethanol-induced conditioned taste aversion when compared to wild-type mice (Blednov et al., 2005, 2012; Erickson et al., 2019). Whereas a substantial body of research has implicated TLR4 in the pathogenesis of ethanol-related neuroimmune activation and neural cell loss (*c.f.*, Crews, Qin, Sheedy, Vetreno, & Zou, 2013; Crews & Vetreno, 2016), as reviewed by Erickson and colleagues, manipulation of TLR4 receptor activation by genetic and pharmacological methods does not seem to significantly impact ethanol drinking behavior in some models (Erickson et al., 2019; Harris et al., 2017). Instead, it seems that alterations in TLR3 are more consistent in impacting ethanol intake among rodents (*e.g.*, McCarthy et al., 2018; Warden et al., 2019a, 2019b), with TLR2 manipulations also impacting ethanol drinking in mice (Blednov et al., 2017). These findings would indicate that TLR4 receptor interaction and NF-κB signaling are not the only neuroimmune contributors to ethanol-associated behaviors and neurodegeneration, and that other immune-related pathways may be more relevant for distinct ethanol responses (*i.e.*, binge-related cell loss *versus* ethanol drinking and acceptance).

Similarly, administration of a microglial inhibitor—minocycline—was shown to significantly reduce ethanol drinking (Agrawal, Hewetson, George, Syapin, & Bergeson, 2011). Moreover, inhibition of microglia with a colony stimulating factor receptor 1 inhibitor (PLX5622) attenuated escalation of dependence-related alcohol drinking in adult mice (Warden et al., 2020). In contrast, activation of the immune system by LPS led to increased ethanol intake in mice (Blednov et al., 2011). Additionally, alterations in cytokine signaling *via* administration of minocycline or IL-1Ra reduced responsiveness to acute ethanol-induced sedation and motor impairment during ethanol intoxication (Wu et al., 2011). Other manipulations of neuroimmune function have had less striking results on ethanol-induced behavioral alterations. For example, intracerebroventricular (i.c.v.) administration of AG490, a JAK/STAT inhibitor that has been shown to block the effects of endogenous IL-6, had a minimal effect on ethanol-induced conditioned taste aversion (Barney, Vore, Gano, Mondello, & Deak, 2019). Reduced social interaction observed during acute ethanol withdrawal was also largely unaffected by i.c.v. administration of an IL-1 receptor antagonist or peripheral administration of indomethacin (Richey, Doremus-Fitzwater, Buck, & Deak, 2012). These seemingly disparate findings for neuroimmune factors in ethanol-related behaviors underscores the need for more focused studies that strike a balance between global knockout models in which specific neuroimmune factors are completely absent (sledgehammer) *versus* more discrete studies with pharmacological injections, which may have limited distribution of drug into neural circuits critical to ethanol-mediated changes in behavior.

Whereas the above studies demonstrated that immune factors can impact ethanol drinking and sensitivity in adulthood, it has also been shown that ethanol itself is a stimulus that potently affects expression of neuroimmune factors (*c.f.*, Crews et al., 2017; Cui et al., 2014; Erickson et al., 2019)—including signaling molecules like cytokines, chemokines, DAMPs, and pathogen-associated molecular patterns (PAMPs)—even in the absence of an immune challenge. Moreover, ethanol significantly impacts many types of inflammatory receptors (*e.g.*, toll-like receptors, chemokine receptors), as well as the activity of CNS glial cells, such as microglia and astrocytes (Cui et al., 2014; Erickson et al., 2019). However, it is important to note that these effects of ethanol on neuroimmune processes appear to be dependent on ethanol dose, route of administration, chronicity of the exposure, and time of post-ethanol examination (*i.e.*, intoxication *versus* acute withdrawal *versus* protracted withdrawal).

3.1 Acute ethanol effects

If the effects of acute ethanol are considered in adult rodents, overall, the data seem to indicate a unique effect of ethanol administration on neuroinflammation (*e.g.*, Deak, Okrainets, & Doremus-Fitzwater, 2012). In particular, data from our laboratory and others have demonstrated a distinct pattern of acute ethanol intoxication effects on several key cytokines in adults, which have been observed across numerous brain regions and following multiple routes of administration and doses of ethanol. When adult rats were challenged with a binge-like dose of ethanol and expression of several neuroimmune factors assessed during peak intoxication, we consistently observed marked and significant changes in expression levels of several neuroinflammatory factors. More specifically, ethanol intoxication increased expression of *Il-6* and Iκ*Bα* in the hippocampus, prefrontal cortex, amygdala, and hypothalamus, whereas reduced expression of *Tnfα* and *Il-1β* mRNA was simultaneously observed in these same regions (Doremus-Fitzwater et al., 2015; Doremus-Fitzwater, Paniccia, Gano, Vore, & Deak, 2018; Gano, Doremus-Fitzwater, & Deak, 2016, 2017). We have termed these ethanol intoxication-induced patterns of cytokine changes "rapid alterations in neuroimmune gene expression" (RANGE) (Gano et al., 2016; Gano, Doremus-Fitzwater, & Deak, 2017), and we have observed such changes in multiple strains of rats (Doremus-Fitzwater et al., 2020; Gano, Doremus-Fitzwater, & Deak, 2017) and in both sexes (Doremus-Fitzwater et al., 2020; Gano, Doremus-Fitzwater, & Deak, 2017). While examining the first few intoxicating exposures to an ethanol challenge, similar results were recently reported by another group (Terasaki & Schwarz, 2017). Adult male and female rats were challenged with a 4.5 g/kg acute binge-like dose of ethanol, and a strikingly similar pattern of ethanol-induced alterations in cytokine expression was observed by Terasaki & Schwarz, 2017. In that experiment, acute ethanol increased *Il-6* levels in the hippocampus, prefrontal cortex, and perirhinal cortex, yet significantly decreased expression of *Il-1β* in the hippocampus and perirhinal cortex (Terasaki & Schwarz, 2017).

Since adolescence represents the age at which most individuals experience their first intoxicating (and sometimes binge-like) doses of alcohol, we examined whether adolescents might be more or less sensitive to acute ethanol-induced RANGE. To our knowledge, few studies to date have examined the effects of an acute initial exposure to ethanol on adolescent neuroimmune responses, and in many cases direct adolescent-adult comparisons were not included. Using male Sprague-Dawley rats we reported that, when age differences were present, adolescents exhibited a significantly blunted neuroimmune response to an acute ethanol challenge (Doremus-Fitzwater et al., 2015). This attenuated sensitivity during adolescence was apparent following both i.p. and intragastric (i.g.) challenges with ethanol, where blood ethanol concentrations were in the supra-binge range

(150–225 mg/dl). A similarly reduced neuroimmune response was observed among adolescent mice that were examined 1 day after their final exposure to ethanol. In that experiment, adolescent and adult mice were administered 6 g/kg of ethanol *via* gavage (twice per day for 10 days) and then several key brain regions examined for expression of inflammatory-related genes and astrocytic staining using GFAP (Kane et al., 2014). The authors reported that this pattern of ethanol administration significantly increased several inflammation genes in adults (*e.g., CCL2, IL-6*) but that the neuroinflammatory gene activation was not observed in adolescents. Together, these findings seem to suggest that induction of neuroimmune genes is not yet mature among adolescents. It should be noted, however, that in one study from our laboratory in which adolescent (Long Evans) rats were examined after prenatal ethanol exposure (or control conditions), we found similar ethanol-induced RANGE patterns in all brain regions examined, regardless of whether they were challenged during adolescence or adulthood (Doremus-Fitzwater et al., 2020). While our design did not allow a direct statistical comparison between adolescents and adults, both age groups exhibited comparable alterations in expression of immune factors. Other, significant differences in that cohort of rats (*i.e.*, rats were timed pregnant rats ordered from a vendor, rather than bred on-site) might further explain this discrepancy. Another recent study examined the expression of proinflammatory cytokines in microglia extracted from the hippocampus of rats exposed to 4 days of adolescent intermittent binge ethanol exposures (Peng & Nixon, 2021). During the intoxication phase after their final ethanol dose at approximately P40, IL-6 mRNA expression was elevated, yet other cytokines were decreased or unchanged—results similar to the RANGE pattern observed by adolescents in our laboratory. Additionally, another study investigated the consequences of an abbreviated adolescent binge exposure to acute ethanol (once per day for 3 days) and compared this to a more protracted period of chronic intermittent ethanol exposure (Bellozi et al., 2019). Overall, this group discovered that a shorter duration of adolescent exposure to acute ethanol resulted in few significant changes in the expression levels of many inflammatory cytokines—a result that was dramatically different from the proinflammatory effects observed after more chronic adolescent exposure (see below). While somewhat equivocal, taken together, these studies do seem to indicate a similar pattern of cytokine changes in adolescent brain in response to acute ethanol, but with the magnitude of these changes being notably muted among adolescents compared to adults, likely due to a functional immaturity of the neuroimmune system.

Despite these intriguing findings, there is still a lot that we do not know about adolescent neuroimmune responses to acute alcohol challenge. For instance, future studies should examine the functional sensitivity of the adolescent brain to bioactive cytokines and other neuroimmune signaling factors, to further probe the functionality of the adolescent neuroimmune system. In addition, studies from our laboratory have largely focused on early adolescence (P28-P42) when making comparisons to adults, and several other key adolescent-specific issues require investigation, including the potential role of puberty and its associated hormonal surges in neuroimmune maturation, sex differences in the timing of puberty, and potential differences in adolescent neuroimmune maturation between early and late adolescence (Spear, 2015).

3.2 Chronic ethanol effects in adolescence

When considering the impact of chronic ethanol exposure in adulthood, ethanol generally appears to have a stimulatory effect on neuroimmune processes that has been suggested to precipitate neuroinflammatory brain damage, behavioral and cognitive deficits, as well as impair host defense (Alfonso-Loeches, Pascual-Lucas, Blanco, Sanchez-Vera, & Guerri, 2011; Crews et al., 2017; Deak et al., 2012; Montesinos, Alfonso-Loeches, & Guerri, 2016; Szabo & Saha, 2015). Although some studies have performed systematic comparisons of adolescent *versus* adult ethanol exposure, the last decade has witnessed a profound increase in studies examining adolescent intermittent ethanol (AIE) exposure to binge-like doses of ethanol. Though the timing, doses, and routes of administration vary somewhat across laboratories, most AIE models span the majority of early (P28-P42) and late (P42-P56) adolescence. For instance, a typical AIE model may involve 16 binge-like ethanol i.g. intubations (5 g/kg) spanning P25-P57, delivered on a 2-days on, 2-days off schedule (Fernandez & Savage, 2017; Macht, Crews, & Vetreno, 2020; Vetreno & Crews, 2012; Vetreno, Lawrimore, Rowsey, & Crews, 2018). Similarly, an AIE model used in our laboratory involves 12 i.g. ethanol intubations (4 g/kg) administered in a 3-days on, 2-days off pattern that spans roughly P30-P50. After a 2–3 week washout period, neurobehavioral assessments are then conducted in early adulthood (Vore, Doremus-Fitzwater, Gano, & Deak, 2017). Other models of AIE utilize vapor exposure to induce high-dose, binge-like levels of ethanol administration. In some instances, the vapor exposure spans the entirety of peri-adolescence (~P22 to P48 or up to P57) and occurs daily for shorter (*e.g.*, 6 h; Logrip et al., 2013) or longer (*e.g.*, 14 h; Desikan, Wills, & Ehlers, 2014) durations, whereas in other vapor models the exposure spans early to mid-adolescence (*e.g.*, approximately P28 to P44) and involves more intermittently spaced bouts with washout days between exposures (Barker, Bryant, Osborne, & Chandler, 2017; Healey et al., 2020).

Several of these research groups have explored the effects of AIE administration on brain regions known to undergo substantial remodeling during adolescence. For instance, McClain et al. (2011) reported that adolescent binge ethanol exposure led to an upregulation of partially activated microglia in the hippocampus. Additionally, BrdU/Iba-1 co-labeling demonstrated that adolescent binge ethanol led to increased proliferation of new microglia in this brain region, with these microglia remaining up to 30 days after ethanol exposure (McClain, Morris, et al., 2011). Within this same brain structure, AIE was found to impair subsequent neurogenesis in the hippocampus during adulthood (McClain, Hayes, et al., 2011; Morris, Eaves, Smith, & Nixon, 2010; Swartzwelder et al., 2019; Vetreno & Crews, 2015). Crews and colleagues have also reported that a similar adolescent binge ethanol exposure upregulated RAGE/TLR-4 and HMGB1 in the PFC of adolescent rats, as well as gene expression of numerous other neuroimmune factors, and these effects persisted well into adulthood (Vetreno & Crews, 2012; Vetreno, Qin, & Crews, 2013). A recent review of chronic binge-like adolescent ethanol exposure on the PFC reported that AIE upregulated TLR signaling and cytokine mRNA expression levels in adolescents, but not adults, and that alterations in myelin-associated proteins would indicate that this might underlie myelin-associated disruptions induced by ethanol (Pascual, Pla, Minarro, & Guerri, 2014). Notably, many studies examining the impact of adolescent ethanol exposure on neuroimmune processes do not include a direct adolescent to adult comparison. However, in the Pascual et al. (2014) study, a group of adult control and alcohol-exposed animals were also examined. The results suggested that some of the effects of adolescent binge ethanol administration are specific to exposure during adolescence, as ethanol-related changes in neuroinflammatory factors were not observed in adult subjects.

Studies from our laboratory have examined the influence of AIE on both circulating (peripheral) and central expression of inflammation-related genes. For instance, Vore et al. (2017) showed that AIE exposure led to sex-specific changes in LPS-induced cytokine gene induction among circulating lymphocytes. Specifically, AIE-exposed males displayed severely reduced cytokine gene induction, whereas AIE-exposed females exhibited normal cytokine induction. Interestingly, this basic pattern of suppressed response to LPS among AIE-exposed males was not cytokine-specific, suggesting the influence of AIE may occur upstream of cytokine gene regulatory control, likely through modulation of Nf-κB signaling pathways. In a separate set of studies, Gano, Mondello, Doremus-Fitzwater, and Deak (2019), Gano, Vore, Sammakia, and Deak (2019) used large-molecule microdialysis to assess extracellular cytokines in the hippocampus after acute and/or chronic ethanol exposure. Although acute ethanol challenge was largely without effect on hippocampal cytokine release, extracellular cytokines were notably reduced among AIE-exposed males that received an additional i.p. ethanol challenge during adulthood (Gano, Mondello, et al., 2019). This effect was reminiscent of the profound influence of AIE on cytokine induction within circulating leukocytes (Vore et al., 2017), suggesting comparable, long-lasting effects of AIE on both central and peripheral cytokine regulation. Consistent with this, Telles et al. (2017) showed that LPS-evoked fever responses were significantly reduced in AIE-exposed males, an effect that may be due, in part, to disrupted thermogenic activity in AIE-exposed rats (Cruz et al., 2020). Together, these findings are generally consistent with the notion that LPS-induced (*i.e.*, primarily TLR4-mediated) inflammation may be globally compromised after AIE. This is generally consistent with the immunosuppressive effects of ethanol that have been discussed elsewhere (Szabo & Saha, 2015), and the developmental specificity of these effects have not yet been tested.

The effects of AIE on neuroimmune function are not quite that simple, however, and one needs to carefully consider the distinction between (i) long-lasting differences in *basal* cytokine expression/activity; (ii) how AIE influences the *inducibility* of cytokine gene regulatory networks and signaling events; and (iii) the *specificity* of AIE effects to distinct modalities of challenges faced later in life (*e.g.*, are long-lasting AIE effects present in adulthood and evoked by LPS, tissue damage, or stress challenges?). For instance, our more recent work demonstrated that AIE-exposed males displayed sensitization of certain neuroimmune genes in the hippocampus when challenged with ethanol (2.5 g/kg, i.p.) as adults (Vore et al., 2017). While this would seem to conflict with a previous study in our laboratory in which AIE-exposed males displayed reduced cytokine *protein* captured with large-molecule microdialysis (Gano, Vore, et al., 2019), there are at least two plausible explanations for this discrepancy. First, sensitized induction of neuroimmune genes may be a compensatory response to help overcome ethanol-mediated disruptions in translational efficiency and post-translational protein modifications (Osna et al., 2016). Consistent with this, studies examining coherence between cytokine mRNA and protein have shown mixed results after ethanol (Doremus-Fitzwater et al., 2014; Gano, Vore, et al., 2019; Knapp, Harper, Whitman, Zimomra, & Breese, 2016). Second, microdialysis involves substantial tissue damage that occurs on two discrete occasions: at initial cannula implantation, as well as in response to probe insertion prior to the microdialysis phase. Indeed, prior cannulation led to a diminished neuroimmune response to an acute ethanol challenge (Gano, Mondello, et al., 2019), and probe insertion led to a rapid surge in both chemokine and cytokine release (Gano, Vore, et al., 2019). For these reasons, the influence of AIE on later neuroimmune function requires careful consideration of methodology and approach to contextualize the outcomes.

If we take some of these data at face value, however, then certain findings might point toward an intriguing developmental hypothesis that could begin to explain what we know about acute and chronic effects of ethanol during adolescence. Since adolescents appear to display muted neuroimmune responses to ethanol when compared to adults, and AIE exposure (at least in male rodents) seems to elicit signs of muted inflammatory responses when challenged later in adulthood, then perhaps AIE is "locking-in" an adolescent-typical phenotype of neuroimmune function. Arrested development has certainly been shown for other adolescent-typical phenotypes (*e.g.*, Spear & Swartzwelder, 2014), though to our knowledge, it has not been extended to neuroimmune system function. To test this hypothesis, future studies will need to include a comparable ethanol exposure procedure in young adults to demonstrate the adolescent-specific vulnerability/sensitivity, as well as explore the durability of the AIE effect(s) across time (*i.e.*, does the AIE effect fade as individuals move beyond young adulthood?). Moreover, it will be imperative to establish cellular mechanisms to account for the locking-in-like phenotype. Clearly, there is much more work needed in this area.

With this in mind, it is important to note that other forms of neuroimmune plasticity may be independent of adolescent ethanol exposure. More specifically, there are likely significant effects of chronic intermittent ethanol exposure that are independent of the period of development in which they occur, and thus would not reflect adolescent-specific vulnerabilities or sensitivities. For instance, ethanol-induced IL-6 expression in the hippocampus was observed consistently after up to 6 once-daily intubations of ethanol, but demonstrated considerable habituation when intubations were delivered on an every-other-day schedule (Gano et al., 2016). Later studies demonstrated that ethanol-paired cues led to a context-dependent sensitization of the IL-6 response in the hippocampus and amygdala, suggesting that neuroimmune responses can be classically conditioned with as little as 4 CS-US pairings (Gano, Pautassi, Doremus-Fitzwater, & Deak, 2017). Using a drinking model, rats exposed to a prolonged period of voluntary ethanol access using the Intermittent Access, 2-bottle Choice (IA2BC) procedure for approximately 10 weeks showed tolerance to the effects of a 4-g/kg ethanol challenge (i.g.) in the PVN, but not amygdala or hippocampus (Doremus-Fitzwater et al., 2014). Thus, a growing body of literature suggests considerable plasticity of the neuroimmune response to ethanol as the number of ethanol exposures accumulate. Understanding the cellular mechanisms underlying various forms of neuroimmune plasticity will be an important area of focus for future studies.

4. Impact of stress and immune stimuli on adolescent neuroimmune functioning

Stress has emerged as a variable that significantly impacts drug- and alcohol-related behaviors and addiction processes in general (*c.f.*, Goeders, Guerin, & Schmoutz, 2014; Lu, Shepard, Hall, & Shaham, 2003; Sinha, Shaham, & Heilig, 2011; Stewart, 2003). In particular, there appears to be a complicated relationship between alcohol use and stress exposure, with these two influencing each other in a bidirectional manner (Brady & Sonne, 1999; Pohorecky, 1991; Sillaber & Henniger, 2004; Sinha, 2012; Spanagel, Noori, & Heilig, 2014). Indeed the complex, and often times contradictory, effects of stress on ethanol intake, behavioral responsiveness, and CNS function have been previously reviewed in adults (Becker, 2012; Becker, Lopez, & Doremus-Fitzwater, 2011; Pohorecky, 1990). The impact of stress on alcohol drinking behavior has been the focus of extensive research. When taken together, the results from such studies are mixed at best. For instance, whereas numerous stressors including restraint (*e.g.*, Lynch, Kushner, Rawleigh, Fiszdon, & Carroll, 1999), forced-swim (Füllgrabe, Vengeliene, & Spanagel, 2007; Siegmund, Vengeliene, Singer, & Spanagel, 2005), and footshock (Vengeliene et al., 2003; Volpicelli, Ulm, & Hopson, 1990) have been shown to significantly elevate ethanol consumption, there are numerous contradictory reports of these same stressors decreasing or not changing ethanol consumption (*e.g.*, Boyce-Rustay, Janos, & Holmes, 2008; Darnaudery et al., 2007; Ng Cheong Ton, Brown, Michalakeas, & Amit, 1983).

When ethanol exposure precedes a stress challenge, the timing of the stress challenge relative to the last ethanol exposure is critical to the observed outcomes. For instance, chronic ethanol studies have shown long-term exposure to ethanol *via* a liquid diet dampened the HPA axis response to a later stress challenge (*e.g.*, Lee & Rivier, 1993; Rivier, 1995). In contrast, our laboratory reported an exacerbated corticosterone response (Buck, Hueston, Bishop, & Deak, 2011) and increased fear conditioning (Marsland, Parrella, Orlofsky, et al., 2021; Marsland, Parrella, Vore, et al., 2021) when the stress challenge occurred during withdrawal from acute ethanol exposure. Results from another research group reported a similar stress sensitization, but during protracted withdrawal from long-term ethanol exposure (Valdez, Zorrilla, Roberts, & Koob, 2003). Examples such as these provide evidence that ethanol can alter stress responsiveness. Yet, there are also reports that stress history can instead influence a later response to ethanol exposure. Acute stress exposures—such as handling (Peris & Cunningham, 1986) and inescapable footshock (Drugan, Coyle, Healy, & Chen, 1996; Drugan, Wiedholz, Holt, Kent, & Christianson, 2007)—have been shown to exacerbate behavioral sensitivity to an acute ethanol challenge, with longer periods of stress exposure also reportedly affecting sensitivity to the effects of acute ethanol (Blakley & Pohorecky, 2006; Fernandez et al., 2016; Matsumoto, Ojima, & Watanabe, 1997; Pohorecky, 2006). Furthermore, when a history of repeated stress exposure was established using intermittent restraint, the development of ethanol-withdrawal-related anxiogenesis was sensitized in a manner similar to when rats experience repeated cycles of ethanol intoxication and withdrawal (Breese, Knapp, & Overstreet, 2004; Knapp, Overstreet, & Breese, 2007).

Adolescent intermittent ethanol exposure has also emerged as a type of developmental ethanol administration that might impart lasting alterations in responsiveness to later stress challenges, including subsequent ethanol exposure. There are conflicting reports from several different groups that have examined the lasting effects of AIE on the later response to an adult binge ethanol challenge, which would represent a homotypic stress challenge (*i.e.*, repeated exposure to the same stressor). Whereas Logrip et al. (2013) and Vore et al. (2017) failed to observe significant lasting alterations in HPA axis sensitivity to an adult ethanol challenge in males, other groups have demonstrated significant AIE-related changes. For example, AIE that was delivered *via* a vapor inhalation method led to a blunted PVN CRF mRNA response to an ethanol challenge in adulthood (Allen, Rivier, & Lee, 2011). Subsequently, Allen and colleagues reported that the same AIE vapor model resulted in a blunted HPA axis response to an ethanol challenge in adult male rats (Allen et al., 2016). In contrast, Pak and colleagues used a different i.g. model of AIE and found that ethanol challenge later in adulthood was associated with a sensitized HPA axis response to an acute ethanol challenge (Przybycien-Szymanska, Mott, Paul, Gillespie, & Pak, 2011). There is evidence to suggest that the lasting consequences of adolescent AIE may be sex- and/or stressor-specific, as we reported that an stress challenge in adulthood led to an exacerbated HPA axis response in AIE-exposed animals, but only in females (Vore et al., 2017). Pak and others have also observed sex differences in AIE effects, with females generally showing greater sensitivity to AIE-related alterations (*e.g.*, Logrip et al., 2013; Przybycien-Szymanska, Gillespie, & Pak, 2012; Przybycien-Szymanska, Rao, & Pak, 2010). The type of stress challenge imposed on AIE animals during adulthood appears to be important as well, as we reported AIE-induced HPA axis sensitization in females when challenged with restraint stress, but not ethanol (Vore et al., 2017). Yet, when Allen, Lee, Koob, and Rivier (2011) examined a footshock challenge in adulthood, they observed no lasting AIE effects. Similarly, AIE did not impact the HPA axis response to restraint stress plus elevated plus-maze exposure during adulthood in male rats (Torcaso, Asimes, Meagher, & Pak, 2017). Hence, while the lasting effects of AIE seem to vary according to sex, type of stress challenge, dose/route of ethanol administration, and methodology of the AIE model used, studies such as these show that exposure to ethanol during the critical period of adolescence induces long-term alterations in the response of the HPA axis to later perturbations.

Importantly, the immune system has been identified as a potential common mechanism by which each of these variables (stress and ethanol) might impact physiology and behavior, as each independently impacts neuroimmune processes (Deak et al., 2012). As described above, ethanol (both acutely and under conditions of chronic administration), can profoundly alter neuroimmune processes. However, stress exposure alone has also been reported to increase expression of inflammatory related factors in the brain (*e.g.*, Deak et al., 2015; Rossetti et al., 2016), with the hypothalamus (Blandino et al., 2009; Deak et al., 2005; Girotti, Donegan, & Morilak, 2013; Grippo, Francis, Beltz, Felder, & Johnson, 2005), hippocampus (Goshen et al., 2008;Rossetti et al., 2016 ; You et al., 2011), and cortex (Rossetti et al., 2016; You et al., 2011) exhibiting significant stress-induced changes in neuroimmune factors. Furthermore, stress-associated increases in depression-like behavior (Feng et al., 2012; Vitale et al., 2009; *c.f.*, Willner, 2005; Wu et al., 2013) have been shown to be attenuated by genetic (Goshen et al., 2008) or pharmacological (Karson, Demirtas, Bayramgurler, Balci, & Utkan, 2013; Santiago et al., 2014) blockade of neuroimmune processes.

Thus, the interaction between stress and ethanol on neuroimmune function is an area of research that is gaining greater attention. In studies such as these, which have primarily been conducted with adult rodents, the impact of previous ethanol or stress exposure on the subsequent neuroimmune response to specific types of stimuli needs to be examined. For example, when Crews and colleagues examined the HPA axis and neuroimmune response to an acute stress challenge after prior acute or chronic ethanol exposure, they found that previous acute ethanol exacerbated several components of the neuroimmune response to stress, including microglial activation and plasma endotoxin induction (Walter, Vetreno, & Crews, 2017). Previous chronic ethanol exposure, however, only exacerbated the microglial response, but not plasma endotoxin. Experiments from our laboratory also investigated the combined influence of stress and ethanol on neuroimmune processes, but in the opposite direction. In other words, rats experienced either acute stress or a history of chronic escalating stress before being administered a binge-like ethanol challenge (Doremus-Fitzwater et al., 2018). Under these circumstances, acute stress augmented the neuroimmune cytokine response to the ethanol challenge, whereas chronic stress seemed to suppress it (Doremus-Fitzwater et al., 2018).

The above studies were completed with adult rodents and not adolescents, however, and so it is worth noting that there is a dearth of information regarding a possible stress-ethanol cross-sensitization or suppression on neuroimmunity during this critical developmental period. This is surprising given the extensive body of work demonstrating differences in the HPA axis response to stressors among adolescents (*c.f.*, McCormick et al., 2017), with acute stressors generally inducing a significantly prolonged corticosterone response in both male and female adolescent rats when compared to their adult counterparts (*e.g.*, Green, Zeidan, Hodges, & McCormick, 2019; Romeo et al., 2006; Romeo, Lee, Chhua, McPherson, & McEwen, 2004; Romeo, Lee, & McEwen, 2004). Doremus-Fitzwater, Varlinskaya, and Spear (2009) reported a similar prolonged response to restraint stress among adolescent males and females relative to adults, and then further demonstrated that habituation of the corticosterone response to restraint across 5 days of repeated exposure was more pronounced for adults compared to adolescents. Given the lasting impact of adolescent stress on later sensitivity to acute stress and ethanol challenges, as well as on affective and cognitive functioning (*e.g.*, Mathews, Mills, & McCormick, 2008; *c.f.*, McCormick & Green, 2013; McCormick & Mathews, 2010), it is not surprising that research is just beginning to emerge suggesting that the immune system may be implicated in, and affected by, these experiences during adolescence (McCormick et al., 2020). Clearly, this area calls for more research, given the seemingly blunted neuroimmune response to ethanol among adolescents, the involvement of stress in neuroimmune processes, and increased levels of stress that some researchers have argued occurs during adolescence.

5. Conclusions

Adolescence is an ontogenetic phase that is now recognized as a critical window for development of multiple physiological, hormonal, neural, and behavioral processes. A time during development that was once thought of as a mere precursor to adulthood, adolescence is now recognized as a formative transitional period in which the brain is sculpted by an individual's experiences, environment, and biology. Given the nature of the adolescent-to-adult transition, this critical period becomes a time for positive experiences to have lasting advantageous impact on an individual. Yet, this also presents a point of vulnerability, as negative experiences can also impart lasting consequences on CNS development. The prototypical behaviors of adolescence include increases in interactions with same-aged peers, novelty-seeking, sensation-seeking, risk-taking behaviors, impulsivity, and impaired inhibitory behavioral control. Likely a product of such behavioral alterations, adolescents are known to begin experimenting with alcohol and other drugs, with use of these substances often becoming excessive in nature during this developmental phase.

Importantly, alcohol exposure during adolescence significantly alters brain structure, function, and behavior. Adolescent alcohol exposure can change the normal processes of myelination, synaptic pruning, and neurogenesis, as well as normative adolescent-associated alterations in neurotransmitter systems and connectivity in neural circuits. With these wide-spread and sometimes long-lasting changes in brain structure and function, it should not be surprising that adolescent alcohol use can also dramatically alter behavior, including acceptance, and intake, of alcohol and other drugs—both during adolescence and into adulthood. Thus, an understanding of the mechanisms responsible for these consequences of adolescent ethanol exposure remains a large focus of alcohol research. The immune system has emerged as a system that is uniquely influenced by ethanol during adolescence, and that can conversely affect the organism's response to an ethanol challenge. While early adolescents exhibit a muted sensitivity to acute binge-like doses of ethanol on neuroinflammation, long-term intermittent binge administration of ethanol appears to incur enhanced sensitivity to neuronal cell loss, significant impairment in hippocampal neurogenesis, and increased degeneration of myelin in white matter-dense brain regions. At this time, it is unclear what exact mechanisms are responsible for these differential ontogenetic responses to acute *versus* chronic high-dose ethanol. But, no matter the mechanisms, taken together, studies such as these suggest that the neuroimmune system represents a novel and promising target for preventing and ameliorating consequences of binge drinking on the adolescent brain. Early ontogenetic exposure to ethanol confers a heightened risk for problematic drinking behavior later in life, as does excessive use. These are conditions of alcohol use in which immune system involvement appears to be particularly pronounced. Hence, neuroimmune interactions with alcohol exposure represents a mechanistic pathway by which adolescent alcohol might contribute to increased risk for AUD diagnosis in adolescence and beyond. A better understanding of adolescent ethanol-neuroimmune interactions will hopefully provide a novel target by which the development of problematic drinking behavior can be identified and circumvented before excessive patterns of use begin and alcohol use disorders emerge.

Acknowledgments

Supported by Supported by National Institute of Health grant P50AA017823 (TD) and the Center for Development and Behavioral Neuroscience (CDBN) at Binghamton University. Any opinions, findings, and conclusions or recommendations expressed in this material are those of the author(s) and do not necessarily reflect the views of the above stated funding agencies.

Conflicts of interest

Neither author has any conflicts of interest to disclose.

References

Abrahao, K. P., Salinas, A. G., & Lovinger, D. M. (2017). Alcohol and the brain: Neuronal molecular targets, synapses, and circuits. *Neuron*, *96*(6), 1223–1238. https://doi.org/10.1016/j.neuron.2017.10.032.

Agrawal, R. G., Hewetson, A., George, C. M., Syapin, P. J., & Bergeson, S. E. (2011). Minocycline reduces ethanol drinking. *Brain, Behavior, and Immunity*, *25*(Suppl. 1), S165–S169. https://doi.org/10.1016/j.bbi.2011.03.002. S0889-1591(11)00070-5 [pii].

Alfonso-Loeches, S., Pascual-Lucas, M., Blanco, A. M., Sanchez-Vera, I., & Guerri, C. (2011). Pivotal role of TLR4 receptors in alcohol-induced neuroinflammation and brain damage. *Journal of Neuroscience*, *30*(24), 8285–8295. https://doi.org/30/24/8285 [pii] https://doi.org/10.1523/JNEUROSCI.0976-10.2010.

Allen, C. D., Grigoleit, J. S., Hong, J., Bae, S., Vaughan, J., & Lee, S. (2016). Exposure to alcohol during adolescence exerts long-term effects on stress response and the adult brain stress circuits. *Neuroscience, 339*, 64–71. https://doi.org/10.1016/j.neuroscience.2016.09.031.

Allen, C. D., Lee, S., Koob, G. F., & Rivier, C. (2011). Immediate and prolonged effects of alcohol exposure on the activity of the hypothalamic-pituitary-adrenal axis in adult and adolescent rats. *Brain, Behavior, and Immunity, 25*(Suppl. 1), S50–S60. https://doi.org/10.1016/j.bbi.2011.01.016.

Allen, C. D., Rivier, C. L., & Lee, S. Y. (2011). Adolescent alcohol exposure alters the central brain circuits known to regulate the stress response. *Neuroscience, 182*, 162–168. https://doi.org/10.1016/j.neuroscience.2011.03.003.

Alliot, F., Godin, I., & Pessac, B. (1999). Microglia derive from progenitors, originating from the yolk sac, and which proliferate in the brain. *Brain Research. Developmental Brain Research, 117*(2), 145–152. https://doi.org/10.1016/s0165-3806(99)00113-3.

Aloisi, F. (2001). Immune function of microglia. *Glia, 36*(2), 165–179. https://doi.org/10.1002/glia.1106.

Anderson, R. I., Varlinskaya, E. I., & Spear, L. P. (2010). Ethanol-induced conditioned taste aversion in male Sprague-dawley rats: Impact of age and stress. *Alcoholism: Clinical and Experimental Research, 34*(12), 2106–2115. https://doi.org/10.1111/j.1530-0277.2010.01307.x.

Antel, J. P., Becher, B., Ludwin, S. K., Prat, A., & Quintana, F. J. (2020). Glial cells as regulators of neuroimmune interactions in the central nervous system. *Journal of Immunology, 204*(2), 251–255. https://doi.org/10.4049/jimmunol.1900908.

Arakawa, H., Arakawa, K., & Deak, T. (2009). Acute illness induces the release of aversive odor cues from adult, but not prepubertal, male rats and suppresses social investigation by conspecifics. *Behavioral Neuroscience, 123*(5), 964–978. https://doi.org/2009-18253-020 [pii] https://doi.org/10.1037/a0017114.

Arcuri, C., Mecca, C., Bianchi, R., Giambanco, I., & Donato, R. (2017). The pathophysiological role of microglia in dynamic surveillance, phagocytosis and structural Rrmodeling of the developing CNS. *Frontiers in Molecular Neuroscience, 10*, 191. https://doi.org/10.3389/fnmol.2017.00191.

Banks, W. A. (2016). From blood-brain barrier to blood-brain interface: New opportunities for CNS drug delivery. *Nature Reviews Drug Discovery, 15*(4), 275–292. https://doi.org/10.1038/nrd.2015.21.

Barker, J. M., Bryant, K. G., Osborne, J. I., & Chandler, L. J. (2017). Age and sex interact to mediate the effects of intermittent, high-dose ethanol exposure on behavioral flexibility. *Frontiers in Pharmacology, 8*, 450. https://doi.org/10.3389/fphar.2017.00450.

Barney, T. M., Vore, A. S., Gano, A., Mondello, J. E., & Deak, T. (2019). The influence of central interleukin-6 on behavioral changes associated with acute alcohol intoxication in adult male rats. *Alcohol, 79*, 37–45. https://doi.org/10.1016/j.alcohol.2018.11.004.

Bava, S., & Tapert, S. F. (2010). Adolescent brain development and the risk for alcohol and other drug problems. *Neuropsychology Review, 20*(4), 398–413. https://doi.org/10.1007/s11065-010-9146-6.

Becker, H. C. (2012). Effects of alcohol dependence and withdrawal on stress responsiveness and alcohol consumption. *Alcohol Research: Current Reviews, 34*(4), 448–458. https://www.ncbi.nlm.nih.gov/pubmed/23584111.

Becker, H. C., Lopez, M. F., & Doremus-Fitzwater, T. L. (2011). Effects of stress on alcohol drinking: A review of animal studies. *Psychopharmacology, 218*(1), 131–156. https://doi.org/10.1007/s00213-011-2443-9.

Bell, R. L., Hauser, S. R., McClintick, J. N., Rahman, S., Edenberg, H. J., Szumlinski, K. K., et al. (2016). Ethanol-associated changes in glutamate reward neurocircuitry: A minireview of clinical and preclinical genetic findings. *Progress in Molecular Biology and Translational Science, 137*, 41–85. https://doi.org/10.1016/bs.pmbts.2015.10.018.

Bellozi, P. M. Q., Pelicao, R., Santos, M. C., Lima, I. V. A., Saliba, S. W., Vieira, E. L. M., et al. (2019). URB597 ameliorates the deleterious effects induced by binge alcohol consumption in adolescent rats. *Neuroscience Letters*, *711*, 134408. https://doi.org/10.1016/j.neulet.2019.134408.

Blakemore, S. J. (2012). Imaging brain development: The adolescent brain. *NeuroImage*, *61*(2), 397–406. https://doi.org/10.1016/j.neuroimage.2011.11.080.

Blakley, G., & Pohorecky, L. A. (2006). Psychosocial stress alters ethanol's effect on open field behaviors. *Pharmacology, Biochemistry and Behavior*, *84*(1), 51–61. https://doi.org/10.1016/j.pbb.2006.04.005.

Blandino, P., Jr., Barnum, C. J., & Deak, T. (2006). The involvement of norepinephrine and microglia in hypothalamic and splenic IL-1beta responses to stress. *Journal of Neuroimmunology*, *173*(1–2), 87–95. https://doi.org/S0165-5728(05)00506-0 [pii] https://doi.org/10.1016/j.jneuroim.2005.11.021.

Blandino, P., Jr., Barnum, C. J., Solomon, L. G., Larish, Y., Lankow, B. S., & Deak, T. (2009). Gene expression changes in the hypothalamus provide evidence for regionally-selective changes in IL-1 and microglial markers after acute stress. *Brain, Behavior, and Immunity*, *23*(7), 958–968. https://doi.org/10.1016/j.bbi.2009.04.013. S0889-1591(09)00170-6 [pii].

Blednov, Y. A., Benavidez, J. M., Black, M., Mayfield, J., & Harris, R. A. (2015). Role of interleukin-1 receptor signaling in the behavioral effects of ethanol and benzodiazepines. *Neuropharmacology*, *95*, 309–320. https://doi.org/10.1016/j.neuropharm.2015.03.015.

Blednov, Y. A., Bergeson, S. E., Walker, D., Ferreira, V. M., Kuziel, W. A., & Harris, R. A. (2005). Perturbation of chemokine networks by gene deletion alters the reinforcing actions of ethanol. *Behavioural Brain Research*, *165*(1), 110–125. https://doi.org/10.1016/j.bbr.2005.06.026.

Blednov, Y. A., Black, M., Chernis, J., Da Costa, A., Mayfield, J., & Harris, R. A. (2017). Ethanol consumption in mice lacking CD14, TLR2, TLR4, or MyD88. *Alcoholism: Clinical and Experimental Research*, *41*(3), 516–530. https://doi.org/10.1111/acer.13316.

Blednov, Y. A., Borghese, C. M., McCracken, M. L., Benavidez, J. M., Geil, C. R., Osterndorff-Kahanek, E., et al. (2011). Loss of ethanol conditioned taste aversion and motor stimulation in knockin mice with ethanol-insensitive alpha2-containing GABA(A) receptors. *Journal of Pharmacology and Experimental Therapeutics*, *336*(1), 145–154. https://doi.org/10.1124/jpet.110.171645.

Blednov, Y. A., Ponomarev, I., Geil, C., Bergeson, S., Koob, G. F., & Harris, R. A. (2012). Neuroimmune regulation of alcohol consumption: Behavioral validation of genes obtained from genomic studies. *Addiction Biology*, *17*(1), 108–120. https://doi.org/10.1111/j.1369-1600.2010.00284.x.

Bonomo, Y. (2005). Early onset of drinking increases alcohol use in adulthood. *Evidence-Based Mental Health*, *8*(4), 98. https://doi.org/10.1136/ebmh.8.4.98.

Bonomo, Y. A., Bowes, G., Coffey, C., Carlin, J. B., & Patton, G. C. (2004). Teenage drinking and the onset of alcohol dependence: A cohort study over seven years. *Addiction*, *99*(12), 1520–1528. https://doi.org/10.1111/j.1360-0443.2004.00846.x.

Boyce-Rustay, J. M., Janos, A. L., & Holmes, A. (2008). Effects of chronic swim stress on EtOH-related behaviors in C57BL/6J, DBA/2J and BALB/cByJ mice. *Behavioural Brain Research*, *186*(1), 133–137. https://doi.org/S0166-4328(07)00375-0 [pii] https://doi.org/10.1016/j.bbr.2007.07.031.

Brady, K. T., & Sonne, S. C. (1999). The role of stress in alcohol use, alcoholism treatment, and relapse. *Alcohol Research and Health*, *23*(4), 263–271. http://www.ncbi.nlm.nih.gov/entrez/query.fcgi?cmd=Retrieve&db=PubMed&dopt=Citation&list_uids=10890823.

Breese, G. R., Knapp, D. J., & Overstreet, D. H. (2004). Stress sensitization of ethanol withdrawal-induced reduction in social interaction: Inhibition by CRF-1 and benzodiazepine receptor antagonists and a 5-HT1A-receptor agonist. *Neuropsychopharmacology*, *29*(3), 470–482. https://doi.org/10.1038/sj.npp.13002821300282 (pii).

Brenhouse, H. C., & Schwarz, J. M. (2016). Immunoadolescence: Neuroimmune development and adolescent behavior. *Neuroscience and Biobehavioral Reviews*, *70*, 288–299. https://doi.org/10.1016/j.neubiorev.2016.05.035.

Broadwater, M. A., Liu, W., Crews, F. T., & Spear, L. P. (2014). Persistent loss of hippocampal neurogenesis and increased cell death following adolescent, but not adult, chronic ethanol exposure. *Developmental Neuroscience*, *36*(3–4), 297–305. https://doi.org/10.1159/000362874.

Buck, H. M., Hueston, C. M., Bishop, C., & Deak, T. (2011). Enhancement of the hypothalamic-pituitary-adrenal axis but not cytokine responses to stress challenges imposed during withdrawal from acute alcohol exposure in Sprague-Dawley rats. *Psychopharmacology*, *218*(1), 203–215. https://doi.org/10.1007/s00213-011-2388-z.

Casey, B. J., Duhoux, S., & Malter Cohen, M. (2010). Adolescence: What do transmission, transition, and translation have to do with it? *Neuron*, *67*(5), 749–760. https://doi.org/10.1016/j.neuron.2010.08.033.

Casey, B. J., Getz, S., & Galvan, A. (2008). The adolescent brain. *Developmental Review*, *28*(1), 62–77. https://doi.org/10.1016/j.dr.2007.08.003.

Casey, B. J., Heller, A. S., Gee, D. G., & Cohen, A. O. (2019). Development of the emotional brain. *Neuroscience Letters*, *693*, 29–34. https://doi.org/10.1016/j.neulet.2017.11.055.

Clancy, B., Finlay, B. L., Darlington, R. B., & Anand, K. J. (2007). Extrapolating brain development from experimental species to humans. *Neurotoxicology*, *28*(5), 931–937. https://doi.org/10.1016/j.neuro.2007.01.014.

Colombo, E., & Farina, C. (2016). Astrocytes: Key regulators of neuroinflammation. *Trends in Immunology*, *37*(9), 608–620. https://doi.org/10.1016/j.it.2016.06.006.

Constantinidis, C., & Luna, B. (2019). Neural substrates of inhibitory control maturation in adolescence. *Trends in Neurosciences*, *42*(9), 604–616. https://doi.org/10.1016/j.tins.2019.07.004.

Conti, B., Tabarean, I., Sanchez-Alavez, M., Davis, C., Brownell, S., Behrens, M., et al. (2007). Cytokine receptors in the brain. *NeuroImmune Biology*, *6*, 19–38. https://doi.org/10.1016/S1567-7443(07)10002-8.

Crews, F. T., Bechara, R., Brown, L. A., Guidot, D. M., Mandrekar, P., Oak, S., et al. (2006). Cytokines and alcohol. *Alcoholism: Clinical and Experimental Research*, *30*(4), 720–730. https://doi.org/ACER84 [pii] https://doi.org/10.1111/j.1530-0277.2006.00084.x.

Crews, F. T., Lawrimore, C. J., Walter, T. J., & Coleman, L. G., Jr. (2017). The role of neuroimmune signaling in alcoholism. *Neuropharmacology*, *122*, 56–73. https://doi.org/10.1016/j.neuropharm.2017.01.031.

Crews, F. T., Mdzinarishvili, A., Kim, D., He, J., & Nixon, K. (2006). Neurogenesis in adolescent brain is potently inhibited by ethanol. *Neuroscience*, *137*(2), 437–445. https://doi.org/10.1016/j.neuroscience.2005.08.090.

Crews, F. T., Qin, L., Sheedy, D., Vetreno, R. P., & Zou, J. (2013). High mobility group box 1/toll-like receptor danger signaling increases brain neuroimmune activation in alcohol dependence. *Biological Psychiatry*, *73*(7), 602–612. https://doi.org/10.1016/j.biopsych.2012.09.030.

Crews, F. T., & Vetreno, R. P. (2016). Mechanisms of neuroimmune gene induction in alcoholism. *Psychopharmacology*, *233*(9), 1543–1557. https://doi.org/10.1007/s00213-015-3906-1.

Crews, F. T., Zou, J., & Qin, L. (2011). Induction of innate immune genes in brain create the neurobiology of addiction. *Brain, Behavior, and Immunity*, *25*(Suppl. 1), S4–S12. https://doi.org/S0889-1591(11)00071-7 [pii] https://doi.org/10.1016/j.bbi.2011.03.003.

Cruz, J. V., Maba, I. K., Correia, D., Kaziuk, F. D., Cadena, S., & Zampronio, A. R. (2020). Intermittent binge-like ethanol exposure during adolescence attenuates the febrile response by reducing brown adipose tissue thermogenesis in rats. *Drug and Alcohol Dependence*, *209*, 107904. https://doi.org/10.1016/j.drugalcdep.2020.107904.

Cui, C., Grandison, L., & Noronha, A. (2013). *Neural-immune interactions in brain function and alcohol related disorders*. NY, NY: Springer.
Cui, C., Grandison, L., & Noronha, A. (2011). Neuroimmune mechanisms of brain function and alcohol related disorders. *Brain, Behavior, and Immunity*, *25*(Suppl. 1), S1–S3. https://doi.org/10.1016/j.bbi.2011.03.005.
Cui, C., Shurtleff, D., & Harris, R. A. (2014). Neuroimmune mechanisms of alcohol and drug addiction. *International Review of Neurobiology*, *118*, 1–12. https://doi.org/10.1016/B978-0-12-801284-0.00001-4.
Cutolo, M., Sulli, A., Capellino, S., Villaggio, B., Montagna, P., Seriolo, B., et al. (2004). Sex hormones influence on the immune system: Basic and clinical aspects in autoimmunity. *Lupus*, *13*(9), 635–638. https://doi.org/10.1191/0961203304lu1094oa.
Dantzer, R. (2018). Neuroimmune interactions: From the brain to the immune system and vice versa. *Physiological Reviews*, *98*(1), 477–504. https://doi.org/10.1152/physrev.00039.2016.
Darnaudery, M., Louvart, H., Defrance, L., Leonhardt, M., Morley-Fletcher, S., Gruber, S. H., et al. (2007). Impact of an intense stress on ethanol consumption in female rats characterized by their pre-stress preference: Modulation by prenatal stress. *Brain Research*, *1131*(1), 181–186. https://doi.org/10.1016/j.brainres.2006.11.005.
Davis, C. N., Slutske, W. S., Piasecki, T. M., Martin, N. G., & Lynskey, M. T. (2020). Comparing the potential causal influence of two indicators of early alcohol use on later alcohol use disorder symptoms. *Journal of Abnormal Psychology*, *129*(3), 256–265. https://doi.org/10.1037/abn0000474.
Dawson, D. A., Goldstein, R. B., Chou, S. P., Ruan, W. J., & Grant, B. F. (2008). Age at first drink and the first incidence of adult-onset DSM-IV alcohol use disorders. *Alcoholism: Clinical and Experimental Research*, *32*(12), 2149–2160. https://doi.org/10.1111/j.1530-0277.2008.00806.x.
Deak, T., Bordner, K. A., McElderry, N. K., Barnum, C. J., Blandino, P., Jr., Deak, M. M., et al. (2005). Stress-induced increases in hypothalamic IL-1: A systematic analysis of multiple stressor paradigms. *Brain Research Bulletin*, *64*(6), 541–556. https://doi.org/S0361-9230(04)00255-2 [pii] https://doi.org/10.1016/j.brainresbull.2004.11.003.
Deak, T., Okrainets, A., & Doremus-Fitzwater, T. L. (2012). Mechanisms of stress-dependent neuroinflammation and their implications for understanding consequences of alcohol exposure. In C. Cui, L. Grandison, & A. Noronha (Eds.), *Neural-immune interactions in brain function and alcohol related disorders* (pp. 133–166). Springer. https://doi.org/10.1007/978-1-4614-4729-0_5.
Deak, T., Quinn, M., Cidlowski, J. A., Victoria, N. C., Murphy, A. Z., & Sheridan, J. F. (2015). Neuroimmune mechanisms of stress: Sex differences, developmental plasticity, and implications for pharmacotherapy of stress-related disease. *Stress*, *18*(4), 367–380. https://doi.org/10.3109/10253890.2015.1053451.
Desikan, A., Wills, D. N., & Ehlers, C. L. (2014). Ontogeny and adolescent alcohol exposure in Wistar rats: Open field conflict, light/dark box and forced swim test. *Pharmacology, Biochemistry and Behavior*, *122*, 279–285. https://doi.org/10.1016/j.pbb.2014.04.011.
DeWit, D. J., Adlaf, E. M., Offord, D. R., & Ogborne, A. C. (2000). Age at first alcohol use: A risk factor for the development of alcohol disorders. *American Journal of Psychiatry*, *157*(5), 745–750. http://www.ncbi.nlm.nih.gov/entrez/query.fcgi?cmd=Retrieve&db=PubMed&dopt=Citation&list_uids=10784467.
Ding, Z. B., Song, L. J., Wang, Q., Kumar, G., Yan, Y. Q., & Ma, C. G. (2021). Astrocytes: A double-edged sword in neurodegenerative diseases. *Neural Regeneration Research*, *16*(9), 1702–1710. https://doi.org/10.4103/1673-5374.306064.
Doremus, T. L., Brunell, S. C., Rajendran, P., & Spear, L. P. (2005). Factors influencing elevated ethanol consumption in adolescent relative to adult rats. *Alcoholism: Clinical and Experimental Research*, *29*(10), 1796–1808. http://www.ncbi.nlm.nih.gov/entrez/query.fcgi?cmd=Retrieve&db=PubMed&dopt=Citation&list_uids=16269909.

Doremus, T. L., Brunell, S. C., Varlinskaya, E. I., & Spear, L. P. (2003). Anxiogenic effects during withdrawal from acute ethanol in adolescent and adult rats. *Pharmacology, Biochemistry and Behavior, 75*(2), 411–418. http://www.ncbi.nlm.nih.gov/entrez/query.fcgi?cmd=Retrieve&db=PubMed&dopt=Citation&list_uids=12873633.

Doremus-Fitzwater, T. L., Buck, H. M., Bordner, K., Richey, L., Jones, M. E., & Deak, T. (2014). Intoxication- and withdrawal-dependent expression of central and peripheral cytokines following initial ethanol exposure. *Alcoholism: Clinical and Experimental Research, 38*(8), 2186–2198. https://doi.org/10.1111/acer.12481.

Doremus-Fitzwater, T. L., Gano, A., Paniccia, J. E., & Deak, T. (2015). Male adolescent rats display blunted cytokine responses in the CNS after acute ethanol or lipopolysaccharide exposure. *Physiology and Behavior, 148*, 131–144. https://doi.org/10.1016/j.physbeh.2015.02.032.

Doremus-Fitzwater, T. L., Paniccia, J. E., Gano, A., Vore, A. S., & Deak, T. (2018). Differential effects of acute versus chronic stress on ethanol sensitivity: Evidence for interactions on both behavioral and neuroimmune outcomes. *Brain, Behavior, and Immunity, 70*, 141–156. https://doi.org/10.1016/j.bbi.2018.02.009.

Doremus-Fitzwater, T. L., & Spear, L. P. (2016). Reward-centricity and attenuated aversions: An adolescent phenotype emerging from studies in laboratory animals. *Neuroscience and Biobehavioral Reviews, 70*, 121–134. https://doi.org/10.1016/j.neubiorev.2016.08.015.

Doremus-Fitzwater, T. L., Varlinskaya, E. I., & Spear, L. P. (2009). Social and non-social anxiety in adolescent and adult rats after repeated restraint. *Physiology & Behavior, 97*(3–4), 484–494. https://doi.org/10.1016/j.physbeh.2009.03.025.

Doremus-Fitzwater, T. L., Youngentob, S. L., Youngentob, L., Gano, A., Vore, A. S., & Deak, T. (2020). Lingering effects of prenatal alcohol exposure on basal and ethanol-evoked expression of inflammatory-related genes in the CNS of adolescent and adult rats. *Frontiers in Behavioral Neuroscience, 14*, 82. https://doi.org/10.3389/fnbeh.2020.00082.

Drugan, R. C., Coyle, T. S., Healy, D. J., & Chen, S. (1996). Stress controllability influences the ataxic properties of both ethanol and midazolam in the rat. *Behavioral Neuroscience, 110*(2), 360–367. https://doi.org/10.1037//0735-7044.110.2.360.

Drugan, R. C., Wiedholz, L. M., Holt, A., Kent, S., & Christianson, J. P. (2007). Environmental and immune stressors enhance alcohol-induced motor ataxia in rat. *Pharmacology, Biochemistry and Behavior, 86*(1), 125–131. https://doi.org/10.1016/j.pbb.2006.12.016.

Ehlers, C. L., Slutske, W. S., Gilder, D. A., Lau, P., & Wilhelmsen, K. C. (2006). Age at first intoxication and alcohol use disorders in Southwest California Indians. *Alcoholism: Clinical and Experimental Research, 30*(11), 1856–1865. https://doi.org/10.1111/j.1530-0277.2006.00222.x.

Engelhardt, B., Vajkoczy, P., & Weller, R. O. (2017). The movers and shapers in immune privilege of the CNS. *Nature Immunology, 18*(2), 123–131. https://doi.org/10.1038/ni.3666.

Erickson, E. K., Grantham, E. K., Warden, A. S., & Harris, R. A. (2019). Neuroimmune signaling in alcohol use disorder. *Pharmacology, Biochemistry and Behavior, 177*, 34–60. https://doi.org/10.1016/j.pbb.2018.12.007.

Faden, V. B. (2006). Trends in initiation of alcohol use in the United States 1975 to 2003. *Alcoholism: Clinical and Experimental Research, 30*(6), 1011–1022. https://doi.org/10.1111/j.1530-0277.2006.00115.x.

Feng, S. F., Shi, T. Y., Fan, Y., Wang, W. N., Chen, Y. C., & Tan, Q. R. (2012). Long-lasting effects of chronic rTMS to treat chronic rodent model of depression. *Behavioural Brain Research, 232*(1), 245–251. https://doi.org/10.1016/j.bbr.2012.04.019.

Fernandez, G. M., & Savage, L. M. (2017). Adolescent binge ethanol exposure alters specific forebrain cholinergic cell populations and leads to selective functional deficits in the prefrontal cortex. *Neuroscience, 361*, 129–143. https://doi.org/10.1016/j.neuroscience.2017.08.013.

Fernandez, M. S., Fabio, M. C., Miranda-Morales, R. S., Virgolini, M. B., De Giovanni, L. N., Hansen, C., et al. (2016). Age-related effects of chronic restraint stress on ethanol drinking, ethanol-induced sedation, and on basal and stress-induced anxiety response. *Alcohol, 51*, 89–100. https://doi.org/10.1016/j.alcohol.2015.11.009.

Franco, R., & Fernandez-Suarez, D. (2015). Alternatively activated microglia and macrophages in the central nervous system. *Progress in Neurobiology, 131*, 65–86. https://doi.org/10.1016/j.pneurobio.2015.05.003.

Füllgrabe, M. W., Vengeliene, V., & Spanagel, R. (2007). Influence of age at drinking onset on the alcohol deprivation effect and stress-induced drinking in female rats. *Pharmacology, Biochemistry, and Behavior, 86*(2), 320–326. https://doi.org/10.1016/j.pbb.2006.10.00.

Gano, A., Doremus-Fitzwater, T. L., & Deak, T. (2016). Sustained alterations in neuroimmune gene expression after daily, but not intermittent, alcohol exposure. *Brain Research, 1646*, 62–72. https://doi.org/10.1016/j.brainres.2016.05.027.

Gano, A., Doremus-Fitzwater, T. L., & Deak, T. (2017). A cross-sectional comparison of ethanol-related cytokine expression in the hippocampus of young and aged Fischer 344 rats. *Neurobiology of Aging, 54*, 40–53. https://doi.org/10.1016/j.neurobiolaging.2017.01.025.

Gano, A., Mondello, J. E., Doremus-Fitzwater, T. L., & Deak, T. (2019). Rapid alterations in neuroimmune gene expression after acute ethanol: Timecourse, sex differences and sensitivity to cranial surgery. *Journal of Neuroimmunology, 337*, 577083. https://doi.org/10.1016/j.jneuroim.2019.577083.

Gano, A., Pautassi, R. M., Doremus-Fitzwater, T. L., & Deak, T. (2017). Conditioned effects of ethanol on the immune system. *Experimental Biology and Medicine, 242*(7), 718–730. https://doi.org/10.1177/1535370217694097.

Gano, A., Vore, A. S., Sammakia, M. N., & Deak, T. (2019). Assessment of extracellular cytokines in the hippocampus of the awake behaving rat using large-molecule microdialysis combined with multiplex arrays after acute and chronic ethanol exposure. *Alcoholism: Clinical and Experimental Research, 43*(4), 640–654. https://doi.org/10.1111/acer.13963.

Gao, B., Ahmad, M. F., Nagy, L. E., & Tsukamoto, H. (2019). Inflammatory pathways in alcoholic steatohepatitis. *Journal of Hepatology, 70*(2), 249–259. https://doi.org/10.1016/j.jhep.2018.10.023.

Garcia-Burgos, D., Gonzalez, F., Manrique, T., & Gallo, M. (2009). Patterns of ethanol intake in preadolescent, adolescent, and adult Wistar rats under acquisition, maintenance, and relapse-like conditions. *Alcoholism: Clinical and Experimental Research, 33*(4), 722–728. https://doi.org/10.1111/j.1530-0277.2008.00889.x.

Giedd, J. N., Blumenthal, J., Jeffries, N. O., Castellanos, F. X., Liu, H., Zijdenbos, A., et al. (1999). Brain development during childhood and adolescence: A longitudinal MRI study. *Nature Neuroscience, 2*(10), 861–863. https://doi.org/10.1038/13158.

Giedd, J. N., Clasen, L. S., Lenroot, R., Greenstein, D., Wallace, G. L., Ordaz, S., et al. (2006). Puberty-related influences on brain development. *Molecular and Cellular Endocrinology, 254–255*, 154–162. https://doi.org/10.1016/j.mce.2006.04.016.

Ginhoux, F., Greter, M., Leboeuf, M., Nandi, S., See, P., Gokhan, S., et al. (2010). Fate mapping analysis reveals that adult microglia derive from primitive macrophages. *Science, 330*(6005), 841–845. https://doi.org/10.1126/science.1194637.

Girotti, M., Donegan, J. J., & Morilak, D. A. (2013). Influence of hypothalamic IL-6/gp130 receptor signaling on the HPA axis response to chronic stress. *Psychoneuroendocrinology, 38*(7), 1158–1169. https://doi.org/10.1016/j.psyneuen.2012.11.004.

Goeders, N. E., Guerin, G. F., & Schmoutz, C. D. (2014). The combination of metyrapone and oxazepam for the treatment of cocaine and other drug addictions. *Advances in Pharmacology*, *69*, 419–479. https://doi.org/10.1016/B978-0-12-420118-7.00011-1.

Goshen, I., Kreisel, T., Ben-Menachem-Zidon, O., Licht, T., Weidenfeld, J., Ben-Hur, T., et al. (2008). Brain interleukin-1 mediates chronic stress-induced depression in mice via adrenocortical activation and hippocampal neurogenesis suppression. *Molecular Psychiatry*, *13*(7), 717–728. https://doi.org/10.1038/sj.mp.4002055.

Green, M. R., Zeidan, M., Hodges, T. E., & McCormick, C. M. (2019). Age-dependent regulation by androgens of gene expression in the anterior hypothalamus and stress-induced release of adrenal hormones in adolescent and adult male rats. *Journal of Neuroendocrinology*, *31*(6), e12714. https://doi.org/10.1111/jne.12714.

Grippo, A. J., Francis, J., Beltz, T. G., Felder, R. B., & Johnson, A. K. (2005). Neuroendocrine and cytokine profile of chronic mild stress-induced anhedonia. *Physiology & Behavior*, *84*(5), 697–706. https://doi.org/10.1016/j.physbeh.2005.02.011.

Hammond, T. R., Dufort, C., Dissing-Olesen, L., Giera, S., Young, A., Wysoker, A., et al. (2019). Single-cell RNA sequencing of microglia throughout the mouse lifespan and in the injured brain reveals complex cell-state changes. *Immunity*, *50*(1), 253–271.e256. https://doi.org/10.1016/j.immuni.2018.11.004.

Harris, R. A., Bajo, M., Bell, R. L., Blednov, Y. A., Varodayan, F. P., Truitt, J. M., et al. (2017). Genetic and pharmacologic manipulation of TLR4 has minimal impact on ethanol consumption in rodents. *The Journal of Neuroscience*, *37*(5), 1139–1155. https://doi.org/10.1523/JNEUROSCI.2002-16.2016.

Healey, K. L., Landin, J. D., Dubester, K., Kibble, S., Marquardt, K., Brutman, J. N., et al. (2020). Effects of ethanol on plasma ghrelin levels in the rat during early and late adolescence. *Alcohol*, *85*, 111–118. https://doi.org/10.1016/j.alcohol.2019.12.006.

Hingson, R. W., Heeren, T., & Winter, M. R. (2006). Age at drinking onset and alcohol dependence: Age at onset, duration, and severity. *Archives of Pediatrics and Adolescent Medicine*, *160*(7), 739–746. https://doi.org/10.1001/archpedi.160.7.739.

Homo-Delarche, F., Fitzpatrick, F., Christeff, N., Nunez, E. A., Bach, J. F., & Dardenne, M. (1991). Sex steroids, glucocorticoids, stress and autoimmunity. *Journal of Steroid Biochemistry and Molecular Biology*, *40*(4–6), 619–637. https://doi.org/10.1016/0960-0760(91)90285-d.

Jensen, C. J., Massie, A., & De Keyser, J. (2013). Immune players in the CNS: The astrocyte. *Journal of Neuroimmune Pharmacology*, *8*(4), 824–839. https://doi.org/10.1007/s11481-013-9480-6.

Johnston, L. D., Miech, R. A., O'Malley, P. M., Bachman, J. G., Schulenberg, J. E., & Patrick, J. E. (2020). *Monitoring the future national survey results on drug use 1975–2019: Overview, key findings on adolescent drug use*. Ann Arbor: Institute for Social Research, The University of Michigan.

Kalia, M. (1992). Early ontogeny of the vagus nerve: An analysis of the medulla oblongata and cervical spinal cord of the postnatal rat. *Neurochemistry International*, *20*(1), 119–128. https://doi.org/10.1016/0197-0186(92)90135-e.

Kane, C. J., Phelan, K. D., Douglas, J. C., Wagoner, G., Johnson, J. W., Xu, J., et al. (2014). Effects of ethanol on immune response in the brain: Region-specific changes in adolescent versus adult mice. *Alcoholism: Clinical and Experimental Research*, *38*(2), 384–391. https://doi.org/10.1111/acer.12244.

Kang, S. S., & McGavern, D. B. (2008). Lymphocytic choriomeningitis infection of the central nervous system. *Frontiers in Bioscience*, *13*, 4529–4543. https://doi.org/10.2741/3021.

Kanny, D., Naimi, T. S., Liu, Y., Lu, H., & Brewer, R. D. (2018). Annual total binge drinks consumed by U.S. adults, 2015. *American Journal of Preventive Medicine*, *54*(4), 486–496. https://doi.org/10.1016/j.amepre.2017.12.021.

Karson, A., Demirtas, T., Bayramgurler, D., Balci, F., & Utkan, T. (2013). Chronic administration of infliximab (TNF-alpha inhibitor) decreases depression and anxiety-like behaviour in rat model of chronic mild stress. *Basic and Clinical Pharmacology and Toxicology*, *112*(5), 335–340. https://doi.org/10.1111/bcpt.12037.

Kettenmann, H., Hanisch, U. K., Noda, M., & Verkhratsky, A. (2011). Physiology of microglia. *Physiological Reviews*, *91*(2), 461–553. https://doi.org/10.1152/physrev.00011.2010.

Keshavan, M. S., Giedd, J., Lau, J. Y., Lewis, D. A., & Paus, T. (2014). Changes in the adolescent brain and the pathophysiology of psychotic disorders. *Lancet Psychiatry*, *1*(7), 549–558. https://doi.org/10.1016/S2215-0366(14)00081-9.

Kitabgi, P., Melik-Parsadaniantz, S., & Rostene, W. (2006). Chemokines: A new peptide family of neuromodulators. In A. J. Kastin (Ed.), *Handbook of biologically active peptides* (pp. 559–565). Academic Press. https://doi.org/10.1016/B978-012369442-3/50083-0.

Knapp, D. J., Harper, K. M., Whitman, B. A., Zimomra, Z., & Breese, G. R. (2016). Stress and withdrawal from chronic ethanol induce selective changes in neuroimmune mRNAs in differing brain sites. *Brain Sciences*, *6*(3). https://doi.org/10.3390/brainsci6030025.

Knapp, D. J., Overstreet, D. H., & Breese, G. R. (2007). Baclofen blocks expression and sensitization of anxiety-like behavior in an animal model of repeated stress and ethanol withdrawal. *Alcoholism: Clinical and Experimental Research*, *31*(4), 582–595. https://doi.org/10.1111/j.1530-0277.2007.00342.x.

Lee, S., & Rivier, C. (1993). Effect of exposure to an alcohol diet for 10 days on the ability of interleukin-1 beta to release ACTH and corticosterone in the adult ovariectomized female rat. *Alcoholism: Clinical and Experimental Research*, *17*(5), 1009–1013. https://doi.org/10.1111/j.1530-0277.1993.tb05656.x.

Lenz, K. M., & McCarthy, M. M. (2015). A starring role for microglia in brain sex differences. *The Neuroscientist*, *21*(3), 306–321. https://doi.org/10.1177/1073858414536468.

Licinio, J. (1997). Central nervous system cytokines and their relevance for neurotoxicity and apoptosis. *Journal of Neural Transmission. Supplementum*, *49*, 169–175. https://doi.org/10.1007/978-3-7091-6844-8_18.

Licinio, J., & Frost, P. (2000). The neuroimmune-endocrine axis: Pathophysiological implications for the central nervous system cytokines and hypothalamus-pituitary-adrenal hormone dynamics. *Brazilian Journal of Medical and Biological Research*, *33*(10), 1141–1148. https://doi.org/S0100-879X2000001000003 [pii].

Lively, S., & Schlichter, L. C. (2013). The microglial activation state regulates migration and roles of matrix-dissolving enzymes for invasion. *Journal of Neuroinflammation*, *10*, 75. https://doi.org/10.1186/1742-2094-10-75.

Logrip, M. L., Rivier, C., Lau, C., Im, S., Vaughan, J., & Lee, S. (2013). Adolescent alcohol exposure alters the rat adult hypothalamic-pituitary-adrenal axis responsiveness in a sex-specific manner. *Neuroscience*, *235*, 174–186. https://doi.org/10.1016/j.neuroscience.2012.12.069.

Lu, L., Shepard, J. D., Hall, F. S., & Shaham, Y. (2003). Effect of environmental stressors on opiate and psychostimulant reinforcement, reinstatement and discrimination in rats: A review. *Neuroscience and Biobehavioral Reviews*, *27*(5), 457–491. https://doi.org/10.1016/s0149-7634(03)00073-3.

Lynch, W. J., Kushner, M. G., Rawleigh, J. M., Fiszdon, J., & Carroll, M. E. (1999). The effects of restraint stress on voluntary ethanol consumption in rats. *Experimental and Clinical Psychopharmacology*, 7(4), 318–323. http://www.ncbi.nlm.nih.gov/entrez/query.fcgi?cmd=Retrieve&db=PubMed&dopt=Citation&list_uids=10609966.

Macht, V., Crews, F. T., & Vetreno, R. P. (2020). Neuroimmune and epigenetic mechanisms underlying persistent loss of hippocampal neurogenesis following adolescent intermittent ethanol exposure. *Current Opinion in Pharmacology*, *50*, 9–16. https://doi.org/10.1016/j.coph.2019.10.007.

Maier, S. F., & Watkins, L. R. (1998). Cytokines for psychologists: Implications of bidirectional immune-to-brain communication for understanding behavior, mood, and cognition. *Psychological Review*, *105*(1), 83–107. http://www.ncbi.nlm.nih.gov/entrez/query.fcgi?cmd=Retrieve&db=PubMed&dopt=Citation&list_uids=9450372.

Maldonado, A. M., Finkbeiner, L. M., Alipour, K. K., & Kirstein, C. L. (2008). Voluntary ethanol consumption differs in adolescent and adult male rats using a modified sucrose-fading paradigm. *Alcoholism: Clinical and Experimental Research*, *32*(9), 1574–1582. https://doi.org/10.1111/j.1530-0277.2008.00733.x.

Markwiese, B. J., Acheson, S. K., Levin, E. D., Wilson, W. A., & Swartzwelder, H. S. (1998). Differential effects of ethanol on memory in adolescent and adult rats. *Alcoholism: Clinical and Experimental Research*, *22*(2), 416–421. https://www.ncbi.nlm.nih.gov/pubmed/9581648.

Marshall, W. A. (1975). Growth and sexual maturation in normal puberty. *Clinics in Endocrinology and Metabolism*, *4*(1), 3–25. https://doi.org/10.1016/s0300-595x(75)80032-6.

Marsland, P., Parrella, A., Orlofsky, M., Lovelock, D. F., Vore, A. S., Varlinskaya, E. I., et al. (2021). Neuroendocrine and neuroimmune responses in male and female rats: Evidence for functional immaturity of the neuroimmune system during early adolescence. *European Journal of Neuroscience*. https://doi.org/10.1111/ejn.15118.

Marsland, P., Parrella, A., Vore, A. S., Barney, T. M., Varlinskaya, E. I., & Deak, T. (2021). Male, but not female, Sprague Dawley rats display enhanced fear learning following acute ethanol withdrawal (hangover). *Pharmacology, Biochemistry, and Behavior*, *208*, 173229. https://doi.org/10.1016/j.pbb.2021.173229.

Masten, A. S., Faden, V. B., Zucker, R. A., & Spear, L. P. (2009). A developmental perspective on underage alcohol use. *Alcohol Research and Health*, *32*(1), 3–15. https://www.ncbi.nlm.nih.gov/pubmed/23104443.

Mathews, I. Z., Mills, R. G., & McCormick, C. M. (2008). Chronic social stress in adolescence influenced both amphetamine conditioned place preference and locomotor sensitization. *Developmental Psychobiology*, *50*(5), 451–459. https://doi.org/10.1002/dev.20299.

Matsumoto, K., Ojima, K., & Watanabe, H. (1997). Central corticotropin-releasing factor and benzodiazepine receptor systems are involved in the social isolation stress-induced decrease in ethanol sleep in mice. *Brain Research*, *753*(2), 318–321. https://doi.org/10.1016/s0006-8993(97)00080-2.

Matullo, C. M., O'Regan, K. J., Curtis, M., & Rall, G. F. (2011). CNS recruitment of CD8 + T lymphocytes specific for a peripheral virus infection triggers neuropathogenesis during polymicrobial challenge. *PLoS Pathogens*, 7(12), e1002462. https://doi.org/10.1371/journal.ppat.1002462.

Mayfield, J., Ferguson, L., & Harris, R. A. (2013). Neuroimmune signaling: A key component of alcohol abuse. *Current Opinion in Neurobiology*, *23*(4), 513–520. https://doi.org/10.1016/j.conb.2013.01.024.

McCarthy, G. M., Farris, S. P., Blednov, Y. A., Harris, R. A., & Mayfield, R. D. (2018). Microglial-specific transcriptome changes following chronic alcohol consumption. *Neuropharmacology*, *128*, 416–424. https://doi.org/10.1016/j.neuropharm.2017.10.035.

McClain, J. A., Hayes, D. M., Morris, S. A., & Nixon, K. (2011). Adolescent binge alcohol exposure alters hippocampal progenitor cell proliferation in rats: Effects on cell cycle kinetics. *Journal of Comparative Neurology*, *519*(13), 2697–2710. https://doi.org/10.1002/cne.22647.

McClain, J. A., Morris, S. A., Deeny, M. A., Marshall, S. A., Hayes, D. M., Kiser, Z. M., et al. (2011). Adolescent binge alcohol exposure induces long-lasting partial activation of microglia. *Brain, Behavior, and Immunity*, *25*(Suppl. 1), S120–S128. https://doi.org/10.1016/j.bbi.2011.01.006.

McCormick, C. M., & Green, M. R. (2013). From the stressed adolescent to the anxious and depressed adult: Investigations in rodent models. *Neuroscience, 249*, 242–257. https://doi.org/10.1016/j.neuroscience.2012.08.063.
McCormick, C. M., Green, M. R., & Simone, J. J. (2017). Translational relevance of rodent models of hypothalamic-pituitary-adrenal function and stressors in adolescence. *Neurobiology of Stress, 6*, 31–43. https://doi.org/10.1016/j.ynstr.2016.08.003.
McCormick, C. M., & Mathews, I. Z. (2010). Adolescent development, hypothalamic-pituitary-adrenal function, and programming of adult learning and memory. *Progress in Neuro-Psychopharmacology and Biological Psychiatry, 34*(5), 756–765. https://doi.org/10.1016/j.pnpbp.2009.09.019.
McCormick, C. M., Smith, K., Baumbach, J. L., de Lima, A. P. N., Shaver, M., Hodges, T. E., et al. (2020). Adolescent social instability stress leads to immediate and lasting sex-specific changes in the neuroendocrine-immune-gut axis in rats. *Hormones and Behavior, 126*, 104845. https://doi.org/10.1016/j.yhbeh.2020.104845.
Molofsky, A. V., & Deneen, B. (2015). Astrocyte development: A guide for the perplexed. *Glia, 63*(8), 1320–1329. https://doi.org/10.1002/glia.22836.
Montesinos, J., Alfonso-Loeches, S., & Guerri, C. (2016). Impact of the innate immune response in the actions of ethanol on the central nervous system. *Alcoholism: Clinical and Experimental Research, 40*(11), 2260–2270. https://doi.org/10.1111/acer.13208.
Morris, S. A., Eaves, D. W., Smith, A. R., & Nixon, K. (2010). Alcohol inhibition of neurogenesis: A mechanism of hippocampal neurodegeneration in an adolescent alcohol abuse model. *Hippocampus, 20*(5), 596–607. https://doi.org/10.1002/hipo.20665.
Ng Cheong Ton, M. J., Brown, Z., Michalakeas, A., & Amit, Z. (1983). Stress induced suppression of maintenance but not of acquisition of ethanol consumption in rats. *Pharmacology, Biochemistry and Behavior, 18*(1), 141–144. https://doi.org/10.1016/0091-3057(83)90264-2.
National Institute on Alcohol Abuse and Alcoholism (NIAAA). (2021). https://www.niaaa.nih.gov/alcohol-health/overview-alcohol-consumption/moderate-binge-drinking.
Nikodemova, M., Kimyon, R. S., De, I., Small, A. L., Collier, L. S., & Watters, J. J. (2015). Microglial numbers attain adult levels after undergoing a rapid decrease in cell number in the third postnatal week. *Journal of Neuroimmunology, 278*, 280–288. https://doi.org/10.1016/j.jneuroim.2014.11.018.
Nixon, K. (2006). Alcohol and adult neurogenesis: Roles in neurodegeneration and recovery in chronic alcoholism. *Hippocampus, 16*(3), 287–295. https://doi.org/10.1002/hipo.20162.
Nixon, K., & Crews, F. T. (2002). Binge ethanol exposure decreases neurogenesis in adult rat hippocampus. *Journal of Neurochemistry, 83*(5), 1087–1093. https://doi.org/10.1046/j.1471-4159.2002.01214.x.
Norden, D. M., Trojanowski, P. J., Villanueva, E., Navarro, E., & Godbout, J. P. (2016). Sequential activation of microglia and astrocyte cytokine expression precedes increased Iba-1 or GFAP immunoreactivity following systemic immune challenge. *Glia, 64*(2), 300–316. https://doi.org/10.1002/glia.22930.
Obernier, J. A., Bouldin, T. W., & Crews, F. T. (2002). Binge ethanol exposure in adult rats causes necrotic cell death. *Alcoholism: Clinical and Experimental Research, 26*(4), 547–557. https://www.ncbi.nlm.nih.gov/pubmed/11981132.
Olsson, C. A., Romaniuk, H., Salinger, J., Staiger, P. K., Bonomo, Y., Hulbert, C., et al. (2016). Drinking patterns of adolescents who develop alcohol use disorders: Results from the Victorian adolescent health cohort study. *BMJ Open, 6*(2), e010455. https://doi.org/10.1136/bmjopen-2015-010455.
Orecchioni, M., Ghosheh, Y., Pramod, A. B., & Ley, K. (2019). Macrophage polarization: Different gene signatures in M1(LPS+) vs. classically and M2(LPS-) vs. alternatively activated macrophages. *Frontiers in Immunology, 10*, 1084. https://doi.org/10.3389/fimmu.2019.01084.

Orihuela, R., McPherson, C. A., & Harry, G. J. (2016). Microglial M1/M2 polarization and metabolic states. *British Journal of Pharmacology*, *173*(4), 649–665. https://doi.org/10.1111/bph.13139.

Osna, N. A., Carter, W. G., Ganesan, M., Kirpich, I. A., McClain, C. J., Petersen, D. R., et al. (2016). Aberrant post-translational protein modifications in the pathogenesis of - alcohol-induced liver injury. *World Journal of Gastroenterology*, *22*(27), 6192–6200. https://doi.org/10.3748/wjg.v22.i27.6192.

Pascual, M., Balino, P., Alfonso-Loeches, S., Aragon, C. M., & Guerri, C. (2011). Impact of TLR4 on behavioral and cognitive dysfunctions associated with alcohol-induced neuroinflammatory damage. *Brain, Behavior, and Immunity*, *25*(Suppl. 1), S80–S91. https://doi.org/10.1016/j.bbi.2011.02.012.

Pascual, M., Pla, A., Minarro, J., & Guerri, C. (2014). Neuroimmune activation and myelin changes in adolescent rats exposed to high-dose alcohol and associated cognitive dysfunction: A review with reference to human adolescent drinking. *Alcohol and Alcoholism*, *49*(2), 187–192. https://doi.org/10.1093/alcalc/agt164.

Patrick, M. E., Schulenberg, J. E., Martz, M. E., Maggs, J. L., O'Malley, P. M., & Johnston, L. D. (2013). Extreme binge drinking among 12th-grade students in the United States: Prevalence and predictors. *JAMA Pediatrics*, *167*(11), 1019–1025. https://doi.org/10.1001/jamapediatrics.2013.2392.

Patrick, M. E., & Terry-McElrath, Y. M. (2017). High-intensity drinking by underage young adults in the United States. *Addiction*, *112*(1), 82–93. https://doi.org/10.1111/add.13556.

Peng, H., & Nixon, K. (2021). Microglia phenotypes following the induction of alcohol dependence in adolescent rats. *Alcoholism: Clinical and Experimental Research*, *45*(1), 105–116. https://doi.org/10.1111/acer.14504.

Peris, J., & Cunningham, C. L. (1986). Handling-induced enhancement of alcohol's acute physiological effects. *Life Sciences*, *38*(3), 273–279. https://doi.org/10.1016/0024-3205(86)90313-9.

Perkins, A. E., Varlinskaya, E. I., & Deak, T. (2019). From adolescence to late aging: A comprehensive review of social behavior, alcohol, and neuroinflammation across the lifespan. *International Review of Neurobiology*, *148*, 231–303. https://doi.org/10.1016/bs.irn.2019.08.001.

Pohorecky, L. A. (1990). Interaction of ethanol and stress: Research with experimental animals—An update. *Alcohol and Alcoholism*, *25*(2–3), 263–276. http://www.ncbi.nlm.nih.gov/entrez/query.fcgi?cmd=Retrieve&db=PubMed&dopt=Citation&list_uids=1973897.

Pohorecky, L. A. (1991). Stress and alcohol interaction: An update of human research. *Alcoholism: Clinical and Experimental Research*, *15*(3), 438–459. http://www.ncbi.nlm.nih.gov/entrez/query.fcgi?cmd=Retrieve&db=PubMed&dopt=Citation&list_uids=1898494.

Pohorecky, L. A. (2006). Housing and rank status of male long-Evans rats modify ethanol's effect on open-field behaviors. *Psychopharmacology*, *185*(3), 289–297. https://doi.org/10.1007/s00213-005-0257-3.

Przybycien-Szymanska, M. M., Gillespie, R. A., & Pak, T. R. (2012). 17beta-Estradiol is required for the sexually dimorphic effects of repeated binge-pattern alcohol exposure on the HPA axis during adolescence. *PLoS One*, 7(2), e32263. https://doi.org/10.1371/journal.pone.0032263.

Przybycien-Szymanska, M. M., Mott, N. N., Paul, C. R., Gillespie, R. A., & Pak, T. R. (2011). Binge-pattern alcohol exposure during puberty induces long-term changes in HPA axis reactivity. *PLoS One*, *6*(4), e18350. https://doi.org/10.1371/journal.pone.0018350.

Przybycien-Szymanska, M. M., Rao, Y. S., & Pak, T. R. (2010). Binge-pattern alcohol exposure during puberty induces sexually dimorphic changes in genes regulating the HPA axis. *American Journal of Physiology Endocrinology and Metabolism*, *298*(2), E320–E328. https://doi.org/10.1152/ajpendo.00615.2009.

Pyapali, G. K., Turner, D. A., Wilson, W. A., & Swartzwelder, H. S. (1999). Age and dose-dependent effects of ethanol on the induction of hippocampal long-term potentiation. *Alcohol*, *19*(2), 107–111. https://doi.org/10.1016/s0741-8329(99)00021-x.

Quinn, R. (2005). Comparing rat's to human's age: How old is my rat in people years? *Nutrition*, *21*(6), 775–777. https://doi.org/10.1016/j.nut.2005.04.002.

Raghavendra, V., Tanga, F. Y., & DeLeo, J. A. (2004). Complete Freunds adjuvant-induced peripheral inflammation evokes glial activation and proinflammatory cytokine expression in the CNS. *European Journal of Neuroscience*, *20*(2), 467–473. https://doi.org/10.1111/j.1460-9568.2004.03514.x.

Ransohoff, R. M. (2016). A polarizing question: do M1 and M2 microglia exist? *Nature Neuroscience*, *19*(8), 987–991. https://doi.org/10.1038/nn.4338.

Richey, L., Doremus-Fitzwater, T. L., Buck, H. M., & Deak, T. (2012). Acute illness-induced behavioral alterations are similar to those observed during withdrawal from acute alcohol exposure. *Pharmacology, Biochemistry and Behavior*, *103*(2), 284–294. https://doi.org/10.1016/j.pbb.2012.08.004.

Rivier, C. (1995). Adult male rats exposed to an alcohol diet exhibit a blunted adrenocorticotropic hormone response to immune or physical stress: Possible role of nitric oxide. *Alcoholism: Clinical and Experimental Research*, *19*(6), 1474–1479. https://doi.org/10.1111/j.1530-0277.1995.tb01010.x.

Romanovsky, A. A., Ivanov, A. I., Lenczowski, M. J., Kulchitsky, V. A., Van Dam, A. M., Poole, S., et al. (2000). Lipopolysaccharide transport from the peritoneal cavity to the blood: Is it controlled by the vagus nerve? *Autonomic Neuroscience*, *85*(1–3), 133–140. https://doi.org/10.1016/S1566-0702(00)00232-0.

Romeo, H., Tio, D., Rahman, S., Chiappelli, F., & Taylor, A. (2001). The glossopharyngeal nerve as a novel pathway in immune-to-brain communication: Relevance to neuroimmune surveillance of the oral cavity. *Journal of Neuroimmunology*, *115*, 91–100.

Romeo, H., Tio, D., & Taylor, A. (2003). Effects of glossopharyngeal nerve transection on central and peripheral cytokines and serum corticosterone induced by localized inflammation. *Journal of Neuroimmunology*, *136*, 104–111.

Romeo, R. D., Bellani, R., Karatsoreos, I. N., Chhua, N., Vernov, M., Conrad, C. D., et al. (2006). Stress history and pubertal development interact to shape hypothalamic-pituitary-adrenal axis plasticity. *Endocrinology*, *147*(4), 1664–1674. http://referenc.lib.binghamton.edu:2048/login?url=http://search.ebscohost.com/login.aspx?direct=true&db=cmedm&AN=16410296&site=ehost-live.

Romeo, R. D., Lee, S. J., Chhua, N., McPherson, C. R., & McEwen, B. S. (2004). Testosterone cannot activate an adult-like stress response in prepubertal male rats. *Neuroendocrinology*, *79*(3), 125–132. http://referenc.lib.binghamton.edu:2048/login?url=http://search.ebscohost.com/login.aspx?direct=true&db=cmedm&AN=15103225&site=ehost-live.

Romeo, R. D., Lee, S. J., & McEwen, B. S. (2004). Differential stress reactivity in intact and ovariectomized prepubertal and adult female rats. *Neuroendocrinology*, *80*(6), 387–393. http://referenc.lib.binghamton.edu:2048/login?url=http://search.ebscohost.com/login.aspx?direct=true&db=cmedm&AN=15741744&site=ehost-live.

Rossetti, A. C., Papp, M., Gruca, P., Paladini, M. S., Racagni, G., Riva, M. A., et al. (2016). Stress-induced anhedonia is associated with the activation of the inflammatory system in the rat brain: Restorative effect of pharmacological intervention. *Pharmacological Research*, *103*, 1–12. https://doi.org/10.1016/j.phrs.2015.10.022.

Santiago, R. M., Barbiero, J., Martynhak, B. J., Boschen, S. L., da Silva, L. M., Werner, M. F., et al. (2014). Antidepressant-like effect of celecoxib piroxicam in rat models of depression. *Journal of Neural Transmission, 121*(6), 671–682. https://doi.org/10.1007/s00702-014-1159-5.

Savina, A., & Amigorena, S. (2007). Phagocytosis and antigen presentation in dendritic cells. *Immunological Reviews, 219*, 143–156. https://doi.org/10.1111/j.1600-065X.2007.00552.x.

Shi, X., DeLucia, A. L., Bao, J., & Zhang, P. (2019). Alcohol abuse and disorder of granulopoiesis. *Pharmacology and Therapeutics, 198*, 206–219. https://doi.org/10.1016/j.pharmthera.2019.03.001.

Siegmund, S., Vengeliene, V., Singer, M. V., & Spanagel, R. (2005). Influence of age at drinking onset on long-term ethanol self-administration with deprivation and stress phases. *Alcoholism: Clinical and Experimental Research, 29*(7), 1139–1145. https://doi.org/10.1097/01.alc.0000171928.40418.46.

Sillaber, I., & Henniger, M. S. (2004). Stress and alcohol drinking. *Annals of Medicine, 36*(8), 596–605. https://doi.org/10.1080/07853890410018862.

Silveri, M. M. (2012). Adolescent brain development and underage drinking in the United States: Identifying risks of alcohol use in college populations. *Harvard Review of Psychiatry, 20*(4), 189–200. https://doi.org/10.3109/10673229.2012.714642.

Silveri, M. M., Dager, A. D., Cohen-Gilbert, J. E., & Sneider, J. T. (2016). Neurobiological signatures associated with alcohol and drug use in the human adolescent brain. *Neuroscience and Biobehavioral Reviews, 70*, 244–259. https://doi.org/10.1016/j.neubiorev.2016.06.042.

Silveri, M. M., & Spear, L. P. (1998). Decreased sensitivity to the hypnotic effects of ethanol early in ontogeny. *Alcoholism: Clinical and Experimental Research, 22*(3), 670–676. https://doi.org/10.1111/j.1530-0277.1998.tb04310.x.

Sinha, R. (2012). How does stress lead to risk of alcohol relapse? *Alcohol Research: Current Reviews, 34*(4), 432–440. https://www.ncbi.nlm.nih.gov/pubmed/23584109.

Sinha, R., Shaham, Y., & Heilig, M. (2011). Translational and reverse translational research on the role of stress in drug craving and relapse. *Psychopharmacology, 218*(1), 69–82. https://doi.org/10.1007/s00213-011-2263-y.

Sisk, C. L., & Foster, D. L. (2004). The neural basis of puberty and adolescence. *Nature Neuroscience*, 7(10), 1040–1047. https://doi.org/10.1038/nn1326.

Spanagel, R., Noori, H. R., & Heilig, M. (2014). Stress and alcohol interactions: Animal studies and clinical significance. *Trends in Neurosciences, 37*(4), 219–227. https://doi.org/10.1016/j.tins.2014.02.006.

Spear, L. P. (2000). The adolescent brain and age-related behavioral manifestations. *Neuroscience and Biobehavioral Reviews, 24*(4), 417–463. http://www.ncbi.nlm.nih.gov/entrez/query.fcgi?cmd=Retrieve&db=PubMed&dopt=Citation&list_uids=10817843.

Spear, L. P. (2011). Rewards, aversions and affect in adolescence: Emerging convergences across laboratory animal and human data. *Developmental Cognitive Neuroscience, 1*(4), 392–400. https://doi.org/10.1016/j.dcn.2011.08.001.

Spear, L. P. (2013). Adolescent neurodevelopment. *Journal of Adolescent Health, 52*(2 Suppl. 2), S7–13. https://doi.org/10.1016/j.jadohealth.2012.05.006.

Spear, L. P. (2015). Adolescent alcohol exposure: Are there separable vulnerable periods within adolescence? *Physiology & Behavior, 148*, 122–130. https://doi.org/10.1016/j.physbeh.2015.01.027.

Spear, L. P. (2018). Effects of adolescent alcohol consumption on the brain and behaviour. *Nature Reviews Neuroscience, 19*(4), 197–214. https://doi.org/10.1038/nrn.2018.10.

Spear, L. P., & Swartzwelder, H. S. (2014). Adolescent alcohol exposure and persistence of adolescent-typical phenotypes into adulthood: A mini-review. *Neuroscience and Biobehavioral Reviews, 45*, 1–8. https://doi.org/10.1016/j.neubiorev.2014.04.012.

Squeglia, L. M., Jacobus, J., & Tapert, S. F. (2014). The effect of alcohol use on human adolescent brain structures and systems. *Handbook of Clinical Neurology*, *125*, 501–510. https://doi.org/10.1016/B978-0-444-62619-6.00028-8.

Stewart, J. (2003). Stress and relapse to drug seeking: Studies in laboratory animals shed light on mechanisms and sources of long-term vulnerability. *American Journal on Addictions*, *12*(1), 1–17. https://www.ncbi.nlm.nih.gov/pubmed/12623736.

Swartzwelder, H. S., Healey, K. L., Liu, W., Dubester, K., Miller, K. M., & Crews, F. T. (2019). Changes in neuroimmune and neuronal death markers after adolescent alcohol exposure in rats are reversed by donepezil. *Scientific Reports*, *9*(1), 12110. https://doi.org/10.1038/s41598-019-47039-1.

Szabo, G. (1999). Consequences of alcohol consumption on host defence. *Alcohol and Alcoholism*, *34*(6), 830–841. https://doi.org/10.1093/alcalc/34.6.830.

Szabo, G., & Saha, B. (2015). Alcohol's effect on host defense. *Alcohol Research*, *37*(2), 159–170. https://www.ncbi.nlm.nih.gov/pubmed/26695755.

Tay, T. L., Savage, J. C., Hui, C. W., Bisht, K., & Tremblay, M. E. (2017). Microglia across the lifespan: From origin to function in brain development, plasticity and cognition. *Journal of Physiology*, *595*(6), 1929–1945. https://doi.org/10.1113/JP272134.

Telles, T. M., de Oliveira, B. M., Lomba, L. A., Leite-Avalca, M. G., Correia, D., & Zampronio, A. R. (2017). Effects of binge-like ethanol exposure during adolescence on the febrile response in rats. *Alcoholism: Clinical and Experimental Research*, *41*(3), 507–515. https://doi.org/10.1111/acer.13333.

Terasaki, L. S., & Schwarz, J. M. (2017). Impact of prenatal and subsequent adult alcohol exposure on pro-inflammatory cytokine expression in brain regions necessary for simple recognition memory. *Brain Sciences*, 7(10), 125. https://doi.org/10.3390/brainsci7100125.

Torcaso, A., Asimes, A., Meagher, M., & Pak, T. R. (2017). Adolescent binge alcohol exposure increases risk assessment behaviors in male Wistar rats after exposure to an acute psychological stressor in adulthood. *Psychoneuroendocrinology*, *76*, 154–161. https://doi.org/10.1016/j.psyneuen.2016.11.032.

Valdez, G. R., Zorrilla, E. P., Roberts, A. J., & Koob, G. F. (2003). Antagonism of corticotropin-releasing factor attenuates the enhanced responsiveness to stress observed during protracted ethanol abstinence. *Alcohol*, *29*(2), 55–60. https://doi.org/S074183290300020X [pii].

Varlinskaya, E. I., & Spear, L. P. (2002). Acute effects of ethanol on social behavior of adolescent and adult rats: Role of familiarity of the test situation. *Alcoholism: Clinical and Experimental Research*, *26*(10), 1502–1511. http://www.ncbi.nlm.nih.gov/entrez/query.fcgi?cmd=Retrieve&db=PubMed&dopt=Citation&list_uids=12394283.

Varlinskaya, E. I., Vetter-O'Hagen, C. S., & Spear, L. P. (2013). Puberty and gonadal hormones: Role in adolescent-typical behavioral alterations. *Hormones and Behavior*, *64*(2), 343–349. https://doi.org/10.1016/j.yhbeh.2012.11.012.

Vengeliene, V., Siegmund, S., Singer, M. V., Sinclair, J. D., Li, T. K., & Spanagel, R. (2003). A comparative study on alcohol-preferring rat lines: Effects of deprivation and stress phases on voluntary alcohol intake. *Alcoholism: Clinical and Experimental Research*, *27*(7), 1048–1054. https://doi.org/10.1097/01.ALC.0000075829.81211.0C.

Vetreno, R. P., & Crews, F. T. (2012). Adolescent binge drinking increases expression of the danger signal receptor agonist HMGB1 and toll-like receptors in the adult prefrontal cortex. *Neuroscience*, *226*, 475–488. https://doi.org/10.1016/j.neuroscience.2012.08.046.

Vetreno, R. P., & Crews, F. T. (2015). Binge ethanol exposure during adolescence leads to a persistent loss of neurogenesis in the dorsal and ventral hippocampus that is associated with impaired adult cognitive functioning. *Frontiers in Neuroscience*, *9*, 35. https://doi.org/10.3389/fnins.2015.00035.

Vetreno, R. P., Lawrimore, C. J., Rowsey, P. J., & Crews, F. T. (2018). Persistent adult neuroimmune activation and loss of hippocampal neurogenesis following adolescent ethanol exposure: Blockade by exercise and the anti-inflammatory drug indomethacin. *Frontiers in Neuroscience*, *12*, 200. https://doi.org/10.3389/fnins.2018.00200.

Vetreno, R. P., Qin, L., & Crews, F. T. (2013). Increased receptor for advanced glycation end product expression in the human alcoholic prefrontal cortex is linked to adolescent drinking. *Neurobiology of Disease*, *59*, 52–62. https://doi.org/S0969-9961(13)00196-4 [pii] https://doi.org/10.1016/j.nbd.2013.07.002.

Vetter, C. S., Doremus-Fitzwater, T. L., & Spear, L. P. (2007). Time course of elevated ethanol intake in adolescent relative to adult rats under continuous, voluntary-access conditions. *Alcoholism: Clinical and Experimental Research*, *31*(7), 1159–1168. https://doi.org/10.1111/j.1530-0277.2007.00417.x.

Vijayakumar, N., Op de Macks, Z., Shirtcliff, E. A., & Pfeifer, J. H. (2018). Puberty and the human brain: Insights into adolescent development. *Neuroscience and Biobehavioral Reviews*, *92*, 417–436. https://doi.org/10.1016/j.neubiorev.2018.06.004.

Vincenti, I., & Merkler, D. (2021). New advances in immune components mediating viral control in the CNS. *Current Opinion in Virololgy*, *47*, 68–78. https://doi.org/10.1016/j.coviro.2021.02.001.

Vitale, G., Ruggieri, V., Filaferro, M., Frigeri, C., Alboni, S., Tascedda, F., et al. (2009). Chronic treatment with the selective NOP receptor antagonist [Nphe 1, Arg 14, Lys 15]N/OFQ-NH 2 (UFP-101) reverses the behavioural and biochemical effects of unpredictable chronic mild stress in rats. *Psychopharmacology*, *207*(2), 173–189. https://doi.org/10.1007/s00213-009-1646-9.

Volpicelli, J. R., Ulm, R. R., & Hopson, N. (1990). The bidirectional effects of shock on alcohol preference in rats. *Alcoholism: Clinical and Experimental Research*, *14*(6), 913–916. https://doi.org/10.1111/j.1530-0277.1990.tb01837.x.

Vore, A. S., Doremus-Fitzwater, T., Gano, A., & Deak, T. (2017). Adolescent ethanol exposure leads to stimulus-specific changes in cytokine reactivity and hypothalamic-pituitary-adrenal axis sensitivity in adulthood. *Frontiers in Behavioral Neuroscience*, *11*, 78. https://doi.org/10.3389/fnbeh.2017.00078.

Walter, T. J., Vetreno, R. P., & Crews, F. T. (2017). Alcohol and stress activation of microglia and neurons: Brain regional effects. *Alcoholism: Clinical and Experimental Research*, *41*(12), 2066–2081. https://doi.org/10.1111/acer.13511.

Warden, A. S., Azzam, M., DaCosta, A., Mason, S., Blednov, Y. A., Messing, R. O., et al. (2019a). Toll-like receptor 3 activation increases voluntary alcohol intake in C57BL/6J male mice. *Brain, Behavior, and Immunity*, *77*, 55–65. https://doi.org/10.1016/j.bbi.2018.12.004.

Warden, A. S., Azzam, M., DaCosta, A., Mason, S., Blednov, Y. A., Messing, R. O., et al. (2019b). Toll-like receptor 3 dynamics in female C57BL/6J mice: Regulation of alcohol intake. *Brain, Behavior, and Immunity*, *77*, 66–76. https://doi.org/10.1016/j.bbi.2018.12.006.

Warden, A. S., Wolfe, S. A., Khom, S., Varodayan, F. P., Patel, R. R., Steinman, M. Q., et al. (2020). Microglia control escalation of drinking in alcohol-dependent mice: Genomic and synaptic drivers. *Biological Psychiatry*, *88*(12), 910–921. https://doi.org/10.1016/j.biopsych.2020.05.011.

White, A. M., Bae, J. G., Truesdale, M. C., Ahmad, S., Wilson, W. A., & Swartzwelder, H. S. (2002). Chronic-intermittent ethanol exposure during adolescence prevents normal developmental changes in sensitivity to ethanol-induced motor impairments. *Alcoholism: Clinical and Experimental Research*, *26*(7), 960–968. http://referenc.lib.binghamton.edu:2048/login?url=http://search.ebscohost.com/login.aspx?direct=true&db=cmedm&AN=12170104&site=ehost-live.

World Health Organization. (2018). *Global status report on alcohol and health 2018*. World Health Organization. https://apps.who.int/iris/handle/10665/274603. (Accessed 23 August 2021).

Willner, P. (2005). Chronic mild stress (CMS) revisited: Consistency and behavioural-neurobiological concordance in the effects of CMS. *Neuropsychobiology*, *52*(2), 90–110. https://doi.org/10.1159/000087097.

Windle, M., Spear, L. P., Fuligni, A. J., Angold, A., Brown, J. D., Pine, D., et al. (2008). Transitions into underage and problem drinking: Developmental processes and mechanisms between 10 and 15 years of age. *Pediatrics*, *121*(Suppl. 4), S273–S289. https://doi.org/10.1542/peds.2007-2243C.

Wu, T. C., Chen, H. T., Chang, H. Y., Yang, C. Y., Hsiao, M. C., Cheng, M. L., et al. (2013). Mineralocorticoid receptor antagonist spironolactone prevents chronic corticosterone induced depression-like behavior. *Psychoneuroendocrinology*, *38*(6), 871–883. https://doi.org/10.1016/j.psyneuen.2012.09.011.

Wu, Y., Dissing-Olesen, L., MacVicar, B. A., & Stevens, B. (2015). Microglia: Dynamic mediators of synapse development and plasticity. *Trends in Immunology*, *36*(10), 605–613. https://doi.org/10.1016/j.it.2015.08.008.

Wu, Y., Lousberg, E. L., Moldenhauer, L. M., Hayball, J. D., Robertson, S. A., Coller, J. K., et al. (2011). Attenuation of microglial and IL-1 signaling protects mice from acute alcohol-induced sedation and/or motor impairment. *Brain, Behavior, and Immunity*, *25*(Suppl. 1), S155–S164. https://doi.org/10.1016/j.bbi.2011.01.012.

You, Z., Luo, C., Zhang, W., Chen, Y., He, J., Zhao, Q., et al. (2011). Pro- and anti-inflammatory cytokines expression in rat's brain and spleen exposed to chronic mild stress: Involvement in depression. *Behavioural Brain Research*, *225*(1), 135–141. https://doi.org/10.1016/j.bbr.2011.07.006.

Zahr, N. M., & Pfefferbaum, A. (2017). Alcohol's effects on the brain: Neuroimaging results in humans and animal models. *Alcohol Research: Current Reviews*, *38*(2), 183–206. https://www.ncbi.nlm.nih.gov/pubmed/28988573.

CHAPTER SEVEN

Alcohol, inflammation, and blood-brain barrier function in health and disease across development

A.S. Vore and T. Deak*

Behavioral Neuroscience Program, Department of Psychology, Developmental Exposure Alcohol Research Center, Binghamton, NY, United States

*Corresponding author: e-mail address: tdeak@binghamton.edu

Contents

1. Introduction 211
2. Form and function of the BBB: Gatekeeper of neurovascular unit function 214
   2.1 The neurovascular unit 215
   2.2 Molecular passage across the BBB 218
   2.3 The role of the BBB in peripheral to central immune signaling 219
   2.4 Approaches for assessment of the BBB 222
   2.5 Regulation of BBB in development 225
3. Alcohol effects on BBB permeability 227
   3.1 Adolescence is a time of unique vulnerability to the neurotoxic consequences of ethanol 227
   3.2 Acute ethanol alters BBB function 228
   3.3 Adolescent ethanol exposure produces sex-specific changes in ethanol-induced inflammation 231
4. Alterations in BBB permeability exacerbate brain pathology 232
   4.1 Natural aging and inflammation 234
   4.2 Implications for drug/alcohol use and abuse 236

Funding 238

References 238

Abstract

Alcohol is the most commonly used drug of abuse in the world and binge drinking is especially harmful to the brain, though the mechanisms by which alcohol compromises overall brain health remain somewhat elusive. A number of brain diseases and pathological states are accompanied by perturbations in Blood-Brain Barrier (BBB) function, ultimately exacerbating disease progression. The BBB is critical for coordinating activity between the peripheral immune system and the brain. Importantly, BBB integrity is responsive to circulating cytokines and other immune-related signaling molecules,

International Review of Neurobiology, Volume 161
ISSN 0074-7742
https://doi.org/10.1016/bs.irn.2021.06.009

which are powerfully modulated by alcohol exposure. This review will highlight key cellular components of the BBB; discuss mechanisms by which permeability is achieved; offer insight into methodological approaches for assessing BBB integrity; and forecast how alcohol-induced changes in the peripheral and central immune systems might influence BBB function in individuals with a history of binge drinking and ultimately Alcohol Use Disorders (AUD).

Abbreviations

AIE	adolescent intermittent ethanol
ADH	alcohol dehydrogenase
AUD	alcohol use disorder
AD	Alzheimer's disease
Aβ	amyloid beta
BBB	blood brain barrier
BECs	blood ethanol concentrations
BDNF	brain derived neurotrophic factor
CNS	central nervous system
CYP2E1	cytochrome P4502E1
Da	Daltons
DAMP	danger-associated molecular pattern
FITC	fluorescein isothiocyanate
HMGB1	high mobility group box-1
HRP	horseradish peroxidase
IL	interleukin
i.p.	intraperitoneal
i.v.	intravenous
KO	knockout
LPS	lipopolysaccharide
LTP	long-term potentiation
LORR	loss of righting reflex
LRP1	low density lipoprotein receptor-related protein 1
MEOS	microsomal ethanol oxidation system
NIAAA	National Institute of Alcohol Abuse and Alcoholism
NALT	nerve associated lymphoid tissue
NVU	neurovascular unit
NO	nitric oxide
IκBα	nuclear factor of kappa light polypeptide gene enhancer in B-cells inhibitor, alpha
PGP	para-glycoprotein
PD	Parkinson's disease
PNS	peripheral nervous system
PND	postnatal day
RAGE	receptor for advanced glycation end products
SAMHSA	Substance Abuse and Mental Health Services Administration
TJ	tight junction
TNF	tumor necrosis factor
TLR	Toll-like receptor
VEGF	vascular endothelial growth factor

1. Introduction

The BBB functions as one of the most vital barriers to protect the CNS from homeostatic disruption in mammalian species. From birth, the barrier serves to closely regulate active and passive transport of blood-borne substances into and out of the brain. This ensures optimal brain health and facilitates communication between the central nervous system (CNS) and peripheral organ systems that mediate normal bodily function (Saili et al., 2017; Saunders, Liddelow, & Dziegielewska, 2012). In the context of inflammation, insult, stress, or exposure to drugs of abuse, and immune signaling factors as well as the drugs themselves are controlled by gating mechanisms mediated by the BBB (Pimentel, Sivalingam, Doke, & Samikkannu, 2020). The transit of signaling factors from the periphery into the brain confers a degree of "awareness" for the CNS as to what is occurring in the periphery (Fuzzati-Armentero, Cerri, & Blandini, 2019; Reinhold & Rittner, 2017). Dysfunction of this gatekeeping role often culminates in extreme pathology due to excessive influx of immune signaling factors into the CNS, which in turn creates a cyclical pattern of disruption. Thus, damage or alterations to BBB permeability have far-reaching implications for subsequent brain health (Daneman & Prat, 2015; Profaci, Munji, Pulido, & Daneman, 2020).

Ethanol remains one of the most commonly used drugs of abuse in modern society. Owing in part to its cultural/societal acceptance, a reported 85.6% of individuals aged 18 and older self-report consumption of alcohol at some point in their lifetime (Center for Behavioral Health Statistics and, 2020). While there is some data to suggest a small quantity of ethanol consumption can have beneficial impact, 25.8% of individuals 18 and older self-report having engaged in harmful binge consumption within the past month (Center for Behavioral Health Statistics and, 2020). Binge patterns of ethanol consumption and heavy alcohol use are especially harmful to both the brain and organ systems throughout the body. For instance, higher levels of self-reported drinking have been associated with reduced gray matter volumes across the brain (Howell et al., 2013; Kvamme et al., 2016). Binge drinking also strongly predicts a future diagnosis of an Alcohol Use Disorder (AUD), which can compromise overall health across the lifespan (DeWit, Adlaf, Offord, & Ogborne, 2000; Grant & Dawson, 1997).

Chronic alcohol abuse and dependence are associated with pathological increases in inflammation that often leads to organ dysfunction (Leclercq, de

Timary, Delzenne, & Stärkel, 2017; Wang, Zakhari, & Jung, 2010), whereas more moderate doses may produce lower risk levels in some circumstances (Bektas, Sen, & Ferrucci, 2016; Pai et al., 2006; Wang, Pan, Ye, Su, & Li, 2008). One hallmark of alcohol dependence, alcoholic liver disease, is known to involve inflammatory signaling at many different levels of cellular and organ function (Wang, Gao, Zakhari, & Nagy, 2012). For instance, alcohol can mobilize gut-derived endotoxins, increase inflammatory signaling factors such as cytokines, and increase oxidative stress (Fujimoto et al., 2000). Increased TLR4-mediated signaling in Kupffer cells, hepatocytes, stellate cells, and other important cellular components of the liver are functionally changed by alcohol (Mandrekar & Szabo, 2009). Beyond these peripheral changes, there is a growing body of literature that examines the impact of ethanol on the gut and how these changes may have more far-reaching impact on behavior and brain function. For example, gut microbiome dysfunction can modulate vagus nerve signaling, ultimately amplifying the neuroimmune response in the amygdala and contributing to heightened anxiety and other behavioral changes observed during alcohol withdrawal (Gorky & Schwaber, 2016). Finally, both rodent and human data indicate that prolonged alcohol use and abuse produces a persistent immunosuppressive phenotype that compromises host defense. While this phenotype often may be subclinical in magnitude, when it is combined with an additional immune challenge or insult pathological immunosuppression emerges (Szabo & Saha, 2015).

In preclinical studies with rodents, acute, binge-like alcohol exposure produced fluctuations in pro-inflammatory cytokine expression, with unique cytokine signatures apparent during acute intoxication and a separate pattern evident during alcohol hangover/withdrawal. Specifically, IL-6 and IκBα were markedly increased within the first few hours after ethanol exposure, whereas IL-1β and TNFα were only increased after ethanol clearance (Doremus-Fitzwater et al., 2014; Gano, Mondello, Doremus-Fitzwater, & Deak, 2019; Gano, Vore, Sammakia, & Deak, 2019; Walter, Vetreno, & Crews, 2017). Although the magnitude of these changes varies somewhat across the CNS, they are particularly prevalent in limbic structures such as the hippocampus, amygdala, and the paraventricular nucleus of the hypothalamus (Doremus-Fitzwater et al., 2014). Such changes in neuroimmune factor gene expression are both reproducible and highly conserved, with prior studies reporting equivalent responses in males and females (Gano, Mondello, et al., 2019; Gano, Vore, et al., 2019), and multiple rat strains (Gano, Doremus-Fitzwater, & Deak, 2017). Surprisingly, age of alcohol exposure has emerged as a critical demographic variable which determines the magnitude of neuroimmune signal gene expression response to an acute alcohol challenge. Adolescent rats showed severely diminished neuroimmune responses to alcohol, perhaps in part due to a functional immature neuroimmune system (Doremus-Fitzwater, Gano, Paniccia, & Deak, 2015; Marsland et al., 2021). In contrast, late aging (24 month) led to a heightened basal expression of cytokines, which made these aged rats refractory (*i.e.*, less responsive) to ethanol-induced neuroimmune changes (Gano et al., 2017). Thus, age of ethanol exposure appears to be a critical determinant of ethanol-related changes in inflammation.

Given the abundance of data that binge or supra-binge levels of ethanol exposure produce a distinct neuroimmune profile, this review will evaluate the potential influence of ethanol evoked inflammation and cytokine induction might have on BBB function. While few studies have directly assessed ethanol's ability to alter BBB function, there is an established body of work demonstrating that inflammatory signaling is mechanistically linked to changes in BBB function. Indeed, inflammatory agents such as LPS have been used in a number of studies to mimic bacterially derived inflammation. Using LPS as a challenge, alterations in BBB permeability may involve a variety of key BBB regulatory mechanisms, including (i) direct damage to the endothelium; (ii) alterations in tight junction (TJ) expression and/or function that "loosen" adhesion of cells; (iii) increased nitric oxide (NO), a transient gas that promotes vasodilation; and (iv) changes in glial (*e.g.*, astrocytes, pericytes, microglia) function and quantity; and many other BBB regulatory mechanisms (for a more thorough review of this literature see Varatharaj & Galea, 2017). As ethanol impacts many of these same physiological parameters and is known to modulate inflammation, it seems reasonable to hypothesize that binge-like alcohol use and abuse compromises BBB function through inflammation-related mechanisms (Fig. 1). The present review provides a more detailed framework for pursuing this hypothesis with empirical studies.

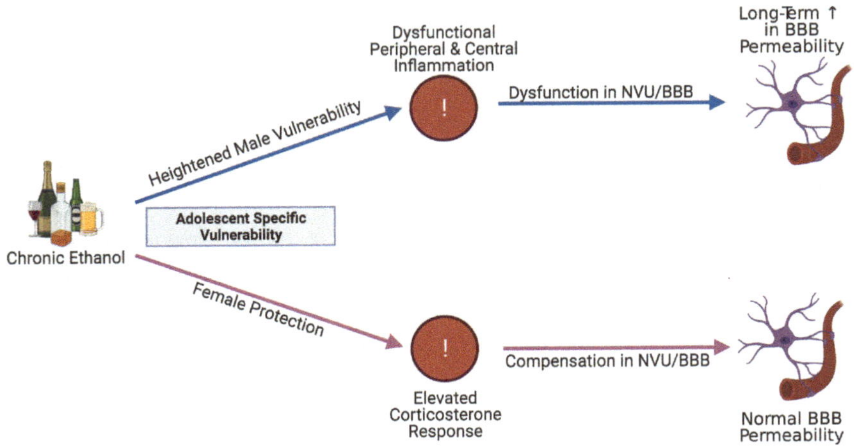

Fig. 1 Hypothetical mechanistic outline by which AIE alters BBB permeability. This figure illustrates the hypothesized mechanism through which chronic ethanol exposure during the important developmental period of adolescence may produce sex-specific changes in inflammation, ultimately resulting in elevated male BBB permeability. *Created with https://biorender.com.*

2. Form and function of the BBB: Gatekeeper of neurovascular unit function

The BBB plays a vital role in regulating passage of cells and biomolecules into the CNS and is conserved in some form across vertebrate species (O'Brown, Pfau, & Gu, 2018). The complex nature of the brain and the processes it governs means that it must gate the passage of infiltrating monocytes and potential toxins and toxicants present in general circulation. At the same time, bidirectional communication is needed to orchestrate effective responses to environmental stimuli. To accomplish this, the BBB functions as a critical gatekeeper of what reaches the parenchyma of the CNS. The BBB is comprised of a variety of cells that work synchronously to provide differing levels of neural protection across the CNS-peripheral nervous system (PNS) and in response to changing physiological circumstances (Zhao, Nelson, Betsholtz, & Zlokovic, 2015). Brain endothelial cells form the tight physical barrier restricting molecules from entering the brain. Cell-to-cell tight junctions and adherens junctions function as physical barriers to large and/or hydrophilic molecules (Zhao et al., 2015).

Tight junctions ensure that only small, lipophilic molecules, estimated to be under 400 Da (Da; or 0.4 kDa) can freely pass (Pardridge, 2015). While this aspect of the BBB helps the CNS maintain persistent exclusivity to larger and potentially harmful entities, the brain requires mechanisms that govern passage of a multitude of macromolecular proteins, amino acids, and other building blocks that can't freely pass through the BBB. Movement of these molecules is accomplished *via* an array of active and passive transport mechanisms. Polar, hydrophilic, small molecules such as amino acids utilize solute carrier membrane transport proteins set within the BBB to gain access to the CNS through facilitated diffusion down the concentration gradient, whereas macromolecule and peptide transit is more frequently handled by more traditional, energy-driven, receptor-mediated transport mechanisms (Banks et al., 2015; Zhao et al., 2015). All of these systems provide important mechanisms for both paracellular and transcellular crossing into and out of the CNS and serve to highlight the gating dynamics as well as the importance of BBB function to overall brain health.

2.1 The neurovascular unit

In some ways, the neurovascular unit (NVU) represents a modification and expansion of the classic "tripartite synapse" concept used to summarize synaptic transmission in the CNS (Muoio, Persson, & Sendeski, 2014). At the heart of the NVU, neurons are exceptionally sensitive to changes in blood oxygen, nutrients, and other blood-borne molecules that are carried through the blood supply. Neuronal communication is energy-intensive and, in response to local changes in neuronal activity other components of the NVU, local blood flow to the NVU is altered to match oxygen and nutrient demand. Thus, form and function of the NVU is designed to be responsive to changing neuronal activity and ultimately to optimize neuronal function.

While the "tripartite synapse" incorporates the most direct cellular elements of synaptic communication (neurons and astrocytes), it does not take into account the many other cellular interactions that chemically influence neuronal health and function (Neuwelt et al., 2011). While not every neurovascular interface is identical, it is most frequently comprised of the endothelium, surrounded by the basement membrane, encapsulated by pericytes, and in close contact with both astrocytic end-feet and microglia (see Fig. 2). Each of these component cells regulate unique aspects of NVU function as

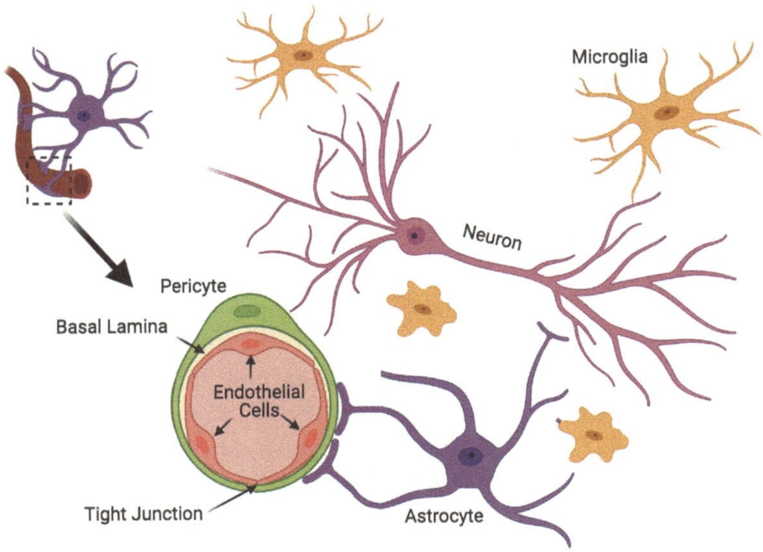

Fig. 2 Schematic representation of the neurovascular unit. This schematic highlights the complex array of cell-to-cell interactions that influence the BBB and molecular access into the CNS. The endothelial cells that form tight junctions are surrounded by the basal lamina as well as pericytes that directly communicate with the endothelial cells. Bidirectional communication occurs at each level between astrocytes, neurons, and microglia as well as astrocytic end-feet connections directly modifying pericyte signaling. *Created with BioRender.com*

a whole, and are poised to significantly influence BBB permeability (McConnell, Kersch, Woltjer, & Neuwelt, 2017). Closely aligned astrocytes and microglia play important roles in buffering the cellular microenvironment, often serving as intermediate signaling cells between blood-borne agents and neurons, and *vice versa.*

Astrocytes function as a direct cellular mediator between neurons and the remainder of the neurovascular unit. Astrocytic end-feet almost completely encapsulate the endothelial cells that comprise the capillaries of the NVU permitting bidirectional communication between both cell types (Fig. 2). Through these connections, the astrocytes release a variety of factors such as prostaglandins and NO that have the capacity to alter blood vessel diameter and therefore blood flow in response to changing neuronal demand (Wolf & Kirchhoff, 2008). Astrocytes also play an active role in the recycling and clearance of neurotransmitters and water from the brain that helps to ensure overall brain function (Sofroniew & Vinters, 2010).

While astrocytic end-feet play an important role in modulating cerebral blood flow, other cells associated with, or encapsulated by, astrocytic end-feet are also critical. The primary contact to the endothelial cells within the NVU is pericytes (Fig. 2). Pericytes have emerged as critical modulators of BBB development and functional integrity, since they specifically aid in the attachment of astrocytic end-feet to the NVU (Armulik et al., 2010). Loss of pericytes is known to promote upregulation of several genes (*i.e.*, Angiopoietin 2, Intercellular Adhesion Molecule 1, and Galectin 3) associated with increased vascular permeability, suggesting a role in BBB maintenance and endothelial cell regulation (Daneman et al., 2009). This can help promote angiogenesis and guide vascular branching within the cerebral vasculature (Bergers & Song, 2005; Eilken et al., 2017). More recently, it has also been noted that pericytes also display a direct role in modulating cerebral blood flow. For example, through contraction of the processes that wrap around the vessel, pericytes directly increase or decrease vessel constriction in response to oxygen/nutrient needs of the brain (McConnell et al., 2017).

Microglia, much like astrocytes, express factors that contribute to both the maintenance and degradation of the BBB. Microglia are considered the "first responders" of the CNS, responding to insult, excessive neuronal activity, tissue damage or infection with rapid expression of chemokines and cytokines that forward-propagate neuroinflammation (Liu & Aguzzi, 2020). Through active monitoring of the NVU, microglia can transition through a number of active states which are accompanied by adaptive changes in the neuroimmune factors they release. One down-stream consequence of microglial activation is impaired BBB function (Haruwaka et al., 2019). A substantial amount of cross-talk occurs between astrocytes and microglia and, through a coordinated response, they orchestrate the NVU's response to damage (Fig. 2). Increases in the total number of active microglia are associated with increased BBB leakiness and multiple diseases associated with BBB dysfunction are correlated with changes in genes unique to microglia such as DAP12 (Sasaki et al., 2015). The final component of the NVU is the basal lamina (Fig. 2). The basal lamina perform an important structural role in the NVU by providing anchor points for the astrocytic end-feet. It functions as an additional extracellular matrix comprised primarily of collagen and laminin. Signaling of both endothelial cells and pericytes guide the creation of the basal lamina and ultimately serve as a physical boundary between the two cell types. Thus, the typical NVU is comprised of multiple cell types in close cellular juxtaposition, with extensive cross-talk between each type.

2.2 Molecular passage across the BBB

The first line of defense forming the BBB is conferred by brain endothelial cells, which form the vascular wall and constrain passage of blood and its cargo into the interstitial tissue. TJs between adjacent endothelial cells prevent the passage of material between endothelial cells and are largely comprised of three families of proteins: occludins, claudins, and other adhesion molecules (Luissint, Artus, Glacial, Ganeshamoorthy, & Couraud, 2012). Studies have shown that interference or knockout of any of these three is associated with increased pathological BBB permeability (Nitta et al., 2003). While the semi-permeable nature of tight junctions varies across the CNS, they are generally considered to be relatively impermeable, allowing only very small (<400 Da), lipid-soluble molecules to cross (Pardridge, 2007).

While small, highly lipophilic molecules may cross freely through TJs, the passage of many endogenous and exogenous molecules across the BBB is governed by active transporters. The variety of membrane transport proteins can then be further divided into uptake/influx (into the brain) and efflux (out of the brain). The parenchyma of the brain is carefully guarded and, generally, for any drug or large molecule, access is extremely limited. Essential nutrients such as amino acids and glucose only gain access to the CNS through uptake systems such as organic anion transporters, organic cation transporters, and multidrug and toxin extrusion proteins (Sanchez-Covarrubias, Slosky, Thompson, Davis, & Ronaldson, 2014). Active transport of these larger polar molecules is facilitated by these membrane proteins. Despite gaining access to the brain, a number of efflux transporters pump out larger, polar molecules that may be present in the CNS. This serves an important regulatory role in preventing over-accumulation of larger molecules, and carefully maintains homeostasis of brain nutrients. Efflux transporters include the Para-glycoprotein (PGP) complex and multidrug resistance-associated proteins (de Boer, van der Sandt, & Gaillard, 2003). The PGP complex specifically plays a vital role in the removal of drugs from the CNS and has remained a focal point in designing new generations of pharmaceuticals that can gain selective access to the brain (for more thorough review see Bendayan, Lee, & Bendayan, 2002; Löscher & Gericke, 2020). The function and quantity of these transporters can be altered in response to a number endogenous and exogenous factors such as neurotransmitters, cytokines, and pathological insult (Avemary et al., 2013; Keaney & Campbell, 2015). Thus, a variety of active transport mechanisms contribute to passage of small molecule and macromolecular complexes across the BBB.

Cytokines interact with the BBB in a number of different ways and many different cytokines transit across the BBB through shared and unique transport mechanisms (see Banks, 2005; Banks, Lynch, & Price, 2009 for excellent reviews). Importantly, cytokine transit is dynamic and can show significant differences in transport rate depending on the animal species, as a result of circadian changes, in region-specific manners, and in response to bodily demands (when challenged). As specific examples, IL-1β is known to cross the BBB into the brain using unidirectional, saturable transport mechanisms (Banks, Plotkin, & Kastin, 1995; Davson & Segal, 1996). Similarly, both IL-6 and TNF-α utilize unidirectional saturable transport mechanisms to transit into the brain; however, it is believed that they use explicitly different transporters (Banks, Kastin, & Gutierrez, 1994; Banks et al., 1995; Gutierrez, Banks, & Kastin, 1993). In contrast, some cytokines, such as neutrophil chemoattracant-1, are able to cross the BBB using diffusion and in some circumstances, such as IL-2, transit is not believed to occur (Pan & Kastin, 2001; Waguespack, Banks, & Kastin, 1994). While saturable transport is a primary mechanism by which many cytokines enter the brain, IL-1β, TNF-α, and IL-6 have all been shown to enter the blood from the brain through cerebrospinal fluid absorption (Banks et al., 1995; Chen, Castro, Chow, & Reichlin, 1997; Chen & Reichlin, 1998). These differences highlight the ability of both the peripheral and central immune system to influence one another.

2.3 The role of the BBB in peripheral to central immune signaling

For an optimal immune response to infection or injury, effective communication must occur between the periphery (and its constituent inflammatory mediators) and the brain (Maier & Watkins, 1998). As such, BBB function is poised to be a key intermediary governing immune-to-brain communication, particularly for immune signaling molecules such as cytokines and chemokines (Banks, 2005; Erickson, Wilson, & Banks, 2020). Studies have shown that TJ permeability is subject to change due to local gene regulation (Balda & Matter, 2009). Inflammatory mediators such as histamines increase permeability in brain endothelial cells in a similar fashion as seen in peripheral capillaries through opening of inter-endothelial cell tight junctions (Abbott, Ronnback, & Hansson, 2006). In response to hypoxia, increased TJ expression and upregulation of PGP, responsible for rapid transit of compounds out of the brain, result in increased barrier stringency (Abbott et al., 2006). These findings suggest that BBB permeability is not a static aspect of physiology, but rather constantly changes in response to physiological conditions.

One of the key components responsible for communication between peripheral and central immune responses are cytokines. Cytokines are produced by microglial cells, endothelial cells, astrocytes, macrophages, and many other types of cells both within the CNS and in the periphery. They are released rapidly during the acute inflammatory process and are able to cross the BBB *via* passive diffusion in CNS regions, only where the BBB is weak or absent, such as the circumventricular organs and the OVLT (Buller, 2001; Roth, Harré, Rummel, Gerstberger, & Hübschle, 2004). Many cytokines are also subject to active transport mechanisms, which specifically transport cytokine proteins across the BBB into the CNS; although, the precise mechanism of cytokine transit across the BBB is largely species and context-dependent (Banks, 2005). Importantly, many cytokines also disrupt the BBB. Pro-inflammatory cytokines and other inflammatory signaling factors cause TJ degradation, thereby increasing permeability of the BBB (Al-Obaidi & Desa, 2018; Coisne & Engelhardt, 2011; Wan, Chen, & Li, 2014). This is, at least partially, caused through disruption of adhesion molecules on endothelial cells. When combined with chemokine signaling, leukocyte recruitment permits extravasation, the movement of white blood cells from the blood capillaries into the CNS often toward the site of insult. Cytokines can also damage the basal lamina by inducing matrix metalloproteinase (MMP) expression by endothelial cells and microglia, which further impairs the structural integrity of the BBB (Gurney, Estrada, & Rosenberg, 2006; Klein & Bischoff, 2011).

In addition to active and passive transport across the BBB, many inflammatory signaling molecules use neural pathways, humoral pathways, and direct interaction with endothelial cells to propagate *de novo* inflammatory signals on the brain side of the BBB (DiSabato, Quan, & Godbout, 2016; Goehler et al., 1999; Laflamme & Rivest, 2001). In this way, inflammatory signals originating in the periphery communicate with the CNS through transcellular communication. Many peripheral cytokines and other inflammatory mediators also bind to peripheral nerves or nerve associated lymphoid tissue (NALT), causing the peripheral inflammatory signal to be transcribed into neural impulses, which in turn may orchestrate central inflammation signaling (Kennedy & Silver, 2016). Thus, most mammalian species have evolved multiple physiological mechanisms for PNS-CNS immune communication that bypasses direct passage of peripherally-derived cytokines into the CNS.

Alterations in BBB function are frequently noted in response to challenges that evoke inflammation or a profound immune response. High levels of proinflammatory cytokines such as IL-1β and IL-6, often released by microglia, lead to increased permeability of the BBB that in times of more extreme pathology allow the infiltration of peripheral macrophages and other immune cells into the CNS (Jin, Silverman, & Vannucci, 2009). Microglia, the prominent and resident immune cell of the CNS, migrate to the damaged area in response to focal injury or conventional immune challenges and, through a variety of different modalities, forward-propagate the expression of chemokines, cytokines, and other signaling molecules crucial for tissue repair (in the case of damage) or host defense (in the case of infection). This increase in inflammatory tone can evoke significant changes in the BBB including increasing local permeability. Proinflammatory cytokines are integral for leukocyte trafficking across the BBB (Blond, Campbell, Butchart, Perry, & Anthony, 2002), one mechanism by which the peripheral immune response may modify central immune function. Minocycline administration, a tetracycline antibiotic that reduces microglial activity, has also been shown to reduce BBB permeability induced by LPS exposure, suggesting a potential role for microglia in inflammation-mediated BBB change (Moretti et al., 2015a, 2015b). LPS exposure in neonatal rats (PND0–PND8) has also been shown to increase BBB permeability, but only once animals reach adulthood and not at P20 (Stolp et al., 2011). These changes suggest a potential developmental vulnerability to some of the inflammation-induced BBB permeability changes. Correlations have also been observed between multiple pathophysiological conditions such as Alzheimer's disease, cancer, and multiple sclerosis, heightened cytokine levels, and the BBB (Moretti et al., 2015a, 2015b; Schenk & de Vries, 2016; Wardill et al., 2016a, 2016b).

Sex differences have been well documented regarding changes in BBB function and data have shown that sex differences exist in many cellular components of the BBB (Torres et al., 2018). Some of the differences noted in BBB permeability are believed to stem from endogenous differences in estrogen and the potential role estrogen may have in protecting against BBB degradation. Female-typical hormones such as estrogens and progestins also have anti-inflammatory properties (Nadkarni & McArthur, 2013), which may short-circuit inflammation-dependent alterations in BBB permeability. In addition, estrogen has been shown to enhance tight junctions, limit lymphocyte extravasation, and protect against BBB degradation resulting from LPS administration (Maggioli et al., 2016). In contrast, some data have shown that certain strains of mice may show heightened female vulnerability to changes in BBB permeability after LPS challenge

(Erickson et al., 2018). Given the number of factors capable of modifying the components of the BBB and NVU, additional studies directly assessing sex differences in BBB regulation are needed.

2.4 Approaches for assessment of the BBB

Several simple and straight-forward approaches for assessment of BBB integrity are used routinely, each with their strengths and limitations. One of the most common methods for quantifying BBB function is through the assessment of individual genes, proteins, or cells that are known to be involved in formation of the BBB. There is no shortage of work examining factors such as claudins, occludins, integrins, and other adhesion molecules (Castro Dias et al., 2019; Liebner et al., 2000). Similarly, direct visualization of endothelial cells, astrocytes, perivascular microglia or pericytes offer additional opportunities to investigate the NVU in relation to BBB integrity (Brown et al., 2020; Kovacs et al., 2017). While these approaches address cellular and structural mechanisms contributing to the BBB, they do not provide a direct *functional* assessment of BBB permeability. In addition, a number of *in vitro* model systems have been used to assess BBB permeability, offering unique opportunities to assess specific BBB regulatory mechanisms (Sivandzade & Cucullo, 2018). Comprehensive *in vitro* modeling of the BBB requires functional incorporation of the various cellular components of the NVU, an incredibly challenging task. However, to truly assess the functional integrity of the BBB, permeability characteristics of the BBB must be probed through the use of dye-based or radioligand-based, *in vivo* assessments.

One of the oldest and most common approaches for measuring BBB permeability is through the use of dyes or labeled compounds to directly quantify brain access. Dating back to Rossner and Temple (1966), Evans Blue (a vital dye that binds to serum albumin and other macromolecules) has been used as a metric for how much albumin is able to cross into the brain. Albumin is a relatively large molecule (~70 kDa) that, under normal conditions, should demonstrate little access to the brain. Due to its function as a vital dye, this allows for a quick and relatively simple quantification of dye extravasation in a variety of tissue compartments following a single *in vivo* exposure. These studies often use i.v. administration of Evans Blue dye followed by quantification or visualization of the amount of dye that accumulates in brain tissue as a metric for BBB permeability, either through tissue slicing and direct visualization; or through gross dissection of brain regions, homogenization of tissue, and quantification of dye using a spectrophotometer. More recently, however, a number of questions have been raised about the validity of the Evans Blue dye assessment. For instance, we now know that Evans Blue does not to bind exclusively to albumin, suggesting the scientific premise underlying this approach may be flawed (see Saunders, Dziegielewska, Mollgard, & Habgood, 2015 for review). Further, 70 kDa is a comparatively large compound relative to many nutrients and humoral signaling molecules commonly distributed in the blood. While it certainly addresses extreme instances of BBB pathology, it does not address smaller molecule permeability that could still have important biological significance. For instance, many pro-inflammatory cytokines range from 14 to 20 kDa in size, suggesting significant passage of these molecules at levels of BBB disruption that would not be effectively quantified with an Evans Blue dye extravasation test.

Another common method to study BBB permeability is the use of horseradish peroxidase after i.v. administration. Horseradish peroxidase is a roughly 45 kDa glycoprotein that, when combined with the proper substrate, can produce a colorimetric or fluorometric derivative allowing it to be easily quantified. As with Evans Blue dye, this allows for simple *in vivo* administration to be quantified in a variety of tissue compartments. The reaction product can be visualized using electron microscopy, allowing even more detailed assessment of BBB permeability. While this technique was shown to work effectively in mice, the method has been questioned because in rats, horseradish peroxidase alone can cause histamine and serotonin release, both of which increase BBB permeability and may potentially confound results of such an analysis (Majno, Palade, & Schoefl, 1961). Moreover, many of the toxic consequences of HRP seem to be dose- and species-specific, making interpretation across experiments and laboratories complicated (Clementi, 1970; Cotran, Karnovsky, & Goth, 1968; Ross, Nuttall, & Wright, 1977).

To more specifically probe BBB integrity across a spectrum of molecular sizes, more recent studies have used fluorophore-tagged dextran molecules (Natarajan, Northrop, & Yamamoto, 2017). Dextrans are complex polysaccharides that can be purchased across a range of sizes (ranging from as small as 4 kDa to as large as 2000 kDa). In this model, transcardial perfusion of Dextran of the target molecular weight is followed by tissue slicing and visualization of fluorescent dye in the interstitial tissue. By probing a range of molecular weights, a more refined and sensitive assessment of subtle, yet biologically important fluctuations in BBB integrity is possible.

Previous work from our laboratory has probed BBB permeability in response to acute ethanol challenge using several approaches. Using the Evans Blue procedure described above, we assessed the sensitivity of the

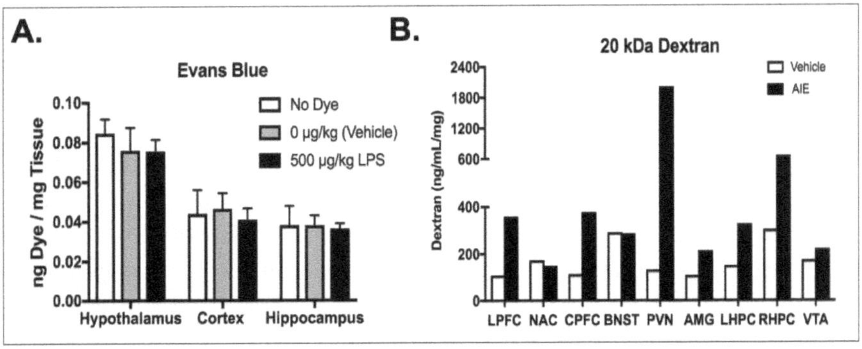

Fig. 3 Comparison of different methods of probing BBB permeability. Adult male, Sprague-Dawley rats received 500 μg/kg i.p. LPS injection and 15 h later BBB permeability was assessed with i.v. 1.0 mL/kg, 2.0% Evans Blue Dye. No differences in dye concentration were observed in gross-dissected brain tissue, evidenced by dye levels being comparable to no dye controls, despite evidence that high dose LPS increased BBB permeability (Banks et al., 2015; Ghosh et al., 2014) (A). In a subsequent pilot, rats (n = 2–3) with a history of adolescent intermittent ethanol (AIE) under basal conditions, revealed significant differences in 20 kDa dextran permeability in a range of brain regions (B).

method using a high dose LPS injection (500 μg/kg) as a presumptive positive control previously documented as producing increased BBB permeability (Banks et al., 2015; Ghosh et al., 2014). We saw no substantial differences in brain dye content relative to no dye or vehicle injected controls (Fig. 3A). These data mirror small pilot studies that were executed to determine whether any observable changes in Evan's Blue dye could be quantified following ethanol exposure. Given our concern over assay sensitivity, we piloted adolescent intermittent ethanol (AIE) exposure evoked changes in BBB permeability using a more sensitive protocol, the FITC-dextran procedure, and observed substantial increases in permeability in adult rats with a history of AIE ethanol exposure at the 20 kDa size (Fig. 3B). Such an assessment would not have been possible without the flexibility of size provided by dextran, and it is highly likely that such a difference would have been missed entirely at the 70 kDa assessment size relevant to Evans Blue.

It is critical to note that 70 kDa passage across the BBB represents a degree of permeability that would imply a highly pathological level of infiltration into the brain. In contrast, the 20 kDa weight range used with dextran is mechanistically relevant as the pro-inflammatory cytokines (IL-6, IL-1β, TNFα) evoked by ethanol are similar in weight. Since these dextrans are available tagged to a variety of fluorophores, they offer an easy method for both visualization of brain tissue as well as quantification through fluorometry. Despite the advantages of this method, as with Evans Blue and HRP, repeated dextran administration has also been shown to be toxic (Edlund, Lofgren, & Vali, 1952) and depending on the structure and size of the polysaccharide branching this may increase toxicity producing a confound. In addition, as dextran is a sugar it is possible that the inherent mechanisms responsible for detecting and responding to blood glucose levels could potentially respond to and facilitate the transfer of smaller dextrans across the BBB. Nevertheless, the FITC-tagged dextran approach offers significant advantages over earlier approaches.

Whereas all methods of assessing BBB integrity have strengths and limitations, all of the methods discussed above are readily available and easy to implement for *in vivo* studies. In selection of a model, one must consider a number of critical issues, including (i) the size of molecule and how it relates to sensitivity of the intended test of BBB integrity; (ii) potential toxicity risks associated with the target compound/dye that will be used; and (iii) potential off-target binding activity of the target compound/dye that may obfuscate conclusions. From our perspective, the ability to procure multiple dextran sizes and probe BBB integrity at different macromolecular weights represents a substantial advantage in study design. This method ensures a high degree of sensitivity in probing BBB permeability to an array of different compounds and minimizes potential type 2 errors that may occur if an experiment using Evans Blue was employed. Overall, careful scrutiny should be employed when evaluating or designing studies to probe BBB integrity to avoid potential type I error (stemming from toxicity of the administered probes) as well as potential type II error (dismissing a potential change in BBB permeability because only one size was tested).

2.5 Regulation of BBB in development

In rodent models, the BBB forms across embryonic development and is largely considered to be stably formed by birth (Blanchette & Daneman, 2015; Engelhardt, 2003). Prenatal BBB development is largely guided by canonical Wnt/beta-catenin signaling and VEGF release, which promotes angiogenesis and normal BBB development (Anderson, Lenar, Quinn, & Rooney, 2011). Genetic disruption of this signaling in early embryonic development leads to both abnormal vessel guidance and morphology (Daneman et al., 2009) and also leads to the vessels displaying abnormal phenotypes that at times miss critical BBB junction regulating proteins such as claudin (Liebner et al., 2008). A number of good reviews outlining specific mechanisms that govern BBB permeability in prenatal development and how exposure to teratogens can alter BBB permeability across the lifespan are available (see Goasdoué, Miller, Colditz, & Björkman, 2017). In contrast, much less work has attempted to directly assess changes in the BBB across postnatal development.

Though few studies have examined BBB permeability across the postnatal and adolescent periods, evidence suggests that the BBB continues to mature. Saunders et al. (2012) hypothesized that the prenatal and newborn BBB was "immature" relative to the adult BBB. While this claim has been challenged by others (Mallard, Ek, & Vexler, 2018), it is possible that maturation of the BBB is non-linear, and instead fluctuates to meet demands uniquely at different developmental epochs. For example, 2-h old rat pups displayed markedly lower BBB permeability in response to pro-inflammatory cytokine injection than post-natal day (PND) 21 rats (Anthony, Bolton, Fearn, & Perry, 1997). Similarly, adult rats displayed increased BBB permeability after acute stroke, an effect that was not observed in neonatal rats (Fernández-López et al., 2012). In addition, functional imaging studies have revealed that increased neural activity almost always corresponded with simultaneous increases in blood flow in adults, and that this coupling was not consistent in neonatal and developing brains, suggesting again that the dynamics that regulate the NVU in the adult brain are not operating in an identical fashion throughout development (Kozberg & Hillman, 2016). Much of the remainder of the evidence supporting potential BBB immaturity during early development comes from changes observed in the components of the NVU across time. While as mentioned previously, endothelial cells, neurons, and even pericytes are in place relatively early in embryonic development, astrocytic end-feet placement tends to occur postnatally in rodents (Rowitch & Kriegstein, 2010). Changes in important regulators of BBB transport, such as endothelial transporters, are present in the newborn rat; these undergo intense periods of growth that ended in a near quadrupling of receptor quantity by the time rats hit early adolescence (Vannucci, 1994). While the role that development may play in "normal" BBB function has yet to be untangled, pathological challenges that occur during these unique stages of brain development can have profound, long-lasting effects on BBB integrity. These parallel developmental differences in CNS and PNS function relative to age-specific physiological demands.

Extending BBB function into a lifespan framework, late aging is associated with natural breakdown of the transport mechanisms governing glucose, amino acid, and hormone transit, as well as lower PGP activity relative to younger individuals (Mooradian, Pinnas, Lung, Yahya, & Meredith, 1994; Toornvliet et al., 2006). These changes functionally mimic an overall decline in BBB permeability, an effect that seems to parallel a marked shift in inflammatory tone later in life. Aging is associated with both an increase in inflammatory pathology such as atherosclerosis as well as an overall increase in systemic and CNS inflammation (Perkins et al., 2021; Walker et al., 2018). This is accompanied by an array of changes in the NVU including alterations in microglial activity (Ronaldson & Davis, 2012), impaired LTP and neurogenesis (Blau et al., 2012), TJ degradation (Elahy et al., 2015), and loss of pericyte coverage (Berthiaume et al., 2018). All of these changes represent different mechanistic ways in which BBB function could change and ultimately, seem to interact negatively with many of the pathologies that often worsen substantially in late aging (Alzheimer's disease (AD), vascular dementia, ischemic stroke).

3. Alcohol effects on BBB permeability

3.1 Adolescence is a time of unique vulnerability to the neurotoxic consequences of ethanol

Alcohol use has been estimated to contribute to 3.3 million deaths per year globally and nearly $223.5 billion in monetary expenses to the United States alone (Centers for Disease Control and Prevention, 2014). Nearly 75% of this is attributed to binge drinking, defined by NIAAA as ethanol consumption resulting in Blood Ethanol Concentrations (BECs) of 0.08 mg/dL or higher (Centers for Disease Control and Prevention, 2014). This type of alcohol consumption is particularly common among adolescents (White, Kraus, & Swartzwelder, 2006). Although substantial progress has been made in curtailing adolescent binge drinking, the highest prevalence of binge drinking across age groups continues to occur among adolescents (Chung, Creswell, Bachrach, Clark, & Martin, 2018). These patterns of intake tend to peak in late adolescence and emerging adulthood (21–25 years of age), with consumption then tapering off throughout adulthood and into late aging. Even more concerning is the emergence of excessive drinking, termed "High Intensity" drinking, where BECs often approach 2–3 times the NIAAA definition of binge (Patrick & Azar, 2018). Such patterns of drinking are particularly hazardous and high frequency binge drinking is strongly correlated with a later AUD diagnosis (Crews et al., 2006). Adolescents display altered sensitivity to many of the positive and negative effects of ethanol. This often manifests as a heightened susceptibility to the positive effects of ethanol such as subjective perception of reward (Pautassi, Myers, Spear, Molina, & Spear, 2008) and social facilitation (Willey, Varlinskaya, & Spear, 2009) and increased resilience to the negative consequences of ethanol such as social inhibition (Varlinskaya & Spear, 2002), sedation (Silveri & Spear, 1999) and aversion (Vetter-O'Hagen, Varlinskaya, & Spear, 2009). This pattern of consequences is believed to contribute to the increased levels of ethanol consumption observed in adolescents.

Beyond the concerns of early adolescent ethanol use on later alcohol use, the adolescent period displays heightened vulnerability to many long-lasting effects of ethanol use. At least some of the adolescent-like phenotypes discussed above display a "locking-in" like effect that persists into adulthood (Spear & Swartzwelder, 2014). In addition, the neurotoxic effects of ethanol are particularly profound during adolescence. Several studies have highlighted that adolescent ethanol exposure promotes inflammatory brain damage (Pascual, Blanco, Cauli, Minarro, & Guerri, 2007), unique patterns of cortical brain damage (Crews, Braun, Hoplight, Switzer 3rd, & Knapp, 2000), inhibited neurogenesis (Crews et al., 2006) and much more (Crews et al., 2000; Guerri & Pascual, 2010). It is worth noting that most studies do not include comparisons of equivalent alcohol exposures in adults, making it difficult to attribute these findings to a vulnerability that is specific to adolescence.

3.2 Acute ethanol alters BBB function

Several studies have examined the impact of acute and chronic ethanol on BBB function alone or in conjunction with other forms of challenge. Going back to 1990, postmortem human analysis revealed that individuals with a history of alcohol use disorders were documented as having abnormal or dysfunctional BBBs (Pratt, Rooprai, Shaw, & Thomson, 1990; Thomsen, Kaatsch, & Asmus, 1994). Rodent studies from the 1970s have also hinted at a possible role that ethanol may have in modulating BBB permeability in response to stab wounds (Rosengren & Persson, 1979), and after chronic vapor exposure when challenged with starvation as a stressor (Phillips & Cragg, 1982). More recent *in vitro* studies have also shown that acute ethanol produced BBB dysfunction through TJ degradation in brain microvascular

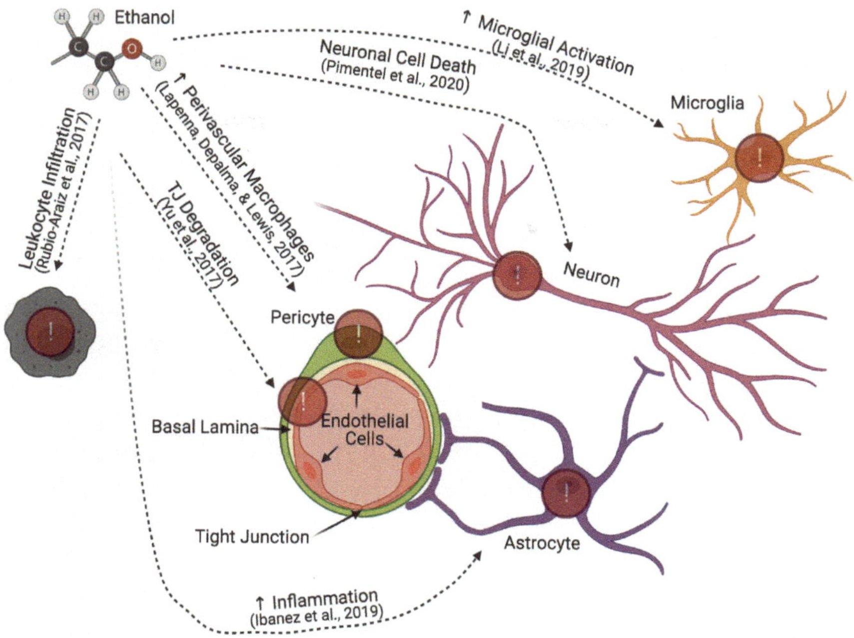

Fig. 4 Evidence of ethanol's ability to disrupt the NVU. Studies indicate potential mechanisms by which ethanol-evoked changes may disrupt BBB permeability are indicated by exclamation points on relevant NVU components. While many of the papers referenced in this schematic highlight the specific role that inflammation plays, there exists substantial correlational evidence that long-term ethanol exposure could alter BBB permeability directly.

endothelial cells (Haorah et al., 2007) (Fig. 4). Several other studies have shown that chronic ethanol exposure degraded many of the proteins associated with normal TJ function (Yu et al., 2017), and enhanced BBB permeability produced by LPS exposure (Singh, Jiang, Gupta, & Benlhabib, 2007). While no consistent mechanism has been identified as being responsible for these changes, ethanol interferes with normal TJ function (Haorah et al., 2007), angiogenesis (Abdul Muneer, Alikunju, Szlachetka, & Haorah, 2012), and endothelial transport receptors (Chang, Huang, Mao, & Mack, 2018).

Another potential mechanism by which ethanol produces substantial changes in BBB permeability is cytokine release. The TLR4 signaling pathway has been associated with BBB dysfunction (Szabo & Lippai, 2014; Wardill et al., 2016b). At binge- or supra-binge levels of ethanol exposure, ethanol increased the functional release of high mobility group box-1 (HMGB1), a danger-associated molecular pattern (DAMP) that serves as an endogenous ligand at TLR4 (Crews & Vetreno, 2018; Ge et al., 2014; Wang et al., 2015; Zou & Crews, 2014;). For this reason, high dose ethanol seems to recapitulate at least some of the effects observed with LPS, which also binds to TLR4 (Alfonso-Loeches, Pascual-Lucas, Blanco, Sanchez-Vera, & Guerri, 2010; Pandey, 2012). Both human and animal studies have supported a role of ethanol-induced TLR4 signaling in BBB dysfunction. For instance, TLR4 KO mice displayed nearly none of the conventional neuroinflammatory consequences of ethanol and showed decreased BBB permeability (Alfonso-Loeches, Urena-Peralta, Morillo-Bargues, Gomez-Pinedo, & Guerri, 2016). TLR4 KO mice exposed to binge drinking paradigms showed reduced levels of several proteins that are used as biomarkers of BBB permeability and hippocampal BBB impairment (Rubio-Araiz et al., 2017). The same studies have also found increased TJ degradation in the brains of human alcohol abusers (Rubio-Araiz et al., 2017). Alcohol preferring P-rats displayed increased LPS-induced BBB abnormalities in comparison to normal rats (Singh et al., 2007). Thus, accumulating evidence posits TLR4 signaling as a potential mechanism for ethanol-induced disruption of the BBB, yet few studies have linked ethanol evoked TLR4 signaling with quantifiable differences in BBB permeability.

A final mechanism by which ethanol is known to affect the BBB is through altered blood flow. Even relatively small doses of ethanol substantially increase systolic, diastolic, and mean blood flow velocities (Stendel, Irnich, al Hassan, Heidenreich, & Pietilae, 2006). This is believed to occur through dilation of the cerebral arteries, and significantly increased blood volume and access to the brain (Gazzieri et al., 2006). In contrast, reduced cerebral blood flow was observed during ethanol withdrawal in ethanol dependent individuals (Mathew & Wilson, 1986). These changes can significantly modify stroke vulnerability and outcome. Consumption of small quantities of ethanol protected against ischemic stroke, whereas heavier alcohol consumption patterns exacerbated injury, presumably by increasing vulnerability to hemorrhage (Li, Li, Xu, & Sun, 2020). In contrast, ethanol substantially increased vulnerability to hemorrhagic stroke, and not only aggravated hemorrhagic volume, but also increased microglial activation and overall inflammation (Liew et al., 2016). These changes result in the hemorrhagic stroke being more severe, and significantly increased mortality and degree of impairment following stroke resolution (Daniel & Bereczki, 2004).

3.3 Adolescent ethanol exposure produces sex-specific changes in ethanol-induced inflammation

Acute ethanol exposure during adolescence is known to evoke a significant neuroimmune signaling response and these neuroimmune signaling changes persist into adulthood. Human data supports that individuals with a history of AUD show increases in many immune signaling molecules such as HMGB1 that correlate with lifetime alcohol exposure (Crews, Qin, Sheedy, Vetreno, & Zou, 2013). In addition, both human data as well as rodent studies have indicated a persistent increase in TLR expression, including TLR4, following AIE (Crews, Lawrimore, Walter, & Coleman Jr, 2017). Long-term changes in histone acetylation as well as brain BDNF levels have been observed following AIE and this correlates strongly with changes in neuroimmune gene expression and ultimately decreased neurogenesis (Broadwater & Spear, 2014; Sakharkar et al., 2016). While these data highlight the potential for AIE to produce lasting changes in immune gene expression, fewer studies have assessed whether the sex of the animal would modify these results and if these changes also reflect lasting differences in peripheral immune activation.

In previous work, male rats with a history of AIE showed a unique (not observed in their female counterparts), robust suppression of adult peripheral cytokine gene expression in response to either an LPS or a restraint stress challenge in adulthood (Vore, Doremus-Fitzwater, Gano, & Deak, 2017). Follow-up work reported that male rats with a history of AIE also displayed a suppressed neuroimmune response to adult ethanol challenge when brain cytokine protein was measured using large molecule microdialysis (Gano, Mondello, et al., 2019; Gano, Vore, et al., 2019). When male and female rats with a history of AIE were challenged with an identical dose of ethanol in adulthood, both sexes displayed elevated hippocampal IL-6 and IκBα, however, only male rats showed this increase in the amygdala (Vore, Barney, Gano, Varlinskaya, & Deak, 2021). Collectively, this work highlights that male rats may be uniquely susceptible to long-lasting changes in inflammation after AIE, with females being resistant to such changes.

In Vore et al. (2021), male rats with a history of AIE also displayed decreased VEGF-A (a known BBB permeability inducing factor) gene expression in response to an adult ethanol challenge that was not observed in female rats. Using the identical experimental manipulation, male rats also showed altered hippocampal ethanol kinetics, achieving significantly higher brain ethanol concentrations more rapidly than adolescent vehicle-exposed counterparts (Gano, Mondello, et al., 2019; Gano, Vore, et al., 2019). The shift in ethanol transit into the CNS among AIE-exposed males indicates the potential for a change in the neurobehavioral response to ethanol. When rats exposed to the same AIE and adult ethanol challenge were assessed for Loss of Righting Reflex (LORR) sensitivity, male rats with a history of AIE showed an almost 50% reduction in sleep time that was not observed in female comparators. Male rats showed no lingering signs of tolerance or altered ethanol metabolism when BEC curves were assessed after a challenge with 0.75, 1.5, or 3.0 g/kg ethanol. The data discussed above highlight critical sex differences in adolescent ethanol sensitivity that require further exploration and could accompany BBB disruption and portend significant changes in later-life pathology.

4. Alterations in BBB permeability exacerbate brain pathology

An abundance of data supports the notion that alterations in BBB permeability are associated with more severe instances of brain pathology. Small differences in access to and from the brain can have widespread impact on brain health in the long-term. The link between vascular dysfunction and Alzheimer's disease (AD) has been studied both as a mechanistic component of the disorder as well as how altered BBB function may worsen disease pathology. Vascular dementia is the second most common form of dementia following AD and perhaps best represents the substantial increase in pathological changes in cerebrovasculature can create (Gorelick et al., 2011). Cerebrovascular disease can result in cognitive impairment ranging from mild to full dementia, similar to what is observed in AD patients. Due to the important role the cerebral vasculature plays in supplying neurons as well as other components of the NVU with oxygen and nutrients, even small vascular alterations can contribute to cognitive impairment. These include factors such as microbleeds, microinfarcts, arteriosclerosis, and vascular stiffening and all of these are connected to reductions in global cerebral perfusion as well as different severities of dementia and cognitive impairment (see Iadecola, 2013 for relevant review). The summated impact of these pathogenic factors may ultimately lead to white matter damage, and increased BBB permeability, leading to increased brain cytokine levels, microglial activation, and oxidative stress. This highlights how even small perturbations in the homeostasis governed by the BBB can lead to substantial physiological and behavioral dysfunction.

One of the hallmark characteristics of Alzheimer's Disease is the steady accumulation of harmful amyloid beta plaques as the individual ages. These Aβ plaques are then believed to ultimately create the cognitive disturbance and many other neurological symptoms observed in AD (Butterfield et al., 2007; Cleary et al., 2005; Lesné et al., 2006). The BBB plays a fundamental role in both amyloid beta production as well as clearance out of the brain. Recent work in both rodent and human models has shown that peripheral amyloid beta contributes to the ultimate accumulation of brain Aβ (Eisele et al., 2010; Sagare, Bell, & Zlokovic, 2012). In addition, peripheral expression of a variety of Aβ sequestering agents such as anti-amyloid beta antibodies have been shown to help slow brain Aβ accumulation by reducing peripheral Aβ contribution (DeMattos et al., 2001). Much of the shuttling of Aβ into the brain and the continued propagation of Aβ is mediated through the receptor for advanced glycation end products (RAGE) (Deane et al., 2003). RAGE expression directly embedded within the brain endothelium permits direct transit of Aβ as well as peripheral monocytes into the brain (another marker of heightened peripheral inflammation) and ultimately helps to propagate Aβ toxicity (Giri et al., 2000). Conversely, low density lipoprotein receptor-related protein 1 (LRP1) also located in the brain endothelium plays a vital role in the shuttling of brain Aβ back into the blood (Deane et al., 2004; Jaeger et al., 2009). Altered LRP1 expression in brain microvessels was associated with the accumulation of Aβ during both natural aging as well as pathological AD (Donahue et al., 2006; Shibata et al., 2000). Clearly, the BBB and brain vasculature as a whole is inextricably linked with both the genesis of AD as well as the rate of its progression and ultimate severity. This has led to more recent studies that highlight a vascular hypothesis of AD wherein vascular injury ultimately leads to BBB dysfunction that produces the aberrant accumulation of Aβ plaques, which ultimately contributes to tau pathology (Zlokovic, 2011).

Similar dysfunctions in BBB integrity have been reportedly associated with Parkinson's pathology and potentially its prognosis (Hirano et al., 2008; Kortekaas et al., 2005; Lee & Pienaar, 2014). Parkinson's disease (PD) is the most common movement disorder in the elderly and is generally considered the second most common degenerative neurological disorder in senescence (Desai, Monahan, Carvey, & Hendey, 2007). It is characterized by the pathological loss of dopamine neurons and the presence of Lewy bodies that ultimately produce significant motor dysfunction such as tremor, bradykinesia, and instability. While the exact mechanisms underlying this dysfunction have yet to be confirmed, animal models have consistently demonstrated elevated inflammatory tone and increased inflammatory cytokine presence in the brain of PD animal models (Barcia, Fernández Barreiro, Poza, & Herrero, 2003). While it was originally believed that PD patients did not show altered BBB function, more recent work using sensitive, modern methods of probing BBB integrity such as FITC-labeled albumin have shown significantly increased BBB permeability in the striatum and other components of the nigrostriatal pathway (Carvey, Punati, & Newman, 2006). These differences were observed both in animals following 6-hydroxydopamine lesions (Carvey et al., 2006) as well as MPTP exposure (Zhao, Ling, Newman, Bhatia, & Carvey, 2007). The latter authors also noted that minocycline treatment (known to inhibit microglial activation) has been shown to prevent the changes in leakage associated with MPTP-induced Parkinsonian symptoms (Zhao et al., 2007). These changes, coupled with an overall increase in microglia numbers in areas where BBB leakage and neuronal degeneration was observed, further suggests a role for inflammation and subsequent BBB damage in PD pathology (Teismann & Schulz, 2004). This was further supported by data showing that TNF-α KO mice showed neither microglial dysfunction nor the increased BBB permeability following MPTP exposure (Zhao et al., 2007). Finally, substantial increases in vascular density/angiogenesis were observed in the substantia nigra of PD patients (Faucheux, Bonnet, Agid, & Hirsch, 1999). Increased angiogenesis occurred in close proximity to areas of high neuronal death and correlated well with heightened levels of VEGF (Barcia et al., 2005). The authors speculated that in response to neuronal degeneration, increased VEGF may be the body's attempt to supplement blood and nutrient flow as well as removal of cellular debris. This activity may help in the short-term but ultimately allows further peripheral macrophage infiltration and exacerbates neuronal degeneration and disease progression (Desai et al., 2007).

4.1 Natural aging and inflammation

Both PD and AD are common neurodegenerative disorders that frequently occur in mid-to-late aging. As discussed above, both disorders are characterized by pathological inflammation that co-occurs with, or perhaps ultimately produces, changes in BBB permeability. This allows the unrestricted transit of toxins or harmful molecules into the brain. This potentially triggers the onset of pathology, which ultimately can initiate a cycle of increased inflammation, disrupted BBB activity, and decreased efflux of toxins producing a more profound pathology. While these disorders represent extreme instances of pathology, natural aging is also characterized by well documented changes in inflammatory tone as well as increased BBB leakage (Perkins et al., 2018). As "normal" individuals age a chronic, low-grade inflammation dubbed "inflammaging" gradually develops (Franceschi, Garagnani, Parini, Giuliani, & Santoro, 2018). This persistent activation of the innate immune system is believed to reflect natural changes that occur during senescence. Ultimately, the inflammaging phenotype develops which is believed to contribute to increased vulnerability to bacterial, viral, and other forms of infection seen in aging (Dall'Olio et al., 2013).

While the exact cause of inflammaging has yet to be identified, it is likely due to the summation of several different CNS changes. The first is the gradual accumulation of cellular debris and damaged molecules that can occur during aging, owing in part to aberrant production and slowing clearance of the same (Franceschi & Campisi, 2014). A variety of different compounds are recognized by different sensors as "danger signals" that trigger an innate immune reaction to prompt resolution of this dysfunction. However, as the body ages and these signals begin to accumulate, it can overwhelm CNS clearance mechanisms and become a chronic condition (Franceschi et al., 2000). Another potential contributor to this chronic immune activation could be increased leakiness of the gut microbiome as an individual ages. While the gut normally traps microbes and their products, over time heightened leakiness may trigger the release of these products into surrounding tissue producing cytokine release and subsequent innate immune activation (Biagi, Candela, Fairweather-Tait, Franceschi, & Brigidi, 2012). It is also likely that natural senescence in cells could contribute to increased inflammatory tone. Normally, the development of senescence in cells serves an important biological function in preventing the rapid proliferation of aged cells that can contribute to cancer; however, the accumulation of senescent cells that release pro-inflammatory cytokines could serve to drive age-related pathology. In mice, the removal of senescent cells has been shown to ameliorate age-related pathology (Coppé, Desprez, Krtolica, & Campisi, 2010). It is also plausible that inflammaging represents a natural decline in immune function. As individuals age, adaptive immunity naturally declines and a possible compensatory mechanism is that innate immunity undergoes a mild increase, again triggering a mild, persistent increase in inflammatory tone (Shaw, Joshi, Greenwood, Panda, & Lord, 2010). Complementing the presence of inflammaging and the gradual increase of inflammatory signaling molecules across aging, "normal" aging is associated with increased BBB leakiness (Zhang et al., 2019), an effect that could be worsened by a lifetime history of alcohol consumption. The combination of increased BBB permeability and heightened inflammation is a natural progression and likely contributes to the significant increase in overall disease vulnerability noted in older individuals.

4.2 Implications for drug/alcohol use and abuse

Several post-mortem studies have documented BBB pathology in individuals with a history of AUD (Pratt et al., 1990; Thomsen et al., 1994). *In vitro* models of the BBB have shown that ethanol and acetaldehyde augmented BBB permeability (Haorah, Schall, Ramirez, & Persidsky, 2008). In addition, neuroimaging studies revealed significantly increased BBB permeability in the hippocampus of social drinkers (Ivanidze et al., 2019). Other studies examining water distribution using diffusion tensor imaging showed abnormal brain diffusivity, reflective of BBB leakage, in the fronto-temporal regions of the brains of individuals with current AUD in contrast to healthy comparators (Monnig et al., 2013). The same study documented persistent damage in the parietal regions of individuals with at least 1 year of abstinence (Monnig et al., 2013). These changes may enhance transit of cytokines and other inflammatory signaling molecules into the brain of individuals with a history of AUD. Subsequently, it could substantially alter the ability of ethanol and other drugs of abuse to cross the BBB. Much as heroin is considered to have a higher degree of subjective reinforcement than morphine due to its ability to more rapidly transit into the brain (Schaefer, Tome, & Davis, 2017), even small changes in brain ethanol kinetics (such as those observed in Gano, Mondello, et al., 2019; Gano, Vore, et al., 2019) could substantially alter ethanol's reinforcing property. Similarly, changes in BBB permeability may significantly alter acetaldehyde levels. Normally, blood acetaldehyde levels do not reflect what is observed in the brain as it is screened out by the BBB (Tabakoff, Anderson, & Ritzmann, 1976). While it is believed that *de novo* acetaldehyde is produced through the metabolism of CNS ethanol by regional ADH (Martínez et al., 2001), increased access of peripheral acetaldehyde into the brain may produce biologically active levels capable of producing reinforcement as observed when acetaldehyde was administered directly into the brain (Amit & Smith, 1985; Brown, Amit, & Rockman, 1979; Myers, Ng, & Singer, 1984; Quertemont, 2004; Rodd-Henricks et al., 2002). In addition, when chronic or high levels of ethanol are consumed, the microsomal ethanol oxidation system (MEOS) and induction of CYP2E1 is also utilized to oxidize ethanol into acetaldehyde (Crabb & Liangpunsakul, 2007). Unlike with ADH, with chronic exposure MEOS enzymes increase in number and are not inhibited by the presence of acetaldehyde (Crabb & Liangpunsakul, 2007). AIE could readily induce CYP2E1 activity in a regionally specific manner given the presence of MEOS systems across the brain (Zimatkin & Deitrich, 1997) that may render specific brain areas vulnerable to acetaldehyde accumulation and subsequent inflammation and behavioral changes.

While this review has focused predominantly upon alcohol, a number of other common drugs of abuse also produce alterations in BBB permeability. Continuous administration of cocaine produces an almost 50% increase in BBB permeability (Sharma, Muresanu, Sharma, & Patnaik, 2009) believed to be modulated through shared mechanistic factors such as increased TNF-α (Sharma et al., 2009); TJ degradation (Dietrich, 2009); and altered astrocyte (Wang et al., 2021) and microglia function (Buch et al., 2012). Similar changes have been shown to occur following methamphetamine administration (Northrop & Yamamoto, 2015). Methamphetamine evoked changes in oxidative stress (Ramirez et al., 2009), changes in astrocytic end-feet contacts (Northrop & Yamamoto, 2012), and potentially methamphetamine-associated excitotoxicity in pericytes (Montiel-Eulefi et al., 2012) have all been identified as potential mechanisms through which methamphetamine may alter BBB integrity. Morphine and other opiates show similar alterations in BBB homeostasis (Kousik, Napier, & Carvey, 2012) and also inhibit PGP expression, reducing morphine removal from the brain and increasing the analgesic and potentially reinforcing consequences of such drugs (Seleman et al., 2014). Finally, chronic nicotine exposure also disrupts TJ proteins (Kousik et al., 2012) and direct binding to nicotinic receptors localized on brain endothelial cells induces NO release and can further compromise BBB function (Mazzone et al., 2010). The present review has highlighted many shared mechanisms by which drugs of abuse with completely different pharmacodynamic properties can result in the same changes of BBB integrity. Polysubstance abuse is common in adolescents and heavy cannabis use is associated with a higher risk for future illicit drug use (Patton et al., 2007). It is possible that adolescent consumption of one drug may alter BBB permeability, subsequently modifying the subjective response to future drug exposures (both to the same drug and to others). Beyond changes in drug access to the brain, alterations in BBB permeability may render drug abusing individuals more vulnerable to comorbid disease states such as HIV infection (Fitting et al., 2010) and hepatitis that are often observed in higher percentages of substance abusing individuals in contrast to individuals not currently abusing drugs (Nath, 2010).

Funding

This work was supported by the National Institutes of Health grants P50AA017823 (TD), T32AA025606 (ASV), and F31AA027959 (ASV). Any opinions, findings, and conclusions or recommendations expressed in this material are those of the author(s) and do not necessarily reflect the views of the above stated funding agencies. The authors have no conflicts of interest to declare.

References

Abbott, N. J., Ronnback, L., & Hansson, E. (2006). Astrocyte-endothelial interactions at the blood-brain barrier. *Nature Reviews. Neuroscience*, *7*, 41–53.

Abdul Muneer, P. M., Alikunju, S., Szlachetka, A. M., & Haorah, J. (2012). The mechanisms of cerebral vascular dysfunction and neuroinflammation by MMP-mediated degradation of VEGFR-2 in alcohol ingestion. *Arteriosclerosis, Thrombosis, and Vascular Biology*, *32*, 1167–1177.

Al-Obaidi, M. M. J., & Desa, M. N. M. (2018). Mechanisms of blood brain barrier disruption by different types of bacteria, and bacterial-host interactions facilitate the bacterial pathogen invading the brain. *Cellular and Molecular Neurobiology*, *38*, 1349–1368.

Alfonso-Loeches, S., Pascual-Lucas, M., Blanco, A. M., Sanchez-Vera, I., & Guerri, C. (2010). Pivotal role of TLR4 receptors in alcohol-induced neuroinflammation and brain damage. *The Journal of Neuroscience*, *30*, 8285–8295.

Alfonso-Loeches, S., Urena-Peralta, J., Morillo-Bargues, M. J., Gomez-Pinedo, U., & Guerri, C. (2016). Ethanol-induced TLR4/NLRP3 neuroinflammatory response in microglial cells promotes leukocyte infiltration across the BBB. *Neurochemical Research*, *41*, 193–209.

Amit, Z., & Smith, B. R. (1985). A multi-dimensional examination of the positive reinforcing properties of acetaldehyde. *Alcohol*, *2*, 367–370.

Anderson, V. C., Lenar, D. P., Quinn, J. F., & Rooney, W. D. (2011). The blood-brain barrier and microvascular water exchange in Alzheimer's disease. *Cardiovascular Psychiatry and Neurology*, *2011*, 615829.

Anthony, D. C., Bolton, S. J., Fearn, S., & Perry, V. H. (1997). Age-related effects of interleukin-1 beta on polymorphonuclear neutrophil-dependent increases in blood-brain barrier permeability in rats. *Brain*, *120*(Pt. 3), 435–444.

Armulik, A., et al. (2010). Pericytes regulate the blood-brain barrier. *Nature*, *468*, 557–561.

Avemary, J., et al. (2013). Dynamic regulation of P-glycoprotein in human brain capillaries. *Molecular Pharmaceutics*, *10*, 3333–3341.

Balda, M. S., & Matter, K. (2009). Tight junctions and the regulation of gene expression. *Biochimica et Biophysica Acta*, *1788*, 761–767.

Banks, W. A. (2005). Blood-brain barrier transport of cytokines: A mechanism for neuropathology. *Current Pharmaceutical Design*, *11*, 973–984.

Banks, W. A., Kastin, A. J., & Gutierrez, E. G. (1994). Penetration of interleukin-6 across the murine blood-brain barrier. *Neuroscience Letters*, *179*, 53–56.

Banks, W. A., Plotkin, S. R., & Kastin, A. J. (1995). Permeability of the blood-brain barrier to soluble cytokine receptors. *Neuroimmunomodulation*, *2*, 161–165.

Banks, W. A., Lynch, J. L., & Price, T. O. (2009). In A. Siegel, & S. S. Zalcman (Eds.), *Cytokines and the blood–brain barrier BT—The neuroimmunological basis of behavior and mental disorders* (pp. 3–17). Springer US. https://doi.org/10.1007/978-0-387-84851-8_1.

Banks, W. A., et al. (2015). Lipopolysaccharide-induced blood-brain barrier disruption: Roles of cyclooxygenase, oxidative stress, neuroinflammation, and elements of the neurovascular unit. *Journal of Neuroinflammation, 12*, 223.

Barcia, C., Fernández Barreiro, A., Poza, M., & Herrero, M.-T. (2003). Parkinson's disease and inflammatory changes. *Neurotoxicity Research, 5*, 411–418.

Barcia, C., et al. (2005). Changes in vascularization in substantia nigra pars compacta of monkeys rendered parkinsonian. *Journal of Neural Transmission, 112*, 1237–1248.

Bektas, A., Sen, R., & Ferrucci, L. (2016). Does a bit of alcohol turn off inflammation and improve health? *Age and Ageing, 45*, 747–748.

Bendayan, R., Lee, G., & Bendayan, M. (2002). Functional expression and localization of P-glycoprotein at the blood brain barrier. *Microscopy Research and Technique, 57*, 365–380.

Bergers, G., & Song, S. (2005). The role of pericytes in blood-vessel formation and maintenance. *Neuro-Oncology, 7*, 452–464.

Berthiaume, A. A., Grant, R. I., McDowell, K. P., Underly, R. G., Hartmann, D. A., Levy, M., et al. (2018). Dynamic remodeling of pericytes in vivo maintains capillary coverage in the adult mouse brain. *Cell Reports, 22*(1), 8–16.

Biagi, E., Candela, M., Fairweather-Tait, S., Franceschi, C., & Brigidi, P. (2012). Aging of the human metaorganism: The microbial counterpart. *Age (Dordrecht, Netherlands), 34*, 247–267.

Blanchette, M., & Daneman, R. (2015). Formation and maintenance of the BBB. *Mechanisms of Development, 138*(Pt. 1), 8–16.

Blau, C. W., et al. (2012). The age-related deficit in LTP is associated with changes in perfusion and blood-brain barrier permeability. *Neurobiology of Aging, 33*, 1005.e23–35.

Blond, D., Campbell, S. J., Butchart, A. G., Perry, V. H., & Anthony, D. C. (2002). Differential induction of interleukin-1beta and tumour necrosis factor-alpha may account for specific patterns of leukocyte recruitment in the brain. *Brain Research, 958*, 89–99.

Broadwater, M., & Spear, L. P. (2014). Consequences of adolescent or adult ethanol exposure on tone and context fear retention: Effects of an acute ethanol challenge during conditioning. *Alcoholism, Clinical and Experimental Research, 38*, 1454–1460.

Brown, Z. W., Amit, Z., & Rockman, G. E. (1979). Intraventricular self-administration of acetaldehyde, but not ethanol, in naive laboratory rats. *Psychopharmacology, 64*, 271–276.

Brown, J. A., et al. (2020). Advances in blood–brain barrier modeling in microphysiological systems highlight critical differences in opioid transport due to cortisol exposure. *Fluids and Barriers of the CNS, 17*, 38.

Buch, S., et al. (2012). Cocaine and HIV-1 interplay in CNS: Cellular and molecular mechanisms. *Current HIV Research, 10*, 425–428.

Buller, K. M. (2001). Role of circumventricular organs in pro-inflammatory cytokine-induced activation of the hypothalamic-pituitary-adrenal axis. *Clinical and Experimental Pharmacology & Physiology, 28*, 581–589.

Butterfield, D. A., et al. (2007). Elevated levels of 3-nitrotyrosine in brain from subjects with amnestic mild cognitive impairment: Implications for the role of nitration in the progression of Alzheimer's disease. *Brain Research, 1148*, 243–248.

Carvey, P. M., Punati, A., & Newman, M. B. (2006). Progressive dopamine neuron loss in Parkinson's disease: The multiple hit hypothesis. *Cell Transplantation, 15*, 239–250.

Castro Dias, M., et al. (2019). Claudin-3-deficient C57BL/6J mice display intact brain barriers. *Scientific Reports, 9*, 203.

Center for Behavioral Health Statistics and Quality. (2020). 2019 National Survey on Drug Use and Health: Methodological summary and definitions. Rockville, MD: Substance Abuse and Mental Health Services Administration. Retrieved from https://www.samhsa.gov/data/.

Centers for Disease Control and Prevention. (2014). *Alcohol use and your health.* http://www.cdc.gov/alcohol/fact-sheets/alcohol-use.htm.
Chang, S. L., Huang, W., Mao, X., & Mack, M. L. (2018). Ethanol's effects on transient receptor potential channel expression in brain microvascular endothelial cells. *Journal of Neuroimmune Pharmacology, 13*, 498–508.
Chen, G., & Reichlin, S. (1998). Clearance of [125I]-tumor necrosis factor-alpha from the brain into the blood after intracerebroventricular injection in rats. *Neuroimmunomodulation, 5*, 261–269.
Chen, G., Castro, W. L., Chow, H. H., & Reichlin, S. (1997). Clearance of 125I-labeled interleukin-6 from brain into blood following intracerebroventricular injection in rats. *Endocrinology, 138*, 4830–4836.
Chung, T., Creswell, K. G., Bachrach, R., Clark, D. B., & Martin, C. S. (2018). Adolescent binge drinking. *Alcohol Research: Current Reviews, 39*, 5–15.
Cleary, J. P., et al. (2005). Natural oligomers of the amyloid-beta protein specifically disrupt cognitive function. *Nature Neuroscience, 8*, 79–84.
Clementi, F. (1970). Effect of horseradish peroxidase on mice lung capillaries' permeability. *The Journal of Histochemistry and Cytochemistry: Official Journal of the Histochemistry Society, 18*, 887–892.
Coisne, C., & Engelhardt, B. (2011). Tight junctions in brain barriers during central nervous system inflammation. *Antioxidants & Redox Signaling, 15*, 1285–1303.
Coppé, J.-P., Desprez, P.-Y., Krtolica, A., & Campisi, J. (2010). The senescence-associated secretory phenotype: The dark side of tumor suppression. *Annual Review of Pathology, 5*, 99–118.
Cotran, R. S., Karnovsky, M. J., & Goth, A. (1968). Resistance of Wistar-Furth rats to the mast cell-damaging effect of horseradish peroxidase. *The Journal of Histochemistry and Cytochemistry: Official Journal of the Histochemistry Society, 16*, 382–383.
Crabb, D. W., & Liangpunsakul, S. (2007). Acetaldehyde generating enzyme systems: roles of alcohol dehydrogenase, CYP2E1 and catalase, and speculations on the role of other enzymes and processes. *Novartis Foundation Symposium, 285*(4–16), 16–22. https://doi.org/10.1002/9780470511848.ch2.
Crews, F. T., & Vetreno, R. P. (2018). Stress and alcohol priming of brain toll-like receptor signaling in alcohol use disorder. *Alcohol and Alcoholism, 53*, 639–641.
Crews, F. T., Braun, C. J., Hoplight, B., Switzer, R. C., 3rd, & Knapp, D. J. (2000). Binge ethanol consumption causes differential brain damage in young adolescent rats compared with adult rats. *Alcoholism, Clinical and Experimental Research, 24*, 1712–1723.
Crews, F., et al. (2006). BHT blocks NF-κB activation and ethanol-induced brain damage. *Alcoholism, Clinical and Experimental Research, 30*, 1938–1949.
Crews, F. T., Qin, L., Sheedy, D., Vetreno, R. P., & Zou, J. (2013). High mobility group box 1/Toll-like receptor danger signaling increases brain neuroimmune activation in alcohol dependence. *Biological Psychiatry, 73*, 602–612.
Crews, F. T., Lawrimore, C. J., Walter, T. J., & Coleman, L. G., Jr. (2017). The role of neuroimmune signaling in alcoholism. *Neuropharmacology, 122*, 56–73.
Dall'Olio, F., et al. (2013). N-glycomic biomarkers of biological aging and longevity: A link with inflammaging. *Ageing Research Reviews, 12*, 685–698.
Daneman, R., & Prat, A. (2015). The blood-brain barrier. *Cold Spring Harbor Perspectives in Biology*, 7, a020412.
Daneman, R., et al. (2009). Wnt/β-catenin signaling is required for CNS, but not non-CNS, angiogenesis. *Proceedings of the National Academy of Sciences, 106*, 641–646.
Daniel, S., & Bereczki, D. (2004). Alcohol as a risk factor for hemorrhagic stroke. *Ideggyógyászati Szemle, 57*, 247–256.
Davson, H., & Segal, M. (1996). *Physiology of the CSF and blood-brain barriers.* CRC Press.

de Boer, A. G., van der Sandt, I. C. J., & Gaillard, P. J. (2003). The role of drug transporters at the blood-brain barrier. *Annual Review of Pharmacology and Toxicology, 43*, 629–656.

Deane, R., et al. (2003). RAGE mediates amyloid-beta peptide transport across the blood-brain barrier and accumulation in brain. *Nature Medicine, 9*, 907–913.

Deane, R., et al. (2004). LRP/amyloid beta-peptide interaction mediates differential brain efflux of Abeta isoforms. *Neuron, 43*, 333–344.

DeMattos, R. B., et al. (2001). Peripheral anti-Aβ antibody alters CNS and plasma Aβ clearance and decreases brain Aβ burden in a mouse model of Alzheimer's disease. *Proceedings of the National Academy of Sciences, 98*, 8850–8855.

Desai, B. S., Monahan, A. J., Carvey, P. M., & Hendey, B. (2007). Blood-brain barrier pathology in Alzheimer's and Parkinson's disease: Implications for drug therapy. *Cell Transplantation, 16*, 285–299.

DeWit, D. J., Adlaf, E. M., Offord, D. R., & Ogborne, A. C. (2000). Age at first alcohol use: A risk factor for the development of alcohol disorders. *The American Journal of Psychiatry, 157*, 745–750.

Dietrich, J. B. (2009). Alteration of blood-brain barrier function by methamphetamine and cocaine. *Cell and Tissue Research, 336*, 385–392.

DiSabato, D. J., Quan, N., & Godbout, J. P. (2016). Neuroinflammation: The devil is in the details. *Journal of Neurochemistry, 139 Suppl*, 136–153.

Donahue, J. E., et al. (2006). RAGE, LRP-1, and amyloid-beta protein in Alzheimer's disease. *Acta Neuropathologica, 112*, 405–415.

Doremus-Fitzwater, T. L., et al. (2014). Intoxication- and withdrawal-dependent expression of central and peripheral cytokines following initial ethanol exposure. *Alcoholism, Clinical and Experimental Research, 38*, 2186–2198.

Doremus-Fitzwater, T. L., Gano, A., Paniccia, J. E., & Deak, T. (2015). Male adolescent rats display blunted cytokine responses in the CNS after acute ethanol or lipopolysaccharide exposure. *Physiology & Behavior, 148*, 131–144.

Edlund, T., Lofgren, B., & Vali, L. (1952). Toxicity of dextran in rats. *Nature, 170*, 125.

Eilken, H. M., et al. (2017). Pericytes regulate VEGF-induced endothelial sprouting through VEGFR1. *Nature Communications, 8*, 1574.

Eisele, Y. S., et al. (2010). Peripherally applied Abeta-containing inoculates induce cerebral beta-amyloidosis. *Science, 330*, 980–982.

Elahy, M., et al. (2015). Blood-brain barrier dysfunction developed during normal aging is associated with inflammation and loss of tight junctions but not with leukocyte recruitment. *Immunity & Ageing, 12*, 2.

Engelhardt, B. (2003). Development of the blood-brain barrier. *Cell and Tissue Research, 314*, 119–129.

Erickson, M. A., et al. (2018). Genetics and sex influence peripheral and central innate immune responses and blood-brain barrier integrity. *PLoS One, 13*, e0205769.

Erickson, M. A., Wilson, M. L., & Banks, W. A. (2020). In vitro modeling of blood–brain barrier and interface functions in neuroimmune communication. *Fluids and Barriers of the CNS, 17*, 26.

Faucheux, B. A., Bonnet, A.-M., Agid, Y., & Hirsch, E. C. (1999). Blood vessels change in the mesencephalon of patients with Parkinson's disease. *Lancet, 353*, 981–982.

Fernández-López, D., et al. (2012). Blood-brain barrier permeability is increased after acute adult stroke but not neonatal stroke in the rat. *The Journal of Neuroscience, 32*, 9588–9600.

Fitting, S., et al. (2010). Interactive comorbidity between opioid drug abuse and HIV-1 Tat: Chronic exposure augments spine loss and sublethal dendritic pathology in striatal neurons. *The American Journal of Pathology, 177*, 1397–1410.

Franceschi, C., & Campisi, J. (2014). Chronic inflammation (inflammaging) and its potential contribution to age-associated diseases. *The Journals of Gerontology. Series A, Biological Sciences and Medical Sciences, 69*, S4–S9.
Franceschi, C., et al. (2000). Inflamm-aging. An evolutionary perspective on immunosenescence. *Annals of the New York Academy of Sciences, 908*, 244–254.
Franceschi, C., Garagnani, P., Parini, P., Giuliani, C., & Santoro, A. (2018). Inflammaging: A new immune–metabolic viewpoint for age-related diseases. *Nature Reviews. Endocrinology, 14*, 576–590.
Fujimoto, M., et al. (2000). Plasma endotoxin and serum cytokine levels in patients with alcoholic hepatitis: Relation to severity of liver disturbance. *Alcoholism, Clinical and Experimental Research, 24*, 48S–54S.
Fuzzati-Armentero, M. T., Cerri, S., & Blandini, F. (2019). Peripheral-central neuroimmune crosstalk in Parkinson's disease: What do patients and animal models tell us? *Frontiers in Neurology, 10*, 232.
Gano, A., Doremus-Fitzwater, T. L., & Deak, T. (2017). A cross-sectional comparison of ethanol-related cytokine expression in the hippocampus of young and aged Fischer 344 rats. *Neurobiology of Aging, 54*, 40–53.
Gano, A., Mondello, J. E., Doremus-Fitzwater, T. L., & Deak, T. (2019). Rapid alterations in neuroimmune gene expression after acute ethanol: Timecourse, sex differences and sensitivity to cranial surgery. *Journal of Neuroimmunology, 337*, 577083.
Gano, A., Vore, A. S., Sammakia, M. N., & Deak, T. (2019). Assessment of extracellular cytokines in the hippocampus of the awake behaving rat using large-molecule microdialysis combined with multiplex arrays after acute and chronic ethanol exposure. *Alcoholism, Clinical and Experimental Research, 43*, 640–654.
Gazzieri, D., et al. (2006). Ethanol dilates coronary arteries and increases coronary flow via transient receptor potential vanilloid 1 and calcitonin gene-related peptide. *Cardiovascular Research, 70*, 589–599.
Ge, X., et al. (2014). High mobility group box-1 (HMGB1) participates in the pathogenesis of alcoholic liver disease (ALD). *The Journal of Biological Chemistry, 289*, 22672–22691.
Ghosh, A., et al. (2014). Assessment of blood-brain barrier function and the neuroinflammatory response in the rat brain by using cerebral open flow microperfusion (cOFM). *PLoS One, 9*, e98143.
Giri, R., et al. (2000). beta-amyloid-induced migration of monocytes across human brain endothelial cells involves RAGE and PECAM-1. *American Journal of Physiology. Cell Physiology, 279*, C1772–C1781.
Goasdoué, K., Miller, S. M., Colditz, P. B., & Björkman, S. T. (2017). Review: The blood-brain barrier; protecting the developing fetal brain. *Placenta, 54*, 111–116.
Goehler, L. E., et al. (1999). Interleukin-1β in immune cells of the abdominal vagus nerve: A link between the immune and nervous systems? *The Journal of Neuroscience, 19*, 2799–2806.
Gorelick, P. B., et al. (2011). Vascular contributions to cognitive impairment and dementia: A statement for healthcare professionals from the American heart association/American stroke association. *Stroke, 42*, 2672–2713.
Gorky, J., & Schwaber, J. (2016). The role of the gut-brain axis in alcohol use disorders. *Progress in Neuro-Psychopharmacology & Biological Psychiatry, 65*, 234–241.
Grant, B. F., & Dawson, D. A. (1997). Age at onset of alcohol use and its association with DSM-IV alcohol abuse and dependence: Results from the national longitudinal alcohol epidemiologic survey. *Journal of Substance Abuse, 9*, 103–110.
Guerri, C., & Pascual, M. (2010). Mechanisms involved in the neurotoxic, cognitive, and neurobehavioral effects of alcohol consumption during adolescence. *Alcohol, 44*, 15–26.
Gurney, K. J., Estrada, E. Y., & Rosenberg, G. A. (2006). Blood-brain barrier disruption by stromelysin-1 facilitates neutrophil infiltration in neuroinflammation. *Neurobiology of Disease, 23*, 87–96.

Gutierrez, E. G., Banks, W. A., & Kastin, A. J. (1993). Murine tumor necrosis factor alpha is transported from blood to brain in the mouse. *Journal of Neuroimmunology*, *47*, 169–176.

Haorah, J., et al. (2007). Oxidative stress activates protein tyrosine kinase and matrix metalloproteinases leading to blood-brain barrier dysfunction. *Journal of Neurochemistry*, *101*, 566–576.

Haorah, J., Schall, K., Ramirez, S. H., & Persidsky, Y. (2008). Activation of protein tyrosine kinases and matrix metalloproteinases causes blood-brain barrier injury: Novel mechanism for neurodegeneration associated with alcohol abuse. *Glia*, *56*, 78–88.

Haruwaka, K., et al. (2019). Dual microglia effects on blood brain barrier permeability induced by systemic inflammation. *Nature Communications*, *10*, 5816.

Hirano, S., et al. (2008). Dissociation of metabolic and neurovascular responses to levodopa in the treatment of Parkinson's disease. *The Journal of Neuroscience*, *28*, 4201–4209.

Howell, N. A., et al. (2013). Increased ventral striatal volume in college-aged binge drinkers. *PLoS One*, *8*, e74164.

Iadecola, C. (2013). The pathobiology of vascular dementia. *Neuron*, *80*, 844–866.

Ivanidze, J., et al. (2019). Dynamic contrast-enhanced MRI reveals unique blood-brain barrier permeability characteristics in the hippocampus in the normal brain. *AJNR. American Journal of Neuroradiology*, *40*, 408–411.

Jaeger, L. B., et al. (2009). Lipopolysaccharide alters the blood-brain barrier transport of amyloid beta protein: A mechanism for inflammation in the progression of Alzheimer's disease. *Brain, Behavior, and Immunity*, *23*, 507–517.

Jin, Y., Silverman, A. J., & Vannucci, S. J. (2009). Mast cells are early responders after hypoxia-ischemia in immature rat brain. *Stroke*, *40*, 3107–3112.

Keaney, J., & Campbell, M. (2015). The dynamic blood-brain barrier. *The FEBS Journal*, *282*, 4067–4079.

Kennedy, R. H., & Silver, R. (2016). In D. W. Pfaff, & N. D. Volkow (Eds.), *Neuroimmune signaling: cytokines and the central nervous system BT—Neuroscience in the 21st century: From basic to clinical* (pp. 601–641). Springer New York. https://doi.org/10.1007/978-1-4939-3474-4_174.

Klein, T., & Bischoff, R. (2011). Physiology and pathophysiology of matrix metalloproteases. *Amino Acids*, *41*, 271–290.

Kortekaas, R., et al. (2005). Blood-brain barrier dysfunction in parkinsonian midbrain in vivo. *Annals of Neurology*, *57*, 176–179.

Kousik, S., Napier, T. C., & Carvey, P. (2012). The effects of psychostimulant drugs on blood brain barrier function and neuroinflammation. *Frontiers in Pharmacology*, *3*, 121.

Kovacs, Z. I., et al. (2017). Disrupting the blood-brain barrier by focused ultrasound induces sterile inflammation. *Proceedings of the National Academy of Sciences of the United States of America*, *114*, E75–E84.

Kozberg, M., & Hillman, E. (2016). Neurovascular coupling and energy metabolism in the developing brain. *Progress in Brain Research*, *225*, 213–242.

Kvamme, T. L., et al. (2016). Sexually dimorphic brain volume interaction in college-aged binge drinkers. *NeuroImage: Clinical*, *10*, 310–317.

Laflamme, N., & Rivest, S. (2001). Toll-like receptor 4: The missing link of the cerebral innate immune response triggered by circulating gram-negative bacterial cell wall components. *The FASEB Journal*, *15*, 155–163.

Leclercq, S., de Timary, P., Delzenne, N. M., & Stärkel, P. (2017). The link between inflammation, bugs, the intestine and the brain in alcohol dependence. *Translational Psychiatry*, 7, e1048.

Lee, H., & Pienaar, I. S. (2014). Disruption of the blood-brain barrier in Parkinson's disease: Curse or route to a cure? *Frontiers in Bioscience*, *19*, 272–280.

Lesné, S., et al. (2006). A specific amyloid-beta protein assembly in the brain impairs memory. *Nature*, *440*, 352–357.

Li, C., Li, J., Xu, G., & Sun, H. (2020). Influence of chronic ethanol consumption on apoptosis and autophagy following transient focal cerebral ischemia in male mice. *Scientific Reports*, *10*, 6164.
Liebner, S., et al. (2000). Claudin-1 and claudin-5 expression and tight junction morphology are altered in blood vessels of human glioblastoma multiforme. *Acta Neuropathologica*, *100*, 323–331.
Liebner, S., et al. (2008). Wnt/beta-catenin signaling controls development of the blood-brain barrier. *The Journal of Cell Biology*, *183*, 409–417.
Liew, H.-K., et al. (2016). Acute alcohol intoxication aggravates brain injury caused by intracerebral hemorrhage in rats. *Journal of Stroke and Cerebrovascular Diseases*, *25*, 15–25.
Liu, Y., & Aguzzi, A. (2020). NG2 glia are required for maintaining microglia homeostatic state. *Glia*, *68*, 345–355.
Löscher, W., & Gericke, B. (2020). Novel intrinsic mechanisms of active drug extrusion at the blood-brain barrier: Potential targets for enhancing drug delivery to the brain? *Pharmaceutics*, *12*(10), 966. https://doi.org/10.3390/pharmaceutics12100966.
Luissint, A.-C., Artus, C., Glacial, F., Ganeshamoorthy, K., & Couraud, P.-O. (2012). Tight junctions at the blood brain barrier: Physiological architecture and disease-associated dysregulation. *Fluids and Barriers of the CNS*, *9*, 23.
Maggioli, E., et al. (2016). Estrogen protects the blood-brain barrier from inflammation-induced disruption and increased lymphocyte trafficking. *Brain, Behavior, and Immunity*, *51*, 212–222.
Maier, S. F., & Watkins, L. R. (1998). Cytokines for psychologists: Implications of bidirectional immune-to-brain communication for understanding behavior, mood, and cognition. *Psychological Review*, *105*, 83–107.
Majno, G., Palade, G. E., & Schoefl, G. I. (1961). Studies on inflammation. II. The site of action of histamine and serotonin along the vascular tree: A topographic study. *The Journal of Biophysical and Biochemical Cytology*, *11*, 607–626.
Mallard, C., Ek, C. J., & Vexler, Z. S. (2018). The myth of the immature barrier systems in the developing brain: Role in perinatal brain injury. *The Journal of Physiology*, *596*, 5655–5664.
Mandrekar, P., & Szabo, G. (2009). Signalling pathways in alcohol-induced liver inflammation. *Journal of Hepatology*, *50*, 1258–1266.
Marsland, P., et al. (2021). Neuroendocrine and neuroimmune responses in male and female rats: Evidence for functional immaturity of the neuroimmune system during early adolescence. *The European Journal of Neuroscience*. n/a.
Martínez, S. E., et al. (2001). Distribution of alcohol dehydrogenase mRNA in the rat central nervous system. Consequences for brain ethanol and retinoid metabolism. *European Journal of Biochemistry*, *268*, 5045–5056.
Mathew, R. J., & Wilson, W. H. (1986). Regional cerebral blood flow changes associated with ethanol intoxication. *Stroke*, *17*, 1156–1159.
Mazzone, P., et al. (2010). Pathophysiological impact of cigarette smoke exposure on the cerebrovascular system with a focus on the blood-brain barrier: Expanding the awareness of smoking toxicity in an underappreciated area. *International Journal of Environmental Research and Public Health*, 7, 4111–4126.
McConnell, H. L., Kersch, C. N., Woltjer, R. L., & Neuwelt, E. A. (2017). The translational significance of the neurovascular unit. *The Journal of Biological Chemistry*, *292*, 762–770.
Monnig, M. A., et al. (2013). Diffusion tensor imaging of white matter networks in individuals with current and remitted alcohol use disorders and comorbid conditions. *Psychology of Addictive Behaviors: Journal of the Society of Psychologists in Addictive Behaviors*, *27*, 455–465.
Montiel-Eulefi, E., et al. (2012). Neural differentiation of rat aorta pericyte cells. *Cytometry. Part A*, *81A*, 65–71.

Mooradian, A. D., Pinnas, J. L., Lung, C.-C., Yahya, M. D., & Meredith, K. (1994). Diabetes-related changes in the protein composition of rat cerebral microvessels. *Neurochemical Research*, *19*, 123–128.

Moretti, R., et al. (2015a). Blood-brain barrier dysfunction in disorders of the developing brain. *Frontiers in Neuroscience*, *9*, 40.

Moretti, S., et al. (2015b). Exposure of adolescent mice to delta-9-tetrahydrocannabinol induces long-lasting modulation of pro- and anti-inflammatory cytokines in hypothalamus and hippocampus similar to that observed for peripheral macrophages. *Journal of Neuroimmune Pharmacology: The Official Journal of the Society on NeuroImmune Pharmacology*, *10*, 371–379.

Muoio, V., Persson, P. B., & Sendeski, M. M. (2014). The neurovascular unit—Concept review. *Acta Physiologica*, *210*, 790–798.

Myers, W., Ng, K., & Singer, G. (1984). Ethanol preference in rats with a prior history of acetaldehyde self-administration. *Experientia*, *40*, 1008–1010.

Nadkarni, S., & McArthur, S. (2013). Oestrogen and immunomodulation: New mechanisms that impact on peripheral and central immunity. *Current Opinion in Pharmacology*, *13*, 576–581.

Natarajan, R., Northrop, N., & Yamamoto, B. (2017). Fluorescein isothiocyanate (FITC)-dextran extravasation as a measure of blood-brain barrier permeability. *Current Protocols in Neuroscience*, *79*, 9.58.1–9.58.15.

Nath, A. (2010). Human immunodeficiency virus-associated neurocognitive disorder: Pathophysiology in relation to drug addiction. *Annals of the New York Academy of Sciences*, *1187*, 122–128.

Neuwelt, E. A., et al. (2011). Engaging neuroscience to advance translational research in brain barrier biology. *Nature Reviews. Neuroscience*, *12*, 169–182.

Nitta, T., et al. (2003). Size-selective loosening of the blood-brain barrier in claudin-5-deficient mice. *The Journal of Cell Biology*, *161*, 653–660.

Northrop, N. A., & Yamamoto, B. K. (2012). Persistent neuroinflammatory effects of serial exposure to stress and methamphetamine on the blood-brain barrier. *Journal of Neuroimmune Pharmacology: The Official Journal of the Society on NeuroImmune Pharmacology*, *7*, 951–968.

Northrop, N. A., & Yamamoto, B. K. (2015). Methamphetamine effects on blood-brain barrier structure and function. *Frontiers in Neuroscience*, *9*, 69.

O'Brown, N. M., Pfau, S. J., & Gu, C. (2018). Bridging barriers: A comparative look at the blood-brain barrier across organisms. *Genes & Development*, *32*(7–8), 466–478. https://doi.org/10.1101/gad.309823.117.

Pai, J. K., et al. (2006). Moderate alcohol consumption and lower levels of inflammatory markers in US men and women. *Atherosclerosis*, *186*, 113–120.

Pan, W., & Kastin, A. J. (2001). Changing the chemokine gradient: CINC1 crosses the blood-brain barrier. *The Journal of Neuroimmunology*, *115*(1–2), 64–70. https://doi.org/10.1016/s0165-5728(01)00256-9.

Pandey, S. C. (2012). TLR4-MyD88 signalling: A molecular target for alcohol actions. *British Journal of Pharmacology*, *165*, 1316–1318.

Pardridge, W. M. (2007). Blood-brain barrier delivery. *Drug Discovery Today*, *12*, 54–61.

Pardridge, W. M. (2015). Blood-brain barrier endogenous transporters as therapeutic targets: A new model for small molecule CNS drug discovery. *Expert Opinion on Therapeutic Targets*, *19*, 1059–1072.

Pascual, M., Blanco, A. M., Cauli, O., Minarro, J., & Guerri, C. (2007). Intermittent ethanol exposure induces inflammatory brain damage and causes long-term behavioural alterations in adolescent rats. *The European Journal of Neuroscience*, *25*, 541–550.

Patrick, M. E., & Azar, B. (2018). High-intensity drinking. *Alcohol Research: Current Reviews*, *39*, 49–55.

Patton, G. C., et al. (2007). Trajectories of adolescent alcohol and cannabis use into young adulthood. *Addiction, 102*, 607–615.
Pautassi, R. M., Myers, M., Spear, L. P., Molina, J. C., & Spear, N. E. (2008). Adolescent but not adult rats exhibit ethanol-mediated appetitive second-order conditioning. *Alcoholism: Clinical and Experimental Research, 32*, 2016–2027. https://doi.org/10.1111/j.1530-0277.2008.00789.x.
Perkins, A. E., Vore, A. S., Lovelock, D., Varlinskaya, E., & Deak, T. (2018). Late aging alters behavioral sensitivity to ethanol in a sex-specific manner in Fischer 344 rats. *Pharmacology, Biochemistry, and Behavior, 175*, 1–9.
Perkins AE, Piazza, MK, Vore, AS, Deak, MM, Varlinskaya, EI, & Deak, T (2021). Assessment of neuroinflammation in the aging hippocampus using large-molecule microdialysis: Sex differences and role of purinergic receptors. Brain, Behavior, and Immunity, 91, 546–555. doi:https://doi.org/10.1016/j.bbi.2020.11.013.
Phillips, S. C., & Cragg, B. G. (1982). Weakening of the blood-brain barrier by alcohol-related stresses in the rat. *Journal of the Neurological Sciences, 54*, 271–278.
Pimentel, E., Sivalingam, K., Doke, M., & Samikkannu, T. (2020). Effects of drugs of abuse on the blood-brain barrier: A brief overview. *Frontiers in Neuroscience, 14*, 513.
Pratt, O. E., Rooprai, H. K., Shaw, G. K., & Thomson, A. D. (1990). The genesis of alcoholic brain tissue injury. *Alcohol and Alcoholism, 25*, 217–230.
Profaci, C. P., Munji, R. N., Pulido, R. S., & Daneman, R. (2020). The blood-brain barrier in health and disease: Important unanswered questions. *The Journal of Experimental Medicine, 217*, 1–16.
Quertemont, E. (2004). Genetic polymorphism in ethanol metabolism: Acetaldehyde contribution to alcohol abuse and alcoholism. *Molecular Psychiatry, 9*, 570–581.
Ramirez, S. H., et al. (2009). Methamphetamine disrupts blood-brain barrier function by induction of oxidative stress in brain endothelial cells. *Journal of Cerebral Blood Flow and Metabolism: Official Journal of the International Society of Cerebral Blood Flow and Metabolism, 29*, 1933–1945.
Reinhold, A. K., & Rittner, H. L. (2017). Barrier function in the peripheral and central nervous system-a review. *Pflügers Archiv, 469*, 123–134.
Rodd-Henricks, Z. A., et al. (2002). The reinforcing effects of acetaldehyde in the posterior ventral tegmental area of alcohol-preferring rats. *Pharmacology, Biochemistry, and Behavior, 72*, 55–64.
Ronaldson, P. T., & Davis, T. P. (2012). Blood-brain barrier integrity and glial support: Mechanisms that can be targeted for novel therapeutic approaches in stroke. *Current Pharmaceutical Design, 18*, 3624–3644.
Rosengren, I. E., & Persson, L. I. (1979). Influence of ethanol and dexamethasone on blood-brain barrier dysfunction to albumin. *Acta Neurologica Scandinavica, 59*, 119–126.
Ross, M. D., Nuttall, A. L., & Wright, C. G. (1977). Horseradish peroxidase acute ototoxicity and the uptake and movement of the peroxidase in the auditory system of the guinea pig. *Acta Oto-Laryngologica, 84*, 187–201.
Rossner, W., & Temple, K. (1966). Quantitative Bestimmung der Permeabilität der sogenannten Blut-Hirnschranke für Evans-Blau. *Medicina et Pharmacologia Experimentalis. International Journal of Experimental Medicine, 14*, 169–182.
Roth, J., Harré, E.-M., Rummel, C., Gerstberger, R., & Hübschle, T. (2004). Signaling the brain in systemic inflammation: Role of sensory circumventricular organs. *Frontiers in Bioscience, 9*, 290–300.
Rowitch, D. H., & Kriegstein, A. R. (2010). Developmental genetics of vertebrate glial-cell specification. *Nature, 468*, 214–222.
Rubio-Araiz, A., et al. (2017). Disruption of blood–brain barrier integrity in postmortem alcoholic brain: Preclinical evidence of TLR4 involvement from a binge-like drinking model. *Addiction Biology, 22*, 1103–1116.

Sagare, A. P., Bell, R. D., & Zlokovic, B. V. (2012). Neurovascular dysfunction and faulty amyloid β-peptide clearance in Alzheimer disease. *Cold Spring Harbor Perspectives in Medicine*, *2*, a011452.

Saili, K. S., et al. (2017). Blood-brain barrier development: Systems modeling and predictive toxicology. *Birth Defects Research*, *109*, 1680–1710.

Sakharkar, A. J., et al. (2016). A role for histone acetylation mechanisms in adolescent alcohol exposure-induced deficits in hippocampal brain-derived neurotrophic factor expression and neurogenesis markers in adulthood. *Brain Structure & Function*, *221*, 4691–4703.

Sanchez-Covarrubias, L., Slosky, L. M., Thompson, B. J., Davis, T. P., & Ronaldson, P. T. (2014). Transporters at CNS barrier sites: Obstacles or opportunities for drug delivery? *Current Pharmaceutical Design*, *20*, 1422–1449.

Sasaki, A., et al. (2015). Variable expression of microglial DAP12 and TREM2 genes in Nasu-Hakola disease. *Neurogenetics*, *16*, 265–276.

Saunders, N. R., Liddelow, S. A., & Dziegielewska, K. M. (2012). Barrier mechanisms in the developing brain. *Frontiers in Pharmacology*, *3*, 46.

Saunders, N. R., Dziegielewska, K. M., Mollgard, K., & Habgood, M. D. (2015). Markers for blood-brain barrier integrity: How appropriate is Evans blue in the twenty-first century and what are the alternatives? *Frontiers in Neuroscience*, *9*, 385.

Schaefer, C. P., Tome, M. E., & Davis, T. P. (2017). The opioid epidemic: A central role for the blood brain barrier in opioid analgesia and abuse. *Fluids and Barriers of the CNS*, *14*, 32.

Schenk, G. J., & de Vries, H. E. (2016). Altered blood-brain barrier transport in neuro-inflammatory disorders. *Drug Discovery Today: Technologies*, *20*, 5–11.

Seleman, M., Chapy, H., Cisternino, S., Courtin, C., Smirnova, M., Schlatter, J., et al. (2014). Impact of P-glycoprotein at the blood-brain barrier on the uptake of heroin and its main metabolites: Behavioral effects and consequences on the transcriptional responses and reinforcing properties. *Psychopharmacology*, *231*(16), 3139–3149.

Sharma, H. S., Muresanu, D., Sharma, A., & Patnaik, R. (2009). Chapter 11—Cocaine-induced breakdown of the blood–brain barrier and neurotoxicity. In *88. New concepts of psychostimulant induced neurotoxicity* (pp. 297–334). Academic Press.

Shaw, A. C., Joshi, S., Greenwood, H., Panda, A., & Lord, J. M. (2010). Aging of the innate immune system. *Current Opinion in Immunology*, *22*, 507–513.

Shibata, M., et al. (2000). Clearance of Alzheimer's amyloid-ss(1-40) peptide from brain by LDL receptor-related protein-1 at the blood-brain barrier. *The Journal of Clinical Investigation*, *106*, 1489–1499.

Silveri, M. M., & Spear, L. P. (1999). Ontogeny of rapid tolerance to the hypnotic effects of ethanol. *Alcoholism, Clinical and Experimental Research*, *23*, 1180–1184.

Singh, A. K., Jiang, Y., Gupta, S., & Benlhabib, E. (2007). Effects of chronic ethanol drinking on the blood brain barrier and ensuing neuronal toxicity in alcohol-preferring rats subjected to intraperitoneal LPS injection. *Alcohol and Alcoholism*, *42*, 385–399.

Sivandzade, F., & Cucullo, L. (2018). In-vitro blood-brain barrier modeling: A review of modern and fast-advancing technologies. *Journal of Cerebral Blood Flow and Metabolism: Official Journal of the International Society of Cerebral Blood Flow and Metabolism.*, *38*, 1667–1681.

Sofroniew, M. V., & Vinters, H. V. (2010). Astrocytes: Biology and pathology. *Acta Neuropathologica*, *119*, 7–35.

Spear, L. P., & Swartzwelder, H. S. (2014). Adolescent alcohol exposure and persistence of adolescent-typical phenotypes into adulthood: A mini-review. *Neuroscience and Biobehavioral Reviews*, *45*, 1–8.

Stendel, R., Irnich, B., al Hassan, A. A., Heidenreich, J., & Pietilae, T. (2006). The influence of ethanol on blood flow velocity in major cerebral vessels. A prospective and controlled study. *Alcohol*, *38*, 139–146.

Stolp, H. B., et al. (2011). Effects of neonatal systemic inflammation on blood-brain barrier permeability and behaviour in juvenile and adult rats. *Cardiovascular Psychiatry and Neurology*, *2011*, 469046.

Szabo, G., & Lippai, D. (2014). Converging actions of alcohol on liver and brain immune signaling. *International Review of Neurobiology*, *118*, 359–380.

Szabo, G., & Saha, B. (2015). Alcohol's effect on host defense. *Alcohol Research: Current Reviews*, *37*, 159–170.

Tabakoff, B., Anderson, R. A., & Ritzmann, R. F. (1976). Brain acetaldehyde after ethanol administration. *Biochemical Pharmacology*, *25*, 1305–1309.

Teismann, P., & Schulz, J. B. (2004). Cellular pathology of Parkinson's disease: Astrocytes, microglia and inflammation. *Cell and Tissue Research*, *318*, 149–161.

Thomsen, H., Kaatsch, H. J., & Asmus, R. (1994). Magnetic resonance imaging of the brain during alcohol absorption and elimination--a study of the rising tide phenomenon. *Blutalkohol*, *31*(3), 178–185.

Thomsen, H., Kaatsch, H. J., & Asmus, R. (1994). Magnetic resonance imaging of the brain during alcohol absorption and elimination—A study of the 'rising tide phenomenon'. *Blutalkohol*, *31*, 178–185.

Toornvliet, R., et al. (2006). Effect of age on functional P-glycoprotein in the blood-brain barrier measured by use of (R)-[11C]verapamil and positron emission tomography. *Clinical Pharmacology and Therapeutics*, *79*, 540–548.

Torres, L., et al. (2018). Toxoplasma gondii alters NMDAR signaling and induces signs of Alzheimer's disease in wild-type, C57BL/6 mice. *Journal of Neuroinflammation*, *15*, 57.

Vannucci, S. J. (1994). Developmental expression of GLUT1 and GLUT3 glucose transporters in rat brain. *Journal of Neurochemistry*, *62*, 240–246.

Varatharaj, A., & Galea, I. (2017). The blood-brain barrier in systemic inflammation. *Brain, Behavior, and Immunity*, *60*, 1–12.

Varlinskaya, E. I., & Spear, L. P. (2002). Acute effects of ethanol on social behavior of adolescent and adult rats: Role of familiarity of the test situation. *Alcoholism, Clinical and Experimental Research*, *26*, 1502–1511.

Vetter-O'Hagen, C., Varlinskaya, E., & Spear, L. (2009). Sex differences in ethanol intake and sensitivity to aversive effects during adolescence and adulthood. *Alcohol and Alcoholism*, *44*, 547–554.

Vore, A. S., Barney, T. M., Gano, A., Varlinskaya, E. L., & Deak, T. (2021). Adolescent intermittent ethanol (AIE) produces sex specific alterations in adult neuroimmune gene expression and ethanol sensitivity that are independent of ethanol metabolism. *Neuropharmacology*, *195*(108635). https://doi.org/10.1016/j.neuropharm.2021.108635.

Vore, A. S., Doremus-Fitzwater, T., Gano, A., & Deak, T. (2017). Adolescent ethanol exposure leads to stimulus-specific changes in cytokine reactivity and hypothalamic-pituitary-adrenal axis sensitivity in adulthood. *Frontiers in Behavioral Neuroscience*, *11*.

Waguespack, P. J., Banks, W. A., & Kastin, A. J. (1994). Interleukin-2 does not cross the blood-brain barrier by a saturable transport system. *Brain Research Bulletin*, *34*, 103–109.

Walker, K. A., et al. (2018). The association of mid- and late-life systemic inflammation with brain amyloid deposition: The ARIC-PET study. *Journal of Alzheimer's Disease*, *66*, 1041–1052.

Walter, T. J., Vetreno, R. P., & Crews, F. T. (2017). Alcohol and stress activation of microglia and neurons: Brain regional effects. *Alcoholism, Clinical and Experimental Research*, *41*, 2066–2081.

Wan, W., Chen, H., & Li, Y. (2014). The potential mechanisms of Aβ-receptor for advanced glycation end-products interaction disrupting tight junctions of the blood-brain barrier in Alzheimer's disease. *The International Journal of Neuroscience*, *124*, 75–81.

Wang, J., Pan, H.-F., Ye, D.-Q., Su, H., & Li, X.-P. (2008). Moderate alcohol drinking might be protective for systemic lupus erythematosus: A systematic review and meta-analysis. *Clinical Rheumatology*, *27*, 1557–1563.

Wang, H. J., Zakhari, S., & Jung, M. K. (2010). Alcohol, inflammation, and gut-liver-brain interactions in tissue damage and disease development. *World Journal of Gastroenterology*, *16*, 1304–1313.

Wang, H. J., Gao, B., Zakhari, S., & Nagy, L. E. (2012). Inflammation in alcoholic liver disease. *Annual Review of Nutrition*, *32*, 343–368.

Wang, X., et al. (2015). The role of HMGB1 signaling pathway in the development and progression of hepatocellular carcinoma: A review. *International Journal of Molecular Sciences*, *16*, 22527–22540.

Wang, J., et al. (2021). Cocaine triggers astrocyte-mediated synaptogenesis. *Biological Psychiatry*, *89*, 386–397.

Wardill, H. R., et al. (2016a). Cytokine-mediated blood brain barrier disruption as a conduit for cancer/chemotherapy-associated neurotoxicity and cognitive dysfunction. *International Journal of Cancer*, *139*, 2635–2645.

Wardill, H. R., et al. (2016b). TLR4-dependent claudin-1 internalization and secretagogue-mediated chloride secretion regulate irinotecan-induced diarrhea. *Molecular Cancer Therapeutics*, *15*, 2767–2779.

White, A. M., Kraus, C. L., & Swartzwelder, H. (2006). Many college freshmen drink at levels far beyond the binge threshold. *Alcoholism, Clinical and Experimental Research*, *30*, 1006–1010.

Willey, A. R., Varlinskaya, E. I., & Spear, L. P. (2009). Social interactions and 50 kHz ultrasonic vocalizations in adolescent and adult rats. *Behavioural Brain Research*, *202*, 122–129.

Wolf, F., & Kirchhoff, F. (2008). Imaging astrocyte activity. *Science*, *320*, 1597–1599.

Yu, H., et al. (2017). Long-term exposure to ethanol downregulates tight junction proteins through the protein kinase Cα signaling pathway in human cerebral microvascular endothelial cells. *Experimental and Therapeutic Medicine*, *14*, 4789–4796.

Zhang, C. E., et al. (2019). Blood-brain barrier leakage in relation to white matter hyperintensity volume and cognition in small vessel disease and normal aging. *Brain Imaging and Behavior*, *13*, 389–395.

Zhao, C., Ling, Z., Newman, M. B., Bhatia, A., & Carvey, P. M. (2007). TNF-alpha knockout and minocycline treatment attenuates blood-brain barrier leakage in MPTP-treated mice. *Neurobiology of Disease*, *26*, 36–46.

Zhao, Z., Nelson, A. R., Betsholtz, C., & Zlokovic, B. V. (2015). Establishment and dysfunction of the blood-brain barrier. *Cell*, *163*, 1064–1078.

Zimatkin, S. M., & Deitrich, R. A. (1997). Ethanol metabolism in the brain. *Addiction Biology*, *2*(4), 387–400. https://doi.org/10.1080/13556219772444.

Zlokovic, B. V. (2011). Neurovascular pathways to neurodegeneration in Alzheimer's disease and other disorders. *Nature Reviews. Neuroscience*, *12*, 723–738.

Zou, J. Y., & Crews, F. T. (2014). Release of neuronal HMGB1 by ethanol through decreased HDAC activity activates brain neuroimmune signaling. *PLoS One*, *9*, e87915.

CHAPTER EIGHT

Age modifies the effect of ethanol on behavior: Investigations in adolescent, adult and aged rats

D.B. Matthews* and B.M. Imhoff
Department of Psychology, University of Wisconsin – Eau Claire, Eau Claire, WI, United States
*Corresponding author: e-mail address: matthedb@uwec.edu

Contents

1. Effect of acute ethanol administration on behavior 253
2. Effects of chronic ethanol administration and corresponding withdrawal state on behavior 260
3. Conclusion 267
Funding 267
References 267

Abstract

The number of older people is increasing in most if not all countries in the world. In addition, the amount of alcohol consumption in the aged population is increasing and the consumption pattern is often in a binge fashion. However, little is known if the effects of alcohol, either acute or chronic exposure, vary in the older population compared to younger populations. The current mini-review will provide an overview of the effects of acute and chronic ethanol exposure at three different periods of development: adolescent, adult and aged on multiple different commonly studied behaviors. The overall conclusion is that biological age of the subject is a critical factor in understanding the effects of ethanol across the lifespan.

Abbreviations

ANOVA analysis of variance
ARR aerial righting reflex
AUD alcohol use disorder
GABA Gamma aminobutyrice acid
LORR loss of righting reflex
NMDA *N*-methyl-D-aspartic acid
PKCγ protein kinase C gamma

International Review of Neurobiology, Volume 161
ISSN 0074-7742
https://doi.org/10.1016/bs.irn.2021.06.010

Alcohol is the most commonly used and misused drug in the world (Ferreira & Willoughby, 2008). In 2018, the World Health Organization estimated that approximately 2.3 billion people were classified as current alcohol drinkers (Global status report on alcohol and health, 2018). In the United States, over 14 million Americans ages 18 and older were diagnosed with an alcohol use disorder (AUD) and this excessive alcohol use crosses all age ranges including that of adolescents, college students, adults and aged individuals (see Alcohol Use in the United States, NIAAA Alcohol Facts and Statistics, Oct., 2020). Survey research is demonstrating that dangerous alcohol use is increasing in many different demographics including young adult women and older women without children (McKetta & Keyes, 2019). While it has long been known that adolescents use and misuse alcohol at alarming rates (Witt, 2010), it has also recently been revealed that the older population is also using alcohol at alarming rates (Breslow, Castle, Chen, & Graubard, 2017; Calvo et al., 2020; Han, Moore, Ferris, & Palamar, 2019; Han, Moore, Sherman, Keyes, & Palamar, 2017). Given the prevalence of alcohol use and misuse at almost every developmental stage from early teenage years too late in life, it is critical to understand how *biological age*, as a factor, influences the effects of alcohol.

Understanding how alcohol impacts behavior at different life stages requires appreciation that the nervous system is not a static system but instead changes over the lifespan and a drug will produce different effects at different developmental stages (c.f., Matthews, Schneider, Kastner, Scaletty, & Szenay, 2019). For example, research has shown that brain regions mature and change as a function of age (e.g., Damoiseaux, 2017; Fjell et al., 2014; Vinke, Ikram, & Vernooij, 2019) and a neurotransmitter's receptor structure is altered by age (e.g., Yu, Wang, Fritschy, Witte, & Redecker, 2006 for $GABA_A$ receptor subunits and Magnusson, 2012 or Kumar, 2015 for NMDA glutamate receptors). Therefore, we hypothesize that the effects of ethanol will be altered by the age of the subject and that by exploring the effects of ethanol in cohorts of different aged animals we will gain important insights into how brain changes during the aging process impacts ethanol-behavior interactions. The resultant insights will help guide future research into identifying specific brain regions and/or neurotransmitter systems that are critical to understanding how ethanol interacts with the brain across the lifespan.

The overarching purpose of the present mini-review is to explore commonalities and differences in behavioral effects of ethanol across three different age-cohorts in preclinical animal models: adolescent, adult and aged subjects. To better provide the ability to synthesize data, the majority of the work reported will have used rats as experimental subjects with most being Sprague-Dawley animals supplemented with data using Long-Evans Hooded and Fisher 344 rats. In addition, we will systematically review a specific set of behaviors that are impacted by acute ethanol first, then, compare and contrast the data from chronic ethanol exposure for the same behaviors. Given the brevity of the current mini-review, only in selected cases will we mention potential neurobiological mechanisms impacting the behavior reviewed. In summary, we believe it is conclusive that that age of subject is one of the primary determinates for ethanol's effects. Such a direct conclusion will hopefully increase the engagement of additional research into primary mechanisms underlying the differential effects of ethanol by age.

1. Effect of acute ethanol administration on behavior

The behavioral effects of acute ethanol in adult rats (often defined as older than 60–70 days of age and younger than 18 months of age) have been well studied. In addition, the behavioral effects of acute ethanol in adolescent rats (those that are typically older than 28 days of age but less than 50 days of age) have been extensively researched in the last two decades. However, the behavioral effects of acute ethanol in aged rats (roughly defined as those over the age of 18 months) is only now being actively investigated. It is known that some of the reported effects of acute ethanol administration include anxiolytic-like responses, impaired muscle coordination, impaired hippocampal dependent cognition and facilitated striatal dependent cognition, production of hypothermia, and hypnosis (i.e., sleep time). In the subsequent paragraphs, we will outline the effect of ethanol in each of these behaviors.

Anxiolytic Effects: The anxiolytic-like effect of acute ethanol is often measured in the elevated plus maze (c.f., Walf & Frye, 2007 for a description of the task), elevated zero maze (Kulkarni, Singh, & Bishnoi, 2007; Shepherd, Grewal, Fletcher, Bill, & Dourish, 1994), the open field test (Bailey & Crawley, 2009; Hall, 1934) or, more recently, the social interaction test (Bailey & Crawley, 2009). Each of these established tests measures how various manipulations (e.g., light, elevation, novelty) can produce an anxiety-like phenotype that interacts with other natural tendencies (e.g., exploration) in animals. A consistent result found is that moderate doses of ethanol, typically between 1.0 and 2.0 g/kg, often produce a reduction in anxiety-like behavior. For example, this anxiolytic effect is seen by an increase in open arm entries and a corresponding increase in the percentage of open arm time in the elevated plus maze (Bertoglio & Carobrez, 2002; Criswell, Knapp, Overstreet, & Breese, 1994; Ferreira, Takahashi, & Morato, 2000; Prunell, Escorihuela, Fernandez-Teruel, Nunez, & Tobena, 1994). While the anxiolytic effect of ethanol in the plus maze is consistent, it can be abolished by previous experience with the apparatus (Bertoglio & Carobrez, 2002). Furthermore, biological sex may be a mediating factor influencing the anxiolytic-like effect of ethanol in the plus maze, such that females show a greater anxiolytic effect than males at higher doses (2.0 g/kg) of ethanol (Wilson, Burghardt, Ford, Wilkinson, & Primeaux, 2004). However, this sex effect may be time-dependent.

The light-dark box test typically measures anxiety based on the latency for an animal to transition from the initial light component of the environment to the dark component of the environment, number of entries to the light component of the environment and the time spent in the lit side of the apparatus. Like that found with other measures of ethanol's effect on anxiolytic behavior, low doses of ethanol produce a reduction in anxiety-like behavior as measured by increased transitions to the light side and longer times in the light side compared to control treated rats (Gallate, Morley, Ambermoon, & McGregor, 2003; Sharko, Kaigler, Fadel, & Wilson, 2016). Finally, social inhibition is a measure of how interaction with an unfamiliar peer may produce an anxiety-like state in subjects (File & Hyde, 1978). Similar to the conclusion that ethanol produces an anxiolytic-like effect in rats, moderate doses of ethanol will reduce social inhibition in rats (Varlinskaya & Spear, 2002). In general, the data in adult rats strongly supports the hypothesis that acute ethanol administration, in the low to moderate range, produces an anxiolytic-like effect.

In adolescent animals, the effect of acute ethanol on anxiety-like behavior is much less studied. What has been reported suggests that both male and female adolescent animals are less sensitive to the reduction in social inhibition produced by ethanol administration compared to adult animals (Varlinskaya & Spear, 2002). Furthermore, adolescent animals show a mostly dose-dependent anxiolytic-like effect to acute ethanol exposure in the open field, elevated plus maze and light-dark box. Unfortunately, no direct measurement to adult animals were made in this specific study (Acevedo, Nizhnikov, Molina, & Pautassi, 2014). While additional work is needed to understand the anxiolytic effect of acute ethanol in adolescents, a dearth of research has occurred in aged animals. Currently, we are unaware of any systematic investigation of acute ethanol's anxiolytic effect in aged animals. This is a significant need in understanding the effect of ethanol across the lifespan.

Ataxia: The impact of ethanol on muscle coordination, often referred to as ataxia, has been well studied (Novier, Van Skike, Chin, Diaz-Granados, & Matthews, 2012; c.f., Novier, Diaz-Granados, & Matthews, 2015) in adult animals. Typically these experiments use either the accelerating rotarod test (Dunham & Miya, 1957), the aerial righting reflex (ARR), tilting plane test or total movement in the open field. The accelerating rotarod requires animals to walk on a rotating dowl rod that slowly accelerates over a set period of time. The dependent measure is the time and/or speed at which the animal is no longer able to walk on the dowl without falling off. The aerial righting reflex is a more gross measure of body movement. In this task, animals are held in a supine position and released at 5-in increments above a thick, soft foam pad. The dependent measure is the height required for the animal to rotate and right itself in the air thereby landing with at least 3 paws on the foam pad. Acute ethanol exposure in adult animals has been shown to produce ataxia that is dose-dependent on both tasks. Specifically, low doses of ethanol (1.0 g/kg) do not impair rotarod performance while moderate doses of ethanol (1.5–2.0 g/kg) significantly decreased time on the accelerating rotarod (Chuck, McLaughlin, Arizzi-LaFrance, Salamone, & Correa, 2006; Dulman, Auta, Teppen, & Pandey, 2019; Khisti, VanDoren, Matthews, & Morrow, 2004; Lai, Kuo, & Lin, 2007). In agreement with the data from the rotarod, acute ethanol administration impairs performance on the aerial righting reflex in that a greater height is needed for the test subject to right themselves following acute ethanol administration (Khisti et al., 2004; Vogel et al., 1981). Furthermore, moderate levels of acute ethanol impair performance of adult animals in the tilting plane task (Ramirez & Spear, 2010; White et al., 2002) suggesting that ethanol reduces balance and/or body control.

In most research reports it is uniformly reported that acute ethanol exposure produces less ataxia in adolescent animals compared to adult animals. Importantly, this reduction in the ataxic effect produced by acute ethanol administration has been proposed as one potential reason why many adolescents consume alcohol at higher rates than adults (Spear & Swartzwelder, 2014). For example, it has been shown that acute ethanol administration reduces the amount of movement in an open field to a smaller extent in adolescent rats compared to adult rats (Van Skike et al., 2010). In addition, adolescent male rats require significantly less height to regain their righting reflex in the aerial righting reflex test compared to adult rats (Van Skike et al., 2010). In agreement with data from the more established tasks, adolescent rats require a significantly larger ethanol challenge to observe similar impairments in the tilting plane (Ramirez & Spear, 2010; White et al., 2002). However, the dose of ethanol appears important in that at lower doses (1.25 g/kg) ethanol may produce similar ataxic deficits in both adolescent and adult rats (Hollstedt, Olsson, & Rydberg, 1980). Interestingly, it has been shown that there are decreased levels of PKCγ expression in the cerebellum from adolescent animals compared to the cerebellum from adult animals (Van Skike et al., 2010). Considering PKCγ knockout mice are less sensitive to the ataxic effects of ethanol (Harris et al., 1995), it is possible that the lower levels of PKCγ expression in adolescence compared to adulthood is one molecular mechanism underlying the reduced ataxic effects in the younger age group.

In aged animals, the opposite effect is found compared to adolescent animals in that aged animals are profoundly more sensitive to the effects of acute ethanol on ataxia than are adult animals (c.f., Matthews, Schneider, et al., 2019; Matthews, Watson, et al., 2019). Our first experiment that included aged animals involved the simple aerial righting reflex (ARR) task in male rats where the ARR was determined every 10 min for 40 min following an acute administration of 2.0 g/kg ethanol (Van Skike et al., 2010). At this time in our research, we were focused primarily on investigating the effect of acute ethanol on adolescent animals. Unfortunately, our data did not directly reveal a significant difference in ARR between aged and adult animals because we did not collect sufficient data across time following the ethanol administration. However, examination of Table 1 (Van Skike et al., 2010) reveals that while not significantly different in this experiment, aged animals' were not showing recovery after 40 min, while there was initial recovery in adult animals after 40 min. We decided to directly investigate this potential age-related effect by determining the effect of 2.0 g/kg ethanol on ARR in aged and adult rats for times up to 90-min postethanol exposure (Novier, Van Skike, Diaz-Granados, Mittleman, & Matthews, 2013). We found that aged rats were profoundly impaired in ARR compared to adult animals and the impairment lasted throughout the 90-min experiment (Novier et al., 2013). In addition to motor impairments in the ARR, we have also reported that aged rats are significantly more impaired on the accelerating rotarod task compared to both adult and adolescent animals (Novier et al., 2013; Ornelas, Novier, Van Skike, Diaz-Granados, & Matthews, 2015). Importantly, blood ethanol levels were determined in these studies and the blood ethanol levels do not explain the large difference in ethanol-induced ataxia between aged animals and younger animals. The lack of a blood ethanol effect strongly suggests that one or more central nervous system mechanism(s) have been altered during aging to increase the sensitivity of aged animals to acute ethanol exposure.

Cognition: A long line of research from multiple laboratories has demonstrated that acute ethanol administration impairs hippocampal-dependent spatial memory in adult animals in a dose-dependent manner (c.f., Van Skike, Goodlett, & Matthews, 2019). Specifically, acute ethanol impairs spatial reference memory and spatial working memory in the radial arm maze (Hoffmann & Matthews, 2001; Matthews, Simson, & Best, 1995) and in the spatial Morris water maze task (Matthews, Morrow, Tokunaga, & McDaniel, 2002). These effects are replicated in commonly used mouse lines like the C57BL/6J (Berry & Matthews, 2004). In addition to the radial arm maze and Morris water maze, acute ethanol will impair trace conditioning (Weitemier & Ryabinin, 2003) and contextual fear conditioning (Melia, Ryabinin, Corodimas, Wilson, & Ledoux, 1996), two tasks commonly used to study hippocampal function. It has been suggested that the degradation in the spatial specificity of hippocampal place cells by acute ethanol administration is one mechanism causing the impairment in hippocampal-dependent memory (Matthews, Simson, & Best, 1996). However, the impairments in cognition are not due to nonselective effects on learning and memory in that acute ethanol administration does not impair, and in some cases can actually facilitate, striatal dependent cue and/or motor-based learning and memory (c.f., Van Skike et al., 2019). For example, acute ethanol administration does not impair nonspatial reference memory or nonspatial working memory in the radial arm maze tasks (Hoffmann & Matthews, 2001; Matthews et al., 1995), nor the standard nonspatial visible platform in the Morris water maze (Matthews et al., 2002). Furthermore, hippocampal-independent cued fear-conditioning is typically not impaired by acute ethanol administration (Melia et al., 1996). Interestingly, in tasks where the use of a spatial strategy or a nonspatial strategy can be employed by the subject, animals tested following either a saline or low dose ethanol administration (1.0 g/kg ethanol) use a hippocampal-dependent spatial strategy while animals tested under a moderate to high ethanol dose (1.5 or 2.0 g/kg ethanol) use a hippocampal-independent nonspatial strategy (Matthews, Ilgen, White, & Best, 1999).

For several years a debate existed in the field concerning whether an acute ethanol challenge produced a greater, or similar, hippocampal dependent memory impairment in adolescent animals compared to adult animals.

The focus of the debate was mostly laboratory-based and fortunately has appeared to resolve itself recently into a general consistence that adolescent animals display the same level of impairment in hippocampal-dependent spatial memory task as do adult animals (c.f., Novier et al., 2015, section 5.2). Briefly, there appears to be two relevant issues. First, adolescent and adult animals use different response strategies in the initial training trials as it relates to specific learning and memory tasks, such as the Morris water maze. Second, individual laboratories were using different dependent measures and the different measures likely facilitated variable results. When such variables were standardized, adolescent and adult animals demonstrated similar levels of ethanol-induced spatial memory impairments (Chin et al., 2011). In addition, a recent report highlights the need to control for state-dependent effects. Specifically, when state-dependent effects are controlled for in fear conditioning, acute ethanol exposure produces similar impairments between adolescent and adult rats (Hunt & Barnet, 2016). While the literature on ethanol's effects on hippocampal-dependent memory appears more consistent in adolescent and adult rats, little work has been done to investigate if ethanol produces greater or lesser effects on striatal-dependent cognitive function in adolescent animals. Given that acute ethanol can facilitate caudate-dependent cue-response memory in adult rats (Matthews et al., 1999), it would be insightful to investigate if adolescent animals were more, less, or similarly facilitated by acute ethanol exposure on caudate-dependent nonspatial memory tasks.

Our initial hypothesis concerning the effect of acute ethanol administration on hippocampal-dependent learning and memory was that aged animals would be significantly more impaired by acute ethanol compared to adult animals. We based this prediction on the general findings that aged animals are more sensitive to ethanol on a variety of tasks. Our first investigation of this hypothesis using spatial memory in the Morris water maze confirmed that aged male rats are indeed more impaired by acute ethanol compared to adult male rats following doses of either 1.5 g/kg ethanol or 2.0 g/kg ethanol (Novier et al., 2013). However, in retrospect, the results reported by Novier et al. (2013) were inconclusive. Specifically, aged rats were more impaired on the spatial version of the Morris water maze if escape latency to the platform was the dependent measure, but no significant effect of age was found if swim pathlength was the dependent measure. To further investigate these conflicting results, we recently took male rats of various ages and first trained them in the spatial version of the Morris water maze naive and only tested those following an ethanol challenge, unlike in the

Novier et al. (2013) study, that were deemed cognitively spared. That is, we only tested aged animals that could actually learn the spatial task to the same level as adult animals. By selectively testing cognitively spared animals, we learned two important new pieces of information. First, aged animals that are still relatively young (19–24 months) show spatial memory impairments to an acute ethanol challenge at the same level as adult animals (9–14 months). Second, very old-aged animals (29- to 33-month-old-animals) are more impaired by the acute ethanol challenge compared to the adult animals in spatial memory but also the cognitive impairment was not selective in that the very old aged animals are more impaired in the standard nonspatial memory task as well (Matthews, Scaletty, Schreiber, & Trapp, 2020). These data are critical because they reveal that relatively young aged animals (19–24 month old) are not always more sensitive to the effects of acute ethanol administration in every task compared to adult animals.

Hypothermia: Acute ethanol exposure produces hypothermia and the effect is dose-dependent (Abel & York, 1979a, 1979b; Freund, 1973; York, 1982, 1983). However, investigating the hypothermic effects of ethanol in adolescent rats is much less straightforward. For example, if ethanol is administered in a gavage fashion to adolescent rats, tolerance to the effect occurs differently compared to adult animals although the initial magnitude of the tolerance is similar (Swartzwelder, Richardson, Markwiese-Foerch, Wilson, & Little, 1998). Further, if the ethanol dose is administered in a manner that produces faster rises in blood ethanol levels (intraperitoneal injections compared to intragastric gavage), less hypothermia is found in adolescent rats compared to adult rats (Silveri & Spear, 2000). Finally, the testing conditions and handling of adolescent rats can impact ethanol-induced hypothermia (Ristuccia, Hernandez, Wilmouth, & Spear, 2007). In aged rats the impact of acute ethanol administration on hypothermia has been less well studied. Initial studies suggest ethanol produces greater hypothermia in aged rats (York, 1982) and this effect has recently been confirmed. Specifically, we investigated ethanol-induced hypothermia in aged, adult, and adolescent male rats following one of three doses of ethanol administered intraperitoneally. In agreement with previous work, acute ethanol exposure produces hypothermia that is dose dependent. However, we also found that the hypothermic effect was exacerbated in aged animals compared to adult or adolescent animals following the highest dose of ethanol administrated, 3.0 g/kg ethanol, but not following lower doses of ethanol, 1.0 and 2.0 g/kg ethanol (Watson, James, Mittleman, & Matthews, 2020).

Hypnosis: Finally, it has been demonstrated that high doses of ethanol can produce hypnosis (i.e., loss of righting reflex [LORR]). The LORR effect is often measured as the amount of time the subject spends in a supine position following an ethanol injection, i.e., sleep time (Majchrowicz, 1975). Consistent with reports that adolescent rats are less sensitive to the motor impairing effects of ethanol, high dose ethanol exposure also produces less LORR in adolescent rats compared to adult rats (Swartzwelder et al., 1998). Recently, the impact of high dose ethanol on LORR has been investigated in aged rats. Similar to other studies, aged animals are significantly more sensitive to the effects of high dose ethanol in that aged rats have longer sleep times compared to adult or adolescent animals (Ornelas et al., 2015; Perkins, Vore, Lovelock, Varlinskaya, & Deak, 2018).

From the reviewed work, it is apparent that biological age is one of the primary factors impacting the effect of acute ethanol administration on behavior. That immediately leads to the question of, "Why? What is different in various ages that can cause such divergent responses to acute ethanol administration?" In essence what is the mechanism(s) that produce such effects? The most obvious answer is differential ethanol uptake and/or ethanol metabolism producing differential blood ethanol levels that can simply account for the divergent behavioral data. However, such is not the case. For example, when compared to adult animals at the time following administration when relevant behavioral data has been assessed, adolescent animals often have *higher* blood ethanol levels compared to adult animals even though adolescents are *less* sensitive than adult animals to the effects of acute ethanol exposure. As it relates to aged animals, when relevant behavioral data are collected (typically within 90 min of ethanol exposure), the blood ethanol level between aged and adult animals is most often *similar* even though aged animals are significantly *more* sensitive to ethanol compared to adult animals. These data suggest that central nervous system mechanisms in aged animals are producing the increased sensitivity to acute ethanol exposure. These central nervous system mechanisms remain to be elucidated.

2. Effects of chronic ethanol administration and corresponding withdrawal state on behavior

The impact of chronic ethanol exposure on various behaviors has also been extensively investigated in adult and adolescent animals. However, the literature is rather sparse as it relates to the effect of chronic ethanol exposure in aged animals. Nonetheless, important insights can be gleaned from the previous literature and several avenues for additional research can be highlighted based on this lack of studies.

Anxiogenic Effects: As previously reviewed, acute ethanol exposure will produce an anxiolytic-like effect in adult animals. However, withdrawal from chronic ethanol exposure produces an anxiogenic-like effect in adult animals that is independent of task (c.f., Pandey, Kyzar, & Zhang, 2017 for a detailed review of behavior and cellular mechanisms underlying anxiety-like increases in ethanol withdrawn animals). For example, withdrawal following chronic ethanol exposure via liquid diet or ethanol vapor in male and female adult rats results in increased anxiety-like measures in the elevated plus maze (Devaud, Matthews, & Morrow, 1999; Morales, McGinnis, Robinson, Chappell, & McCool, 2018) which can last for several weeks (Valdez et al., 2002). As expected, similar anxiogenic effects during withdrawal following chronic ethanol exposure were found in the zero maze (McGinnis, Parrish, & McCool, 2020) and the open field (Fu et al., 2020). Changes in anxiety-like behavior following chronic ethanol exposure in adolescent animals has also received research attention. Specifically, the Neuroscience of Adolescent Drinking in Adulthood (NADIA) research consortium has detailed potentially long-lasting changes in anxiety-like behavior following chronic ethanol exposure in adolescent animals (c.f., Crews et al., 2019 and relevant chapters in this volume). Briefly, chronic intermittent ethanol exposure during adolescence produces an anxiety-like effect both initially and during adulthood when measured via the social interaction test (Varlinskaya, Truxell, & Spear, 2015) although the effect may occur only in males and not in females (Dannenhoffer et al., 2018). In addition, adolescent chronic intermittent ethanol exposure can increase anxiety responses during adulthood on the elevated plus maze (Kokare, Kyzar, Zhang, Sakharkar, & Pandey, 2017; Pandey, Sakharkar, Tang, & Zhang, 2015; Van Skike, Diaz-Granados, & Matthews, 2015), the light-dark box (Lee, Coehlo, Solton, & Szumlinski, 2017; Vetreno, Yaxley, Paniagua, & Crews, 2016) and the open-field test (Coleman, Liu, Oguz, Styner, & Crews, 2014; Vetreno, Broadwater, Liu, Spear, & Crews, 2014). However, similar to lack of research investigating anxiolytic-like effect in aged animals following acute ethanol, there are also few studies that have investigated the effect of anxiety-like behavior following chronic ethanol exposure in aged animals. Our first investigation into the effect of chronic ethanol exposure on anxiety-like behavior in aged rats compared aged and adult male rats who consumed ethanol via a chronic liquid diet. Unfortunately, our exposure paradigm did not produce similar ethanol levels in aged and adult animals. Specifically, while aged animals consumed more fluid than adult animals, when body weight was taken into account, aged animals consumed less ethanol per body weight and their resultant blood ethanol levels were lower than the adult subjects. These differences precluded us from being able to directly compare the different ages in our behavioral tasks. Nonetheless, we still found that chronic ethanol exposure in aged animals would produce an anxiogenic-like effect on the elevated plus maze (Novier, Ornelas, Diaz-Granados, & Matthews, 2016). In our second attempt we treated aged, adult, and adolescent rats with low to moderate doses of ethanol (1.0 or 2.0 g/kg ethanol) every other day for 20 days via intraperitoneal injection and then tested them on a variety of tasks following the completion of the exposure period. In this particular experiment, low dose chronic intermittent ethanol did not produce an anxiety-like effect on the elevated plus maze (Matthews, Watson, et al., 2019). Hence, it appears that aged animals must be exposed to doses greater than 2.0 g/kg ethanol per day for an extended period of time to produce an increase in anxiety-like behavior. However, it is currently unknown if chronic ethanol exposure in aged animals produces more, less, or similar levels of anxiety-like behavior compared to younger animals following such an exposure pattern.

Ataxia: Chronic ethanol exposure can produce tolerance to ethanol-induced motor ataxia (Khanna, Kalant, Sharma, & Chau, 1991). In addition, tolerance to the motor impairing effects of ethanol following chronic exposure is found at all ages tested; although, the age of the subject significantly impacts the degree of ethanol tolerance. For example, chronic intermittent ethanol exposure in adult rats produces small transient tolerance to the motor impairing effects of acute ethanol in the tilting plane test but similar doses and exposure pattern of chronic ethanol in adolescent animals produced tolerance that was immediate and long lasting (White et al., 2002). One of the few studies in aged animals that has investigated the development of tolerance to the ataxic effects of ethanol used the ARR (Matthews & Mittleman, 2017). In this study, male aged, adult, and adolescent rats were exposed to either 1.5 g/kg or 2.5 g/kg ethanol in a 20-day intermittent pattern (ethanol exposure every other day for a total of 10 ethanol exposures). As expected, aged animals were significantly more impaired (required a higher height to successfully right themselves) than adult or adolescent animals, but tolerance did not develop. Additional research is needed to determine if higher doses of ethanol or longer ethanol exposure periods would produce more or less motoric tolerance in aged animals compared to younger animals.

Cognition: The effect of chronic ethanol exposure on hippocampal-dependent spatial memory has received much research attention and the effect of chronic ethanol is likely exacerbated by thiamine deficiency (c.f., Nunes, Kipp, Reitz, & Savage, 2019). However, chronic ethanol administration by itself may produce hippocampal-dependent spatial memory impairments and reduce behavioral flexibility, a task dependent on the orbitofrontal cortex (Floresco, Block, & Tse, 2008; Ghods-Sharifi, Haluk, & Floresco, 2008) in adult rats. For example, early studies demonstrated that chronic ethanol impaired hippocampal memory (Franke, Kittner, Berger, Wirkner, & Schramek, 1997) perhaps by degrading the cholinergic system (Arendt et al., 1989). Interesting, short-term (2 week) chronic ethanol consumption in adult rats does not impair spatial learning but will reduce behavioral flexibility in rats (Fernandez, Lew, Vedder, & Savage, 2017) and can also blunt the spatial memory impairment produced by an acute ethanol challenge (Boulouard, Lelong, Daoust, & Naassila, 2002). Such a blunting of the memory impairing effects of acute ethanol demonstrate cognitive tolerance following chronic ethanol exposure. However, longer term chronic intermittent ethanol exposure (~4–8 weeks) will produce deficits in spatial memory as assessed by a probe trial in the Morris water maze (Cagetti, Pinna, Guidotti, Baicy, & Olsen, 2004). Importantly, the impairments in hippocampal cognition are not limited to the water maze as shown by data that chronic ethanol impairs hippocampal-dependent spatial recognition memory (Garcia-Moreno & Cimadevilla, 2012).

Chronic ethanol exposure during adolescence produces less consistent effects on hippocampal-dependent spatial learning and memory. For example, if male adolescent rats are treated with high dose ethanol (5.0 g/kg) every other day from postnatal day 30 to postnatal day 48 and trained in the spatial version of the Morris water maze task on nontreatment days, ethanol treatment does not alter spatial learning (Silvers, Tokunaga, Mittleman, & Matthews, 2003). Interestingly though, similar to what is found in adult animals, chronic treatment during adolescence produces tolerance to the spatial memory impairment produced by acute ethanol challenge (Silvers et al., 2003) and this data is mirrored by the effect of acute ethanol on the spontaneous activity of hippocampal pyramidal neurons (Tokunaga, McDaniel, Morrow, & Matthews, 2003; Tokunaga, Silvers, & Matthews, 2006). Noteworthy is the fact that the length of chronic ethanol exposure is critical for the expression of cognitive tolerance in that shorter exposure periods (~4–5 days of ethanol exposure) do not produce cognitive tolerance to an acute ethanol challenge even though this exposure pattern will produce metabolic tolerance (Silvers et al., 2006; Van Skike, Novier, Diaz-Granados, & Matthews, 2012). Furthermore, chronic intermittent ethanol exposure during adolescence does not produce greater performance impairments in a reversal task measuring behavioral flexibility if animals are trained in the task during ethanol treatment (Silvers et al., 2006). However, if animals are trained following chronic ethanol exposure during adolescence, they can be impaired during a reversal task measuring behavioral flexibility (Contreras et al., 2019; Fernandez et al., 2017). These data demonstrate that ethanol exposure during adolescence can produce changes in brain function that are time-, dose- and length-of-time-dependent after ethanol exposure. In addition, it appears that reversal tasks that test behavioral flexibility are more sensitive to chronic ethanol exposure effects than regular spatial learning and memory tasks. Given that the orbitofrontal cortex is critical for behavioral flexibility while the hippocampus is critical for spatial cognition, it seems reasonable to predict that the orbitofrontal cortex is more compromised by chronic ethanol exposure than the hippocampus. Consequently, the orbitofrontal cortex is crucial for investigations of the mechanism(s) underlying altered behavioral flexibility following chronic ethanol exposure.

There have been significantly fewer studies that have investigated the effect of chronic ethanol exposure in aged rats on hippocampal-dependent learning and memory, and the results that exist are somewhat mixed. For example, in support of work in adolescent and adult animals, chronic ethanol exposure in aged rats does not appear to have a dramatic effect on hippocampal-dependent spatial learning if animals are exposed to ethanol via liquid diet (Novier et al., 2016). However, there is an impairment in spatial learning if aged animals are treated with ethanol via intraperitoneal injections for 45 days (Baydas, Yasar, & Tuzcu, 2005), but this impairment is dependent on the length of the ethanol exposure in that a spatial learning impairment is not seen if animals are only treated 10 times in an every other day intermittent fashion (Matthews, Watson, et al., 2019). Given the lack of dramatic effects in aged animals following chronic ethanol exposure in basic spatial learning tasks, future research should investigate deficits in behavioral flexibility via reversal learning trials. These studies are currently underway in our laboratory.

Two studies provide interesting insights into the effect of chronic intermittent ethanol exposure during adolescence on the effect of a later-in-life ethanol challenge. Specifically, these two studies exposed animals to ethanol chronically during adolescence and then their spatial memory was tested later in life. In the first study, White and colleagues treated adolescent or adult rats with ethanol in a chronic intermittent fashion and then tested their spatial working memory following an acute ethanol injection 20 days later. Interestingly, the ethanol challenge produced a larger spatial working memory deficit in the subjects treated with ethanol during adolescence compared to those subjects treated with ethanol chronically during adulthood (White, Ghia, Levin, & Swartzwelder, 2000). We dramatically extended and replicated this finding. Specifically, we treated animals with high dose ethanol (5.0 g/kg) every other day for 20 days during adolescence and then maintained the subjects for 532 days testing them on a variety of tasks every 4 months. Similar to the White et al. (2000) study, we found that subjects treated with ethanol during adolescence had significantly greater spatial memory impairments following an acute ethanol challenge compared to those animals not treated with ethanol during adolescence when tested 100s of days later (Matthews et al., 2017). These data suggest that some long-lasting brain changes occur during chronic treatment in adolescence that is maintained throughout the animal's lifespan compromising cognition following subsequent ethanol exposure. The clinical ramifications of this finding are large considering the number of older adults who began consuming alcohol as an adolescent and continued to consume alcohol later in life.

Hypothermia: While acute ethanol administration reliably decreases core body temperature, i.e., produces hypothermia, exposure to chronic ethanol and resulting tolerance to ethanol's hypothermic effects has been less consistent. For example, an early study did show a reduction in the time course of hypothermia during a chronic ethanol exposure regime (Fidecka, Tamborska, & Langwinski, 1986). However, one study revealed that chronic ethanol exposure in adult or adolescent male rats did not result in tolerance to acute ethanol induced hypothermia in adolescents and only a strong trend toward tolerance in adult rats (Pautassi, Godoy, & Molina, 2015). Another study demonstrated greater tolerance to ethanol-induced hypothermia in adolescent rats but not adult rats (Swartzwelder et al., 1998). We are currently unaware of any studies investigating the development of tolerance to ethanol induced hypothermia in aged rats.

While the data on the development of tolerance in ethanol-induced hypothermia is inconsistent, several factors likely impact this. These include the ambient room temperature whereas greater ambient temperatures minimize the development of tolerance to the hypothermic effects of ethanol (Le, Kalant, & Khanna, 1986; Mackenzie-Taylor & Rech, 1991) and alter the motivational and hypothermic effects of ethanol (Cunningham, Hawks, & Niehus, 1988; Cunningham & Niehus, 1989; Cunningham, Niehus, & Bachtold, 1992). In addition, different ethanol treatment patterns and length of treatment likely influences the development of hypothermic tolerance to ethanol exposure.

Hypnosis: Loss of righting reflex is a common measure investigating the sensitivity to high dose ethanol and tolerance to high dose ethanol. It is currently an open question if the development of tolerance to high dose ethanol as measured by the loss of righting reflex (LORR) is due to actual tolerance to this effect or a by-produce of metabolic tolerance (Broadwater, Varlinskaya, & Spear, 2011). Nonetheless, it has been shown that LORR tolerance will develop during chronic ethanol exposure in adolescence and will also reduce age related increases in LORR to an acute ethanol challenge (Silvers et al., 2003) and this occurs in a dose-dependent manner (Matthews, Tinsley, Diaz-Granados, Tokunaga, & Silvers, 2008). In aged animals, we are only aware of one study that examines tolerance to LORR due to chronic ethanol exposure. In this study Matthews, Watson, et al. (2019), aged, adult, and adolescent rats were treated with low dose ethanol (1.0 or 2.0 g/kg) every other day for 20 days (10 treatments) and then challenged with a 3.0 g/kg ethanol injection. Aged animals had significantly longer LORR times and the animals chronically treated with 2.0 g/kg ethanol most likely develop tolerance to LORR (the relevant interaction in the two-way ANOVA had a significance value of $P < 0.051$). These data suggest that aged animals may develop tolerance significantly faster as it relates to high dose ethanol but further research is needed to determine the validity of this prediction. In addition, if adolescents are exposed to high dose ethanol (5.0 g/kg) and then tested every 4 months until 532 days with high dose ethanol, the LORR times are significantly less than animals not exposed to ethanol during adolescence. In other words, if animals are treated chronically with ethanol during adolescence, tolerance to high dose ethanol lasts across the lifespan (Matthews et al., 2017).

When conducting research on chronic ethanol exposure in different aged animals it is always critical to ensure differential blood ethanol levels are not the cause of differential behavioral and/or neurobiological effects. As discussed earlier in this mini-review, behavioral measures recorded within the first ~90 min of acute ethanol administration produces similar blood ethanol levels suggesting the altered behavioral effects in different aged animals are not due to ethanol uptake and/or clearance. However, as it relates to aged animals, we reported that when plasma samples are collected 180 min following ethanol administration, high dose ethanol (3.0 g/kg) did produce differential blood ethanol levels between adolescent animals and aged animals (Watson et al., 2020). This suggests that caution must be given to the interpretation of behavioral data from studies that compare aged and adolescent animals that were exposed to high doses of ethanol and data points collected at later times.

3. Conclusion

Ethanol is the most used and misused drug in the world and recent data shows that ethanol is used by all age demographics. While ethanol use typically begins during adolescence, current work also demonstrates that the older population consumes ethanol and often does so in a dangerous, binge pattern. Significant advances in understanding how ethanol interacts with adolescences and adults to modify various behavioral effects have been made. This work demonstrates that adolescents are often, but not always, less sensitive to the effects of ethanol. However, the dramatic increase in the aged demographic coupled with the amount of ethanol consumed by this population necessitates additional research on the interaction of ethanol and the older subject. Current work demonstrates that aged subjects are often, but not always, more sensitive to the effects of ethanol. By continuing and expanding these lines of research, insights into how normal aging alters the brain and body will help reveal why adolescents are *less sensitive* than adults to ethanol and why the aged are *more sensitive* than adults to ethanol.

Funding

This work was funded in part by the Mayo Healthcare-University of Wisconsin - Eau Claire Research Consortium.

References

Abel, E. L., & York, J. L. (1979a). Absence of effect of prenatal ethanol on adult emotionality and ethanol consumption in rats. *Journal of Studies on Alcohol*, *40*(7), 547–553.

Abel, E. L., & York, J. L. (1979b). Age-related differences in response to ethanol in the rat. *Psychobiology*, 7(4), 391–395.

Acevedo, M. B., Nizhnikov, M. E., Molina, J. C., & Pautassi, R. M. (2014). Relationship between ethanol-induced activity and anxiolysis in the open field, elevated plus maze, light-dark box, and ethanol intake in adolescent rats. *Behavioural Brain Research*, *265*, 203–215. https://doi.org/10.1016/j.bbr.2014.02.032.

Arendt, T., Allen, Y., Marchbanks, R. M., Schugens, M. M., Sinden, J., Lantos, P. L., et al. (1989). Cholinergic system and memory in the rat: Effects of chronic ethanol, embryonic basal forebrain brain transplants and excitotoxic lesions of cholinergic basal forebrain projection system. *Neuroscience*, *33*(3), 435–462. https://doi.org/10.1016/0306-4522(89)90397-7.

Bailey, K. R., & Crawley, J. N. (2009). Anxiety-related behaviors in mice. In J. J. Buccafusco (Ed.), *Methods of behavior analysis in neuroscience*. Boca Raton (FL): CRC Press.

Baydas, G., Yasar, A., & Tuzcu, M. (2005). Comparison of the impact of melatonin on chronic ethanol-induced learning and memory impairment between young and aged rats. *Journal of Pineal Research*, *39*(4), 346–352. https://doi.org/10.1111/j.1600-079X.2005.00257.x.

Berry, R. B., & Matthews, D. B. (2004). Acute ethanol administration selectively impairs spatial memory in C57BL/6J mice. *Alcohol*, *32*(1), 9–18. https://doi.org/10.1016/j.alcohol.2003.09.005.

Bertoglio, L. J., & Carobrez, A. P. (2002). Anxiolytic effects of ethanol and phenobarbital are abolished in test-experienced rats submitted to the elevated plus maze. *Pharmacology, Biochemistry, and Behavior*, *73*(4), 963–969. https://doi.org/10.1016/s0091-3057(02)00958-9.

Boulouard, M., Lelong, V., Daoust, M., & Naassila, M. (2002). Chronic ethanol consumption induces tolerance to the spatial memory impairing effects of acute ethanol administration in rats. *Behavioural Brain Research*, *136*(1), 239–246. https://doi.org/10.1016/s0166-4328(02)00134-1.

Breslow, R. A., Castle, I. P., Chen, C. M., & Graubard, B. I. (2017). Trends in alcohol consumption among older Americans: National Health Interview Surveys, 1997 to 2014. *Alcoholism, Clinical and Experimental Research*, *41*(5), 976–986. https://doi.org/10.1111/acer.13365.

Broadwater, M., Varlinskaya, E. I., & Spear, L. P. (2011). Chronic intermittent ethanol exposure in early adolescent and adult male rats: Effects on tolerance, social behavior, and ethanol intake. *Alcoholism, Clinical and Experimental Research*, *35*(8), 1392–1403. https://doi.org/10.1111/j.1530-0277.2011.01474.x.

Cagetti, E., Pinna, G., Guidotti, A., Baicy, K., & Olsen, R. W. (2004). Chronic intermittent ethanol (CIE) administration in rats decreases levels of neurosteroids in hippocampus, accompanied by altered behavioral responses to neurosteroids and memory function. *Neuropharmacology*, *46*(4), 570–579. https://doi.org/10.1016/j.neuropharm.2003.10.001.

Calvo, E., Allel, K., Staudinger, U. M., Castillo-Carniglia, A., Medina, J. T., & Keyes, K. M. (2020). Cross-country differences in age trends in alcohol consumption among older adults: A cross-sectional study of individuals aged 50 years and older in 22 countries. *Addiction*. https://doi.org/10.1111/add.15292.

Chin, V. S., Van Skike, C. E., Berry, R. B., Kirk, R. E., Diaz-Granados, J., & Matthews, D. B. (2011). Effect of acute ethanol and acute allopregnanolone on spatial memory in adolescent and adult rats. *Alcohol*, *45*(5), 473–483. https://doi.org/10.1016/j.alcohol.2011.03.001.

Chuck, T. L., McLaughlin, P. J., Arizzi-LaFrance, M. N., Salamone, J. D., & Correa, M. (2006). Comparison between multiple behavioral effects of peripheral ethanol administration in rats: Sedation, ataxia, and bradykinesia. *Life Sciences*, *79*(2), 154–161. https://doi.org/10.1016/j.lfs.2005.12.045.

Coleman, L. G., Jr., Liu, W., Oguz, I., Styner, M., & Crews, F. T. (2014). Adolescent binge ethanol treatment alters adult brain regional volumes, cortical extracellular matrix protein and behavioral flexibility. *Pharmacology, Biochemistry, and Behavior*, *116*, 142–151. https://doi.org/10.1016/j.pbb.2013.11.021.

Contreras, A., Polín, E., Miguéns, M., Pérez-García, C., Pérez, V., Ruiz-Gayo, M., et al. (2019). Intermittent-excessive and chronic-moderate ethanol intake during adolescence impair spatial learning. Memory and cognitive flexibility in the adulthood. *Neuroscience*, *418*, 205–217. https://doi.org/10.1016/j.neuroscience.2019.08.051. Epub 2019 Sep 3. PMID:31491502.

Crews, F. T., Robinson, D. L., Chandler, L. J., Ehlers, C. L., Mulholland, P. J., Pandey, S. C., et al. (2019). Mechanisms of persistent neurobiological changes following adolescent alcohol exposure: NADIA consortium findings. *Alcoholism, Clinical and Experimental Research*, *43*(9), 1806–1822. https://doi.org/10.1111/acer.14154.

Criswell, H. E., Knapp, D. J., Overstreet, D. H., & Breese, G. R. (1994). Effects of ethanol, chlordiazepoxide, and MK-801 on performance in the elevated-plus maze and on locomotor activity. *Alcoholism, Clinical and Experimental Research*, *18*(3), 596–601. https://doi.org/10.1111/j.1530-0277.1994.tb00916.x.

Cunningham, C. L., Hawks, D. M., & Niehus, D. R. (1988). Role of hypothermia in ethanol-induced conditioned taste aversion. *Psychopharmacology*, *95*, 318–322.

Cunningham, C. L., & Niehus, D. M. (1989). Effect of ingestion-contingent hypothermia on ethanol self-administration. *Alcohol*, *6*, 377–380.

Cunningham, C. L., Niehus, J. S., & Bachtold, J. F. (1992). Ambient temperature effects on taste aversion conditioned by ethanol: Contribution of ethanol-induced hypothermia. *Alcoholism, Clinical and Experimental Research*, *16*, 1117–1124.

Damoiseaux, J. S. (2017). Effects of aging on functional and structural brain connectivity. *NeuroImage*, *160*, 32–40. https://doi.org/10.1016/j.neuroimage.2017.01.077.

Dannenhoffer, C. A., Kim, E. U., Saalfield, J., Werner, D. F., Varlinskaya, E. I., & Spear, L. P. (2018). Oxytocin and vasopressin modulation of social anxiety following adolescent intermittent ethanol exposure. *Psychopharmacology*, *235*(10), 3065–3077. https://doi.org/10.1007/s00213-018-5003-8.

Devaud, L. L., Matthews, D. B., & Morrow, A. L. (1999). Gender impacts behavioral and neurochemical adaptations in ethanol-dependent rats. *Pharmacology, Biochemistry, and Behavior*, *64*(4), 841–849.

Dulman, R. S., Auta, J., Teppen, T., & Pandey, S. C. (2019). Acute ethanol produces ataxia and induces Fmr1 expression via histone modifications in the rat cerebellum. *Alcoholism, Clinical and Experimental Research*, *43*(6), 1191–1198. https://doi.org/10.1111/acer.14044.

Dunham, N. W., & Miya, T. S. (1957). A note on simple apparatus for detecting neurological deficit in rats and mice. *Journal of Pharmaceutical Science*, *46*, 208–209. https://doi.org/10.1002/jps.3030460322.

Fernandez, G. M., Lew, B. J., Vedder, L. C., & Savage, L. M. (2017). Chronic intermittent ethanol exposure leads to alterations in brain-derived neurotrophic factor within the frontal cortex and impaired behavioral flexibility in both adolescent and adult rats. *Neuroscience*, *348*, 324–334. https://doi.org/10.1016/j.neuroscience.2017.02.045.

Ferreira, V. M., Takahashi, R. N., & Morato, G. S. (2000). Dexamethasone reverses the ethanol-induced anxiolytic effect in rats. *Pharmacology, Biochemistry, and Behavior*, *66*(3), 585–590. https://doi.org/10.1016/s0091-3057(00)00255-0.

Ferreira, M. P., & Willoughby, D. (2008). Alcohol consumption: The good, the bad, and the indifferent. *Applied Physiology, Nutrition, and Metabolism*, *33*(1), 12–20. https://doi.org/10.1139/H07-175.

Fidecka, S., Tamborska, E., & Langwinski, R. (1986). Ethanol-induced hypothermia. *Cross tolerance with morphine. Acta Physiol Pol*, *37*(4–5), 168–176.

File, S. E., & Hyde, J. R. (1978). Can social interaction be used to measure anxiety? *British Journal of Pharmacology*, *62*(1), 19–24. https://doi.org/10.1111/j.1476-5381.1978.tb07001.x.

Fjell, A. M., McEvoy, L., Holland, D., Dale, A. M., Walhovd, K. B., & Alzheimer's Disease Neuroimaging, I. (2014). What is normal in normal aging? Effects of aging, amyloid and Alzheimer's disease on the cerebral cortex and the hippocampus. *Progress in Neurobiology*, *117*, 20–40. https://doi.org/10.1016/j.pneurobio.2014.02.004.

Floresco, S. B., Block, A. E., & Tse, M. T. (2008). Inactivation of the medial prefrontal cortex of the rat impairs strategy set-shifting, but not reversal learning, using a novel, automated procedure. *Behavioural Brain Research, 190*(1), 85–96. https://doi.org/10.1016/j.bbr.2008.02.008.

Franke, H., Kittner, H., Berger, P., Wirkner, K., & Schramek, J. (1997). The reaction of astrocytes and neurons in the hippocampus of adult rats during chronic ethanol treatment and correlations to behavioral impairments. *Alcohol, 14*(5), 445–454. https://doi.org/10.1016/s0741-8329(96)00209-1.

Freund, G. (1973). Hypothermia after acute ethanol and benzyl alcohol administration. *Life Sciences, 13*(4), 345–349.

Fu, R., Mei, Q., Shiwalkar, N., Zuo, W., Zhang, H., Gregor, D., et al. (2020). Anxiety during alcohol withdrawal involves 5-HT2C receptors and M-channels in the lateral habenula. *Neuropharmacology, 163*, 107863. https://doi.org/10.1016/j.neuropharm.2019.107863.

Gallate, J. E., Morley, K. C., Ambermoon, P., & McGregor, I. S. (2003). The consequences of beer consumption in rats: Acute anxiolytic and ataxic effects and withdrawal-induced anxiety. *Psychopharmacology, 166*(1), 51–60. https://doi.org/10.1007/s00213-002-1291-z.

Garcia-Moreno, L. M., & Cimadevilla, J. M. (2012). Acute and chronic ethanol intake: Effects on spatial and non-spatial memory in rats. *Alcohol, 46*(8), 757–762. https://doi.org/10.1016/j.alcohol.2012.08.001.

Ghods-Sharifi, S., Haluk, D. M., & Floresco, S. B. (2008). Differential effects of inactivation of the orbitofrontal cortex on strategy set-shifting and reversal learning. *Neurobiology of Learning and Memory, 89*(4), 567–573. https://doi.org/10.1016/j.nlm.2007.10.007.

Hall, C. S. (1934). Emotional behavior in the rat. I. Defecation and urination as measures of individual differences in emotionality. *Journal of Comparative Psychology, 18*, 385–403.

Han, B. H., Moore, A. A., Ferris, R., & Palamar, J. J. (2019). Binge drinking among older adults in the United States, 2015 to 2017. *Journal of the American Geriatrics Society, 67*(10), 2139–2144. https://doi.org/10.1111/jgs.16071.

Han, B. H., Moore, A. A., Sherman, S., Keyes, K. M., & Palamar, J. J. (2017). Demographic trends of binge alcohol use and alcohol use disorders among older adults in the United States, 2005-2014. *Drug and Alcohol Dependence, 170*, 198–207. https://doi.org/10.1016/j.drugalcdep.2016.11.003.

Harris, R. A., McQuilkin, S. J., Paylor, R., Abeliovich, A., Tonegawa, S., & Wehner, J. M. (1995). Mutant mice lacking the gamma isoform of protein kinase C show decreased behavioral actions of ethanol and altered function of gamma-aminobutyrate type a receptors. *Proceedings of the National Academy of Sciences of the United States of America, 92*(9), 3658–3662. https://doi.org/10.1073/pnas.92.9.3658.

Hoffmann, S. E., & Matthews, D. B. (2001). Ethanol-induced impairments in spatial working memory are not due to deficits in learning. *Alcoholism, Clinical and Experimental Research, 25*(6), 856–861.

Hollstedt, C., Olsson, O., & Rydberg, U. (1980). Effects of ethanol on the developing rat. II. Coordination as measured by the tilting-plane test. *Medical Biology, 58*(3), 164–168.

Hunt, P. S., & Barnet, R. C. (2016). Adolescent and adult rats differ in the amnesic effects of acute ethanol in two hippocampus-dependent tasks: Trace and contextual fear conditioning. *Behavioural Brain Research, 298*(Pt A), 78–87. https://doi.org/10.1016/j.bbr.2015.06.046.

Khanna, J. M., Kalant, H., Sharma, H., & Chau, A. (1991). Initial sensitivity, acute tolerance and alcohol consumption in Fischer 344 and long Evans rats. *Psychopharmacology, 105*(2), 175–180. https://doi.org/10.1007/BF02244305.

Khisti, R. T., VanDoren, M. J., Matthews, D. B., & Morrow, A. L. (2004). Ethanol-induced elevation of 3 alpha-hydroxy-5 alpha-pregnan-20-one does not modulate motor incoordination in rats. *Alcoholism, Clinical and Experimental Research, 28*(8), 1249–1256.

Kokare, D. M., Kyzar, E. J., Zhang, H., Sakharkar, A. J., & Pandey, S. C. (2017). Adolescent alcohol exposure-induced changes in alpha-melanocyte stimulating hormone and neuropeptide Y pathways via histone acetylation in the brain during adulthood. *The International Journal of Neuropsychopharmacology, 20*(9), 758–768. https://doi.org/10.1093/ijnp/pyx041.

Kulkarni, S. K., Singh, K., & Bishnoi, M. (2007). Elevated zero maze: A paradigm to evaluate antianxiety effects of drugs. *Methods and Findings in Experimental and Clinical Pharmacology, 29*(5), 343–348. https://doi.org/10.1358/mf.2007.29.5.1117557.

Kumar, A. (2015). NMDA receptor function during senescence: Implication on cognitive performance. *Frontiers in Neuroscience, 9*, 473. https://doi.org/10.3389/fnins.2015.00473.

Lai, C. C., Kuo, T. I., & Lin, H. H. (2007). The role of protein kinase a in acute ethanol-induced neurobehavioral actions in rats. *Anesthesia and Analgesia, 105*(1), 89–96. https://doi.org/10.1213/01.ane.0000263030.13249.36.

Le, A. D., Kalant, H., & Khanna, J. M. (1986). Influence of ambient temperature on the development and maintenance of tolerance to ethanol-induced hypothermia. *Pharmacology, Biochemistry, and Behavior, 25*(3), 667–672. https://doi.org/10.1016/0091-3057(86)90157-7.

Lee, K. M., Coehlo, M. A., Solton, N. R., & Szumlinski, K. K. (2017). Negative affect and excessive alcohol intake incubate during protracted withdrawal from binge-drinking in adolescent, but not adult, mice. *Frontiers in Psychology, 8*, 1128. https://doi.org/10.3389/fpsyg.2017.01128.

Mackenzie-Taylor, D., & Rech, R. H. (1991). Cellular and learned tolerances for ethanol hypothermia. *Pharmacology, Biochemistry, and Behavior, 38*(1), 29–36. https://doi.org/10.1016/0091-3057(91)90585-p.

Magnusson, K. R. (2012). Aging of the NMDA receptor: From a mouse's point of view. *Future Neurology, 7*(5), 627–637. https://doi.org/10.2217/fnl.12.54.

Majchrowicz, E. (1975). Induction of physical dependence upon ethanol and the associated behavioral changes in rats. *Psychopharmacologia, 43*(3), 245–254.

Matthews, D. B., Ilgen, M., White, A. M., & Best, P. J. (1999). Acute ethanol administration impairs spatial performance while facilitating nonspatial performance in rats. *Neurobiology of Learning and Memory, 72*(3), 169–179. https://doi.org/10.1006/nlme.1998.3900.

Matthews, D. B., & Mittleman, G. (2017). Age-dependent effects of chronic intermittent ethanol treatment: Gross motor behavior and body weight in aged, adult and adolescent rats. *Neuroscience Letters, 657*, 146–150. https://doi.org/10.1016/j.neulet.2017.08.012.

Matthews, D. B., Morrow, A. L., Tokunaga, S., & McDaniel, J. R. (2002). Acute ethanol administration and acute allopregnanolone administration impair spatial memory in the Morris water task. *Alcoholism, Clinical and Experimental Research, 26*(11), 1747–1751. https://doi.org/10.1097/01.ALC.0000037219.79257.17.

Matthews, D. B., Novier, A., Diaz-Granados, J. L., Van Skike, C. E., Ornelas, L., & Mittleman, G. (2017). Impact of adolescent alcohol use across the lifespan: Long-lasting tolerance to high-dose alcohol coupled with potentiated spatial memory impairments to moderate-dose alcohol. *Alcohol, 61*, 33–42. https://doi.org/10.1016/j.alcohol.2017.01.012.

Matthews, D. B., Scaletty, S., Schreiber, A., & Trapp, S. (2020). Acute ethanol administration produces larger spatial and nonspatial memory impairments in 29-33 month old rats compared to adult and 18-24 month old rats. *Pharmacology, Biochemistry, and Behavior, 199*, 173074. https://doi.org/10.1016/j.pbb.2020.173074.

Matthews, D. B., Schneider, A., Kastner, A., Scaletty, S., & Szenay, R. (2019). I can't drink what I used to: The interaction between ethanol and the aging brain. *International Review of Neurobiology, 148*, 79–99. https://doi.org/10.1016/bs.irn.2019.09.003.

Matthews, D. B., Simson, P. E., & Best, P. J. (1995). Acute ethanol impairs spatial memory but not stimulus/response memory in the rat. *Alcoholism, Clinical and Experimental Research, 19*(4), 902–909.

Matthews, D. B., Simson, P. E., & Best, P. J. (1996). Ethanol alters the spatial processing of hippocampal place cells: A mechanism for impaired navigation when intoxicated. *Alcoholism, Clinical and Experimental Research, 20*(2), 404–407.

Matthews, D. B., Tinsley, K. L., Diaz-Granados, J. L., Tokunaga, S., & Silvers, J. M. (2008). Chronic intermittent exposure to ethanol during adolescence produces tolerance to the hypnotic effects of ethanol in male rats: A dose-dependent analysis. *Alcohol, 42*(8), 617–621. https://doi.org/10.1016/j.alcohol.2008.09.001.

Matthews, D. B., Watson, M. R., James, K., Kastner, A., Schneider, A., & Mittleman, G. (2019). The impact of low to moderate chronic intermittent ethanol exposure on behavioral endpoints in aged, adult, and adolescent rats. *Alcohol, 78*, 33–42. https://doi.org/10.1016/j.alcohol.2018.11.005.

McGinnis, M. M., Parrish, B. C., & McCool, B. A. (2020). Withdrawal from chronic ethanol exposure increases postsynaptic glutamate function of insular cortex projections to the rat basolateral amygdala. *Neuropharmacology, 172*, 108129. https://doi.org/10.1016/j.neuropharm.2020.108129.

McKetta, S., & Keyes, K. M. (2019). Heavy and binge alcohol drinking and parenting status in the United States from 2006 to 2018: An analysis of nationally representative cross-sectional surveys. *PLoS Medicine, 16*(11), e1002954. https://doi.org/10.1371/journal.pmed.1002954.

Melia, K. R., Ryabinin, A. E., Corodimas, K. P., Wilson, M. C., & Ledoux, J. E. (1996). Hippocampal-dependent learning and experience-dependent activation of the hippocampus are preferentially disrupted by ethanol. *Neuroscience, 74*(2), 313–322.

Morales, M., McGinnis, M. M., Robinson, S. L., Chappell, A. M., & McCool, B. A. (2018). Chronic intermittent ethanol exposure modulation of glutamatergic neurotransmission in rat lateral/basolateral amygdala is duration-, input-, and sex-dependent. *Neuroscience, 371*, 277–287. https://doi.org/10.1016/j.neuroscience.2017.12.005.

Novier, A., Diaz-Granados, J. L., & Matthews, D. B. (2015). Alcohol use across the lifespan: An analysis of adolescent and aged rodents and humans. *Pharmacology, Biochemistry, and Behavior, 133*, 65–82. https://doi.org/10.1016/j.pbb.2015.03.015.

Novier, A., Ornelas, L. C., Diaz-Granados, J. L., & Matthews, D. B. (2016). Differences in behavioral responding in adult and aged rats following chronic ethanol exposure. *Alcoholism, Clinical and Experimental Research, 40*(7), 1462–1472. https://doi.org/10.1111/acer.13098.

Novier, A., Van Skike, C. E., Chin, V. S., Diaz-Granados, J. L., & Matthews, D. B. (2012). Low and moderate doses of acute ethanol do not impair spatial cognition but facilitate accelerating rotarod performance in adolescent and adult rats. *Neuroscience Letters, 512*(1), 38–42. https://doi.org/10.1016/j.neulet.2012.01.059.

Novier, A., Van Skike, C. E., Diaz-Granados, J. L., Mittleman, G., & Matthews, D. B. (2013). Acute alcohol produces ataxia and cognitive impairments in aged animals: A comparison between young adult and aged rats. *Alcoholism, Clinical and Experimental Research, 37*(8), 1317–1324. https://doi.org/10.1111/acer.12110.

Nunes, P. T., Kipp, B. T., Reitz, N. L., & Savage, L. M. (2019). Aging with alcohol-related brain damage: Critical brain circuits associated with cognitive dysfunction. *International Review of Neurobiology, 148*, 101–168. https://doi.org/10.1016/bs.irn.2019.09.002.

Ornelas, L. C., Novier, A., Van Skike, C. E., Diaz-Granados, J. L., & Matthews, D. B. (2015). The effects of acute alcohol on motor impairments in adolescent, adult, and aged rats. *Alcohol, 49*(2), 121–126. https://doi.org/10.1016/j.alcohol.2014.12.002.

Pandey, S. C., Kyzar, E. J., & Zhang, H. (2017). Epigenetic basis of the dark side of alcohol addiction. *Neuropharmacology, 122*, 74–84. https://doi.org/10.1016/j.neuropharm.2017.02.002.

Pandey, S. C., Sakharkar, A. J., Tang, L., & Zhang, H. (2015). Potential role of adolescent alcohol exposure-induced amygdaloid histone modifications in anxiety and alcohol intake during adulthood. *Neurobiology of Disease, 82*, 607–619. https://doi.org/10.1016/j.nbd.2015.03.019.

Pautassi, R. M., Godoy, J. C., & Molina, J. C. (2015). Adolescent rats are resistant to the development of ethanol-induced chronic tolerance and ethanol-induced conditioned aversion. *Pharmacol Biochem and Behav, 138*, 58–69. https://doi.org/10.1016/j.pbb.2015.09.012.

Perkins, A. E., Vore, A. S., Lovelock, D., Varlinskaya, E., & Deak, T. (2018). Late aging alters behavioral sensitivity to ethanol in a sex-specific manner in Fischer 344 rats. *Pharmacology, Biochemistry, and Behavior, 175*, 1–9. https://doi.org/10.1016/j.pbb.2018.07.012.

Prunell, M., Escorihuela, R. M., Fernandez-Teruel, A., Nunez, J. F., & Tobena, A. (1994). Differential interactions between ethanol and Ro 15-4513 on two anxiety tests in rats. *Pharmacology, Biochemistry, and Behavior, 47*(1), 147–151. https://doi.org/10.1016/0091-3057(94)90124-4.

Ramirez, R. L., & Spear, L. P. (2010). Ontogeny of ethanol-induced motor impairment following acute ethanol: Assessment via the negative geotaxis reflex in adolescent and adult rats. *Pharmacol Biochem and Behav, 95*(2), 242–248. https://doi.org/10.1016/j.pbb.2010.01.013.

Ristuccia, R. C., Hernandez, M., Wilmouth, C. E., & Spear, L. P. (2007). Differential expression of ethanol-induced hypothermia in adolescent and adult rats induced by pretest familiarization to the handling/injection procedure. *Alcoholism, Clinical and Experimental Research, 31*(4), 575–581. https://doi.org/10.1111/j.1530-0277.2007.00341.x.

Sharko, A. C., Kaigler, K. F., Fadel, J. R., & Wilson, M. A. (2016). Ethanol-induced anxiolysis and neuronal activation in the amygdala and bed nucleus of the stria terminalis. *Alcohol, 50*, 19–25. https://doi.org/10.1016/j.alcohol.2015.11.001.

Shepherd, J. K., Grewal, S. S., Fletcher, A., Bill, D. J., & Dourish, C. T. (1994). Behavioural and pharmacological characterisation of the elevated "zero-maze" as an animal model of anxiety. *Psychopharmacology, 116*(1), 56–64. https://doi.org/10.1007/BF02244871.

Silveri, M. M., & Spear, L. P. (2000). Ontogeny of ethanol elimination and ethanol-induced hypothermia. *Alcohol, 20*(1), 45–53.

Silvers, J. M., Tokunaga, S., Mittleman, G., & Matthews, D. B. (2003). Chronic intermittent injections of high-dose ethanol during adolescence produce metabolic, hypnotic, and cognitive tolerance in rats. *Alcoholism, Clinical and Experimental Research, 27*(10), 1606–1612. https://doi.org/10.1097/01.ALC.0000090141.66526.22.

Silvers, J. M., Tokunaga, S., Mittleman, G., O'Buckley, T., Morrow, A. L., & Matthews, D. B. (2006). Chronic intermittent ethanol exposure during adolescence reduces the effect of ethanol challenge on hippocampal allopregnanolone levels and Morris water maze task performance. *Alcohol, 39*(3), 151–158. https://doi.org/10.1016/j.alcohol.2006.09.001.

Spear, L. P., & Swartzwelder, H. S. (2014). Adolescent alcohol exposure and persistence of adolescent-typical phenotypes into adulthood: A mini-review. *Neuroscience and Biobehavioral Reviews, 45*, 1–8. https://doi.org/10.1016/j.neubiorev.2014.04.012.

Swartzwelder, H. S., Richardson, R. C., Markwiese-Foerch, B., Wilson, W. A., & Little, P. J. (1998). Developmental differences in the acquisition of tolerance to ethanol. *Alcohol, 15*(4), 311–314.

Tokunaga, S., McDaniel, J. R., Morrow, A. L., & Matthews, D. B. (2003). Effect of acute ethanol administration and acute allopregnanolone administration on spontaneous hippocampal pyramidal cell neural activity. *Brain Research, 967*(1–2), 273–280.

Tokunaga, S., Silvers, J. M., & Matthews, D. B. (2006). Chronic intermittent ethanol exposure during adolescence blocks ethanol-induced inhibition of spontaneously active hippocampal pyramidal neurons. *Alcoholism, Clinical and Experimental Research, 30*(1), 1–6. https://doi.org/10.1111/j.1530-0277.2006.00020.x.

Valdez, G. R., Roberts, A. J., Chan, K., Davis, H., Brennan, M., Zorrilla, E. P., et al. (2002). Increased ethanol self-administration and anxiety-like behavior during acute ethanol withdrawal and protracted abstinence: Regulation by corticotropin-releasing factor. *Alcoholism, Clinical and Experimental Research, 26*(10), 1494–1501. https://doi.org/10.1097/01.ALC.0000033120.51856.F0.

Van Skike, C. E., Botta, P., Chin, V. S., Tokunaga, S., McDaniel, J. M., Venard, J., et al. (2010). Behavioral effects of ethanol in cerebellum are age dependent: Potential system and molecular mechanisms. *Alcoholism, Clinical and Experimental Research, 34*(12), 2070–2080. https://doi.org/10.1111/j.1530-0277.2010.01303.x.

Van Skike, C. E., Diaz-Granados, J. L., & Matthews, D. B. (2015). Chronic intermittent ethanol exposure produces persistent anxiety in adolescent and adult rats. *Alcoholism, Clinical and Experimental Research, 39*(2), 262–271. https://doi.org/10.1111/acer.12617.

Van Skike, C. E., Goodlett, C., & Matthews, D. B. (2019). Acute alcohol and cognition: Remembering what it causes us to forget. *Alcohol, 79*, 105–125. https://doi.org/10.1016/j.alcohol.2019.03.006.

Van Skike, C. E., Novier, A., Diaz-Granados, J. L., & Matthews, D. B. (2012). The effect of chronic intermittent ethanol exposure on spatial memory in adolescent rats: The dissociation of metabolic and cognitive tolerances. *Brain Research, 1453*, 34–39. https://doi.org/10.1016/j.brainres.2012.03.006.

Varlinskaya, E. I., & Spear, L. P. (2002). Acute effects of ethanol on social behavior of adolescent and adult rats: Role of familiarity of the test situation. *Alcoholism, Clinical and Experimental Research, 26*(10), 1502–1511. https://doi.org/10.1097/01.ALC.0000034033.95701.E3.

Varlinskaya, E. I., Truxell, E. M., & Spear, L. P. (2015). Ethanol intake under social circumstances or alone in Sprague-dawley rats: Impact of age, sex, social activity, and social anxiety-like behavior. *Alcoholism, Clinical and Experimental Research, 39*(1), 117–125. https://doi.org/10.1111/acer.12604.

Vetreno, R. P., Broadwater, M., Liu, W., Spear, L. P., & Crews, F. T. (2014). Adolescent, but not adult, binge ethanol exposure leads to persistent global reductions of choline acetyltransferase expressing neurons in brain. *PLoS One, 9*(11), e113421. https://doi.org/10.1371/journal.pone.0113421.

Vetreno, R. P., Yaxley, R., Paniagua, B., & Crews, F. T. (2016). Diffusion tensor imaging reveals adolescent binge ethanol-induced brain structural integrity alterations in adult rats that correlate with behavioral dysfunction. *Addiction Biology, 21*(4), 939–953. https://doi.org/10.1111/adb.12232.

Vinke, E. J., Ikram, M. A., & Vernooij, M. W. (2019). Brain aging: More of the same!? *Aging (Albany NY), 11*(3), 849–850. https://doi.org/10.18632/aging.101799.

Vogel, R. A., Frye, G. D., Koepke, K. M., Mailman, R. B., Mueller, R. A., & Breese, G. R. (1981). Differential effects of TRH, amphetamine, naloxone, and fenmetozole on ethanol actions: Attenuation of the effects of punishment and impairment of aerial righting reflex. *Alcoholism, Clinical and Experimental Research, 5*(3), 386–392. https://doi.org/10.1111/j.1530-0277.1981.tb04921.x.

Walf, A. A., & Frye, C. A. (2007). The use of the elevated plus maze as an assay of anxiety-related behavior in rodents. *Nature Protocols, 2*(2), 322–328. https://doi.org/10.1038/nprot.2007.44.

Watson, M. R., James, K., Mittleman, G., & Matthews, D. B. (2020). Impact of acute ethanol exposure on body temperatures in aged, adult and adolescent male rats. *Alcohol, 82*, 81–89. https://doi.org/10.1016/j.alcohol.2019.08.001.

Weitemier, A. Z., & Ryabinin, A. E. (2003). Alcohol-induced memory impairment in trace fear conditioning: A hippocampus-specific effect. *Hippocampus, 13*(3), 305–315. https://doi.org/10.1002/hipo.10063.

White, A. M., Bae, J. G., Truesdale, M. C., Ahmad, S., Wilson, W. A., & Swartzwelder, H. S. (2002). Chronic-intermittent ethanol exposure during adolescence prevents normal developmental changes in sensitivity to ethanol-induced motor impairments. *Alcoholism, Clinical and Experimental Research, 26*(7), 960–968. https://doi.org/10.1097/01.ALC.0000021334.47130.F9.

White, A. M., Ghia, A. J., Levin, E. D., & Swartzwelder, H. S. (2000). Binge pattern ethanol exposure in adolescent and adult rats: Differential impact on subsequent responsiveness to ethanol. *Alcoholism, Clinical and Experimental Research, 24*(8), 1251–1256.

Wilson, M. A., Burghardt, P. R., Ford, K. A., Wilkinson, M. B., & Primeaux, S. D. (2004). Anxiolytic effects of diazepam and ethanol in two behavioral models: Comparison of males and females. *Pharmacology, Biochemistry, and Behavior, 78*(3), 445–458. https://doi.org/10.1016/j.pbb.2004.04.017.

Witt, E. D. (2010). Research on alcohol and adolescent brain development: Opportunities and future directions. *Alcohol, 44*(1), 119–124. https://doi.org/10.1016/j.alcohol.2009.08.011.

York, J. L. (1982). Body water content, ethanol pharmacokinetics, and the responsiveness to ethanol in young and old rats. *Developmental Pharmacology and Therapeutics, 4*(1–2), 106–116.

York, J. L. (1983). Increased responsiveness to ethanol with advancing age in rats. *Pharmacology, Biochemistry, and Behavior, 19*(4), 687–691. https://doi.org/10.1016/0091-3057(83)90346-5.

Yu, Z. Y., Wang, W., Fritschy, J. M., Witte, O. W., & Redecker, C. (2006). Changes in neocortical and hippocampal GABAA receptor subunit distribution during brain maturation and aging. *Brain Research, 1099*(1), 73–81. https://doi.org/10.1016/j.brainres.2006.04.118.

CHAPTER NINE

Early life stress and susceptibility to addiction in adolescence

K.E. Tschetter[a,b,c], L.B. Callahan[a,b,c], S.A. Flynn[b], S. Rahman[d], T.P. Beresford[e,f], and P.J. Ronan[a,b,c,e,*]

[a]Research Service, Sioux Falls VA Healthcare System, Sioux Falls, SD, United States
[b]Department of Psychiatry and Basic Biomedical Sciences, University of South Dakota Sanford School of Medicine, Sioux Falls, SD, United States
[c]Division of Basic Biomedical Sciences, University of South Dakota Sanford School of Medicine, Sioux Falls, SD, United States
[d]Department of Pharmaceutical Sciences, South Dakota State University, Brookings, SD, United States
[e]Laboratory for Clinical and Translational Research in Psychiatry, Rocky Mountain Regional, VA Medical Center, Aurora, CO, United States
[f]Department of Psychiatry, University of Colorado School of Medicine, Aurora, CO, United States
*Corresponding author: e-mail address: patrick.ronan@usd.edu

Contents

1. Introduction 278
2. Adolescence 278
3. Early life stress 279
4. Investigation of molecular mechanisms underlying the ELS-induced addiction susceptibility phenotype in adolescence 280
   4.1 Materials and methods 282
   4.2 Results 286
   4.3 Discussion 291
5. Conclusions 295
Acknowledgments 295
Conflict of interest 295
References 295

Abstract

Early life stress (ELS) is a risk factor for developing a host of psychiatric disorders. Adolescence is a particularly vulnerable period for the onset of these disorders and substance use disorders (SUDs). Here we discuss ELS and its effects in adolescence, especially SUDs, and their correlates with molecular changes to signaling systems in reward and stress neurocircuits. Using a maternal separation (MS) model of neonatal ELS, we studied a range of behaviors that comprise a "drug-seeking" phenotype. We then investigated potential mechanisms underlying the development of this phenotype. Corticotropin releasing factor (CRF) and serotonin (5-HT) are widely believed to be

International Review of Neurobiology, Volume 161
ISSN 0074-7742
https://doi.org/10.1016/bs.irn.2021.08.007

involved in "stress-induced" disorders, including addiction. Here, we show that ELS leads to the development of a drug-seeking phenotype indicative of increased susceptibility to addiction and concomitant sex-dependent upregulation of CRF and 5-HT system components throughout extended brain reward/stress neurocircuits.

Abbreviations

5-HT serotonin
AFR animal facility raised
BNST bed nucleus of the stria teminalis
CeA central nucleus of the amygdala
CIL cocaine-induced locomotion
CRF corticotropin releasing factor
ELS early life stress
EPM elevated plus maze
LC locus coeruleus
MS maternal separation
MS15 15 min daily separation (mild separation)
MS180 180 min daily separation
NH non-handled
PGM playground maze
PND post natal day
PVN paraventricular nucleus of the hypothalamus

1. Introduction

Early life stress (ELS) is a risk factor for the development of a wide range of psychiatric disorders later in life (Enoch, 2011; Graham, Heim, Goodman, Miller, & Nemeroff, 1999; Gutman & Nemeroff, 2003). Adolescence is a particularly vulnerable period for the onset of many of these disorders, especially substance abuse (Merikangas et al., 2010) and ELS increases this vulnerability (Enoch, 2011). The underlying biological basis for these effects is largely unknown. Here we review the maternal separation model of ELS that can be used to interrogate these mechanisms in adolescence, and present novel research findings describing a battery of behavioral tests that can reveal a drug-seeking phenotype in adolescent rodents.

2. Adolescence

Adolescence is a critical period for the development of addictive behaviors. A hallmark of adolescence is a generalized increase in risk-taking and impulsive behaviors such as drug use. Experimentation with drugs of abuse during adolescence can quickly lead to dependence. Substance use disorders (SUDs) affect approximately 10–12% of the adolescent population (Merikangas et al., 2010). Alcohol use disorders (AUDs) and SUDs almost always begin during adolescence. Addictive behaviors that start in adolescence often continue into adulthood (Kuhn, 2015; Spear, 2000) and if one does not develop an addiction as an adolescent, they are unlikely to develop one as an adult (Clark, Kirisci, & Tarter, 1998; Kelley, Schochet, & Landry, 2004). Estimates are that around 37% of adolescents with substance abuse problems have a history of some form of ELS (Roos, Horn, Berkman, Pears, & Fisher, 2018).

3. Early life stress

Stress facilitates the acquisition and maintenance of drug-seeking behaviors and is a primary factor leading to relapse (Piazza & Le Moal, 1998; Ronan & Summers, 2011; Sarnyai, Shaham, & Heinrichs, 2001). A prior history of stress also plays a significant role in susceptibility to develop AUDs and SUDs. Evidence indicates that ELS has long-lasting effects that increase drug use in the absence of acute stressors (Forster, Anderson, Scholl, Lukkes, & Watt, 2018; Moffett et al., 2007). In addition, ELS and/or childhood adversities are highly predictive of increased vulnerability to develop an addiction in both adolescents and adults (Enoch, 2011; McLaughlin et al., 2012). Early adverse events such as abuse, neglect, family violence, and parental death, alone or in combination most strongly predict later substance abuse (Dong et al., 2004; Edwards, Holden, Felitti, & Anda, 2003). Children subjected to foster care or poor mothering are more likely to engage in risky behaviors as adolescents, including experimentation with drugs of abuse (Carpenter, Clyman, Davidson, & Steiner, 2001). Along with substance abuse, ELS increases a risk for a range of other psychiatric disorders in adolescents, such as anxiety, that can also contribute to the development of comorbid addictive disorders (Enoch, 2011; Graham et al., 1999; Green et al., 2010; Gutman & Nemeroff, 2003; Kelley et al., 2004; McLaughlin et al., 2012).

Changes in the sensitivity and responsiveness of brain stress systems induced by exposure to ELS may underlie the increased susceptibility to drug addiction seen later in life. ELS can alter central corticotropin releasing factor CRF systems in adult rodents and thus potentially modify behavioral and neuroendocrine responses throughout life (Ladd, Owens, & Nemeroff, 1996). Maternal separation (MS), a group of widely used models of ELS that cause changes to the CRF system in adult rodents (Aisa, Tordera, Lasheras, Del Rio, & Ramirez, 2007; Callaghan & Richardson, 2011; Litvin et al., 2010; Miczek, Yap, & Covington 3rd, 2008; Plotsky & Meaney, 1993; Plotsky et al., 2005). Variations of MS cause both increased (Desbonnet, Garrett, Daly, McDermott, & Dinan, 2008; Huot, Gonzalez, Ladd, Thrivikraman, & Plotsky, 2004; Ladd, Thrivikraman, Huot, & Plotsky, 2005; Plotsky et al., 2005) and decreased (Greisen, Bolwig, Husum, Nedergaard, & Wortwein, 2005; Vazquez, Eskandari, Phelka, & Lopez, 2003; Workel et al., 2001) expression of CRF system components in adult rats. Models of MS vary, sometimes significantly, between research groups, but the majority of studies have demonstrated that MS causes increased measures of CRF activity in adults. In the most widely used variation of MS, Plotsky's group separated pups from dams for 180 min per day for 12 days and found increased levels of CRF mRNA in the paraventricular nucleus of the hypothalamus (PVN), bed nucleus of the stria teminalis (BNST), central nucleus of the amygdala (CeA), and locus coeruleus (LC) of adult rats (Plotsky & Meaney, 1993; Plotsky et al., 2005). Another group used a model in which they separated pups and dams one time for 24 h and found decreased CRF receptor (CRFR) expression mRNA in the amygdala and ventromedial nucleus of the hypothalamus and an increase of $CRFR_1$ and $CRFR_2$ mRNA in the cortex (Vazquez et al., 2003). In contrast, rats exposed to prenatal stress show increased expression of $CRFR_1$ in the amygdala (Ward, Johnson, Salm, & Birkle, 2000).

4. Investigation of molecular mechanisms underlying the ELS-induced addiction susceptibility phenotype in adolescence

We used a version of the MS animal model to explore the role of ELS on the development of a drug-seeking phenotype in adolescence and explore potential neurobiological mechanisms. The corticotropin releasing factor (CRF) and serotonergic (5-HT) systems have been widely implicated as etiological factors in a host of "stress-induced" psychiatric disorders including addiction (Kirby, Zeeb, & Winstanley, 2011; Ronan and Summers, 2011; Valentino, Lucki, & Van Bockstaele, 2010). Yet, there have been surprisingly few studies investigating the possible roles of CRF and

5-HT systems in ELS-induced increases in vulnerability to addiction during adolescence. Serotonin and CRF signaling are tightly linked to anxiety, depression, and drug addiction (Kirby et al., 2011; Ronan & Summers, 2011; Tiwari, Souza, & Muller, 2009).

The CRF system consists of the neuropeptide CRF and related urocortins, two known receptors, $CRFR_1$ and $CRFR_2$, as well as CRF binding protein (CRF-BP). Key brain regions involved in CRF-mediated stress responses are the paraventricular nucleus of the hypothalamus (PVN), CeA and BNST (c.f., Ronan & Summers, 2011). These regions send direct CRF projections to dopaminergic cells of the ventral tegmental area (VTA; Rodaros, Caruana, Amir, & Stewart, 2007) where CRF has both electrophysiological and behavioral effects (Lodge & Grace, 2005). CRF also plays a key role in nucleus accumbens (NAc) reward-related behavior (Marcinkiewcz et al., 2009; Pecina, Schulkin, & Berridge, 2006). Though commonly thought of as involved mainly in anxiety and fear responses; the PVN, CeA and BNST should also be considered components of an extended brain reward/reinforcement system that play a key role in addiction (Nestler, 2005; Ronan & Summers, 2011).

Serotonin signaling also plays a significant role throughout this extended reward system (Kirby et al., 2011). The $5\text{-}HT_{2A}$ receptor has recently been directly implicated in augmenting CRF signaling and CRF-induced anxiety-like behavior (Magalhaes et al., 2010) as well as impulse control in addiction (Anastasio et al., 2011). Also, ELS leads to enhanced $5\text{-}HT_{2A}$ receptor function (Benekareddy, Goodfellow, Lambe, & Vaidya, 2010). Therefore, it is plausible that the $5\text{-}HT_{2A}$ receptor and CRF signaling are specific mechanisms translating ELS to increased vulnerability for behavioral disorders such as addiction in adolescence.

We used the MS procedure described above to model ELS (Aisa et al., 2007; Callaghan & Richardson, 2011; Litvin et al., 2010; Miczek et al., 2008; Plotsky & Meaney, 1993; Plotsky et al., 2005). We determined the effects of MS on a range of behaviors indicative of a drug-seeking phenotype and on the expression of CRF system components (CRF, $CRFR_1$, $CRFR_2$ and CRF-BP) and $5\text{-}HT_{2A}$ in extended reward-related brain regions. Behavioral traits that predict susceptibility to addiction include increased anxiety, novelty-seeking, risk-taking, impulsivity and sensitivity to the locomotor activating properties of a psychostimulant (Bardo, Donohew, & Harrington, 1996; Cleck & Blendy, 2008; Kelley et al., 2004; Kirby et al., 2011; Stanis, Burns, Sherrill, & Gulley, 2008).

4.1 Materials and methods

4.1.1 Animals

Eight timed pregnant Sprague Dawley rats/per behavioral test (CD-IGS; Charles River Laboratories Wilmington, MA) arrived on gestation day 15. This was the first litter for all dams. They were housed in transparent polypropylene cages (20 × 25 cm) with 2 cm of aspen bedding material with 12:12 h light: dark cycle (lights on at 0700) and *ad libitum* access to food and water at an AAALAC-approved facility at the Sioux Falls VA Healthcare System. They gave birth within one day of each other. The day of birth was considered postnatal day 0 (PND0). Dams and pups were left undisturbed for two days following birth. All animal procedures were conducted according to the Guide for the Care and Use of Laboratory Animals (National Research Council, 2011) and approved by the University of South Dakota Institutional Animal Care and Use Committee (IACUC).

4.1.2 Maternal separation

Litters were randomly assigned to one of four conditions: non-handled (NH), animal facility raised (AFR), 15-min daily separation (MS15; mild separation) or 180 min daily separation (MS180; severe separation). Non-handled animals were only exposed to one very brief handling during a cage change on post-natal day 11 (PND11) which was required by our local IACUC. During this cage change for the NH group a small amount of old bedding was mixed with new bedding, per standard husbandry practice. The AFR animals were handled twice weekly coinciding with cage changes for a total of four times over the course of the separation period. No old bedding material was mixed with new during cage changes for AFR. Pups in the mild and severe groups were separated from the dam for 15 min or 180 min daily, respectively, from PND2-PND14. In addition to daily handling and separation MS15 and MS180 groups received one cage change on PND11. During this cage change some old bedding was mixed with new bedding. All separation and handling took place from 0800 to 0900. Dams were removed from the maternity cage before removing pups and placed in individual holding cages with aspen bedding. These holding cages were kept specific for each dam throughout the course of the experiment. Each litter was then removed and placed as a group into a plastic container (15 × 15 cm) with 3 cm of aspen bedding. Again, each container was only used by one litter. They were placed in an incubator (32 ± 0.5 °C for PND2-PND5 and 30 ± 0.5 °C for PND6-PND14). After the separation period, pups were returned to the maternity cage, rolled in the home cage bedding, and placed back into the nest from which they were removed then the dam was returned. Rat pups were weaned on PND21–22 and housed in groups of 4–6 in same sex/same rearing group cages until behavioral testing.

4.1.3 Behavioral testing

All behavioral testing took place between PND40 and PND43 which correlates with peri-adolescence (approximately ages 14–16 in humans; Andersen, 2003). Separate cohorts were used for each behavioral test. Testing was performed during the first half of the light cycle in a room directly adjacent to the animal housing quarters. Behaviors were video recorded for subsequent analysis using EthoVision 7.0 XT software (Noldus Information Technology, Netherlands). A phenotype comprised of many behaviors is likely more predictive of a drug-seeking phenotype than any one behavior. Individuals with high levels of anxiety, impulsivity, risk-taking and novelty-seeking are at increased risk for drug addiction and many other behavioral disorders (Bardo et al., 1996; Kelley et al., 2004). Also, locomotor responses to cocaine in drug-naïve rodents predicts impulsivity that is related to addictive behaviors (Stanis et al., 2008). Initial cocaine-induced locomotion in an open field is also correlated with increased synaptic strength in the VTA that has been proposed as a neuroadaptation that increases susceptibility to cocaine abuse (Wanat & Bonci, 2008). Locomotor activity in drug naïve animals perhaps reflects the degree of motivational significance of cocaine (Borgland, Malenka, & Bonci, 2004).

4.1.4 Elevated plus maze

The elevated plus maze (EPM) is a widely used procedure to assess anxiety-like behavior (Carobrez & Bertoglio, 2005; Harro, Tonissaar, & Eller, 2001; Ueno et al., 2002). The EPM consists of two open arms and two closed arms (each 50 × 10 cm) in a plus configuration elevated 1 m above the floor (Stoelting Model 60240, Wood Dale, IL). The closed arms have walls 40 cm high. Rats were placed in the center of the maze and allowed to explore freely for 15 min. All sessions were video recorded and analyzed using EthoVision 7.0 XT software. Parameters measured included time spent in open arms and total distance traveled. Each group had an n of 12–14 males and 12–14 females.

4.1.5 Playground maze

The playground maze (PGM) is a behavioral protocol that is used to determine novelty-seeking and risk-taking behavior (Klebaur & Bardo, 1999; Klebaur, Bevins, Segar, & Bardo, 2001; Mellanby, Johansen-Berg, Leyland, & Milward, 1999; Stansfield, Philpot, & Kirstein, 2004). The playground maze is a variation of the novel object recognition test. It is a circular forced open field with a diameter of 100 cm and raised 55 cm off the floor with no raised edges made of ¾" plywood painted flat black and sealed for sanitizing purposes. The maze was cleaned between animals with PDI Sani-Cloth Plus disinfectant wipes, a quaternary/low alcohol formula (14.85%; Thermo Fisher Scientific, Houston, TX) followed by a wipe with water to remove the scent of alcohol. Eight different plastic toys were spaced evenly around the perimeter of the maze and attached using Velcro®. On the first 3 trials on 3 consecutive days (for a total of 9 habituation trials), rats were placed in the middle of the maze facing away from the experimenter. The same eight objects were randomly ordered each day to prevent positional bias. On the test day (Day 4), rats were placed on the maze with the familiar eight toys for 3 min. The rat was then removed and placed in the home cage (while still in the behavioral testing room) for 1 min. A novel object replaced one of the eight familiar objects. The novel object replaced the same familiar object for each rat. The rat was placed back on the maze facing away from the novel object. The rat was then allowed to explore for 3 min. All sessions were video recorded and analyzed using EthoVision software. We measured percent time spent with the novel object/% time spent with all other objects and number of approaches to the novel object. Groups had an n of 10–13 each for males and females.

4.1.6 Cocaine-induced locomotion

We followed standard methods for cocaine-induced locomotion (CIL; Badanich, Maldonado, & Kirstein, 2008; Marin & Planeta, 2004). All sessions were video recorded and analyzed using EthoVision software. Briefly, adolescent rats (PND 40—PND43; $n = 8$–10 males/and 12–15 females/group) were placed in a rectangular Plexiglas open field (45 × 24 × 19 cm) in a sound attenuating chamber (Med. Associates Inc., St. Albans, VT). Rats were habituated to the apparatus for 10 min. Baseline locomotor activity was then determined over the next 10 min. Rats were removed, given an injection of cocaine (10 mg/kg, ip) and total distance traveled (cm) was analyzed for 15 min. Results are reported as a percentage of baseline locomotion.

4.1.7 Brain microdissection and qRT-PCR

We adapted a brain microdissection technique (Palkovits & Brownstein, 1988) for use with RT-PCR Array. Immediately following behavioral testing rats were deeply anesthetized with isoflurane and decapitated. Brains were removed and snap frozen in isopentane cooled to −78 °C on dry ice. Frozen brains were sliced at 300 μm at −13 °C using a cryostat microtome (Leica CM1900; Wetzlar, Germany). Sections were very briefly thaw mounted on slides by warming the underside of the slide with a gloved finger. Using a Peltier cold plate set to −30 °C (Physiotemp BFS-30MP; NJ, USA) and dissecting microscope (Leica EZ4; Heerbrugg, Switzerland) areas of interest were punched with a sterile brain punch ranging from 100 to 160 μm diameter (Fine Science Tools, Inc.; Foster City, CA). Punches were placed in extraction buffer (75 μL) and RNA was isolated using a Qiagen RNeasy Lipid Tissue Mini Kit, (Qiagen 74804; Maryland, MD), and quantified using the Quanti-iT Ribogreen RNA kit (Molecular Probes Inc. R11490; OR, USA). RNA (50 ng) was converted to cDNA prior to RT-PCR using RT^2 First Strand Kit (Qiagen C-03/330401; Maryland, MD). Custom RT-PCR arrays were created by SABiosciences (RT^2 Profiler™ PCR Array; SABiosciences, Qiagen; Maryland, MD). An experimental cocktail was made using the cDNA and 2× SABiosciences RT^2 qPCR Master Mix (Qiagen PA-011-12/330512; Maryland, MD) and loaded onto a 96-well RT^2 Profiler PCR array and run on a Bio-Rad iCycler (Bio-Rad iQ5; CA, USA). For each sample, the PCR experimental cocktail was prepared by mixing 14 μL of diluted cDNA, 161 μL of H_2O, and 175 μL of 2× SABiosciences RT^2 qPCR Master Mix (Qiagen, Frederick, MD). Each well received 25 μL of PCR experimental cocktail. A two-step cycling program was used and threshold cycle (Ct) was calculated for each well using the Bio-Rad iQ5 software program (Bio-Rad iQ5; CA, USA). There were 3 plate controls and 2 housekeeping genes: Rpl13a (NM 173340) and Rplp1 (NM 001007604). Genes specific to this study included CRF (NM 031019), CRF-BP (NM 139183), $CRFR_1$ (NM 030999), $CRFR_2$ (NM 022714), and $5\text{-}HT_{2A}$ (NM 017254).

4.1.8 Statistical analysis

The effects of maternal separation (MS) on behavior (EPM, PGM, and CIL) and mRNA expression were analyzed by analysis of variance (one-way or two-way ANOVA followed by a Dunnett's or Bonferroni post-hoc test where appropriate) for sex, treatment, and behavior, using Sigma Stat

(Systat Software Inc., San Jose, CA). All data are expressed as a mean ± SEM. Statistical significance was set at $\alpha = 0.05$. The RT^2 Profiler PCR Array Data Analysis web-based PCR data analysis program was used to calculate expression levels (SABiosciences, Frederick, MD). Expression level was normalized to that of the housekeeping genes and statistical values were determined using raw ΔC_t values. Values for fold-change in gene expression were determined using the $\Delta\Delta C_t$ method ($2^{-\Delta\Delta Ct}$).

4.2 Results

4.2.1 Behavior

4.2.1.1 Early life stress increases anxiety-like behavior on the elevated plus maze

Anxiety-like behavior (decreased percentage time on the open arms) was assessed with the EPM. Maternal separation (MS) led to increased anxiety-like behavior in both males (MS180) and females (MS15 and MS180; Fig. 1; $F_{(3,96)} = 4.452$, $P < 0.05$). Handling stress (AFR) also caused an increase in anxiety-like behavior in males but had no effect on females.

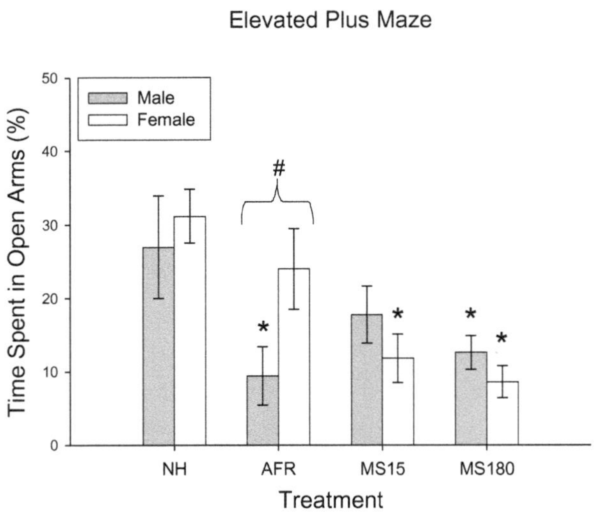

Fig. 1 Effects of early life stress on anxiety-like behavior on the elevated plus maze of adolescent rats. Males ($n = 12$–14/condition) in the animal facility raised (AFR) and 180 min maternal separation (MS180) spent significantly less time in open arms than nonhandled controls (NH). Females (12–14/condition) in the MS15 and MS180 conditions also spent significantly less time in open arms than NH controls. Males in the AFR condition spent less time on open arms compared to female AFR. *Significantly different than NH controls ($P < 0.05$). # Significant difference between sexes. Data are presented as percentage time open v. closed arms (mean ± SEM).

Females in the AFR group spent significantly more time in the open arm of the EPM than AFR males ($F_{(3,96)} = 1.697$, $P < 0.05$). Total locomotion was not significantly different among any groups for males or females.

4.2.1.2 Novelty-seeking and risk-taking on the playground maze is enhanced by early life stress

The playground maze was used to assess novelty-seeking and risk-taking behaviors (Klebaur & Bardo, 1999; Klebaur et al., 2001). Percent time interacting with the novel toy is used as a measure of both novelty-seeking and risk-taking. Both MS15 and MS180 females spent a greater percent time than controls interacting (defined as touching with forepaws or head) with the novel object over the familiar objects ($F_{(3,87)} = 3.463$, < 0.05; Fig. 2A). MS had no effect on this measure in males. However, while ELS increased novelty seeking in females and not males, males had significantly higher baseline novelty seeking. Male controls spent a greater percentage of time with the novel object than their female counterparts ($F_{(3,87)} = 1.696$, $P < 0.05$). An ancillary measure of risk-taking and novelty-seeking is the number of approaches to the novel object. Females in all experimental groups also had significantly more approaches to the novel object than NH controls ($F_{(3,87)} = 3.567$, $P < 0.05$; Fig. 2B). In contrast, only males that had been subjected to mild separation (MS15) approached the novel object significantly more times than controls ($F_{(3,87)} = 1.848$, $P < 0.05$). Female rats in the AFR and MS180 groups approached the novel object significantly more times than male rats.

4.2.1.3 Cocaine-induced locomotion is increased by neonatal stress

In rodents, initial sensitivity to the locomotor activating properties of cocaine predicts impulsivity related to addictive behaviors (Stanis et al., 2008). MS led to an increase in cocaine-induced locomotion in both MS15 and MS180 groups for males but only the MS15 group for females ($F_{(3,65)} = 2.474$, $P < 0.05$; Fig. 3). Animal facility raised animals also had greater cocaine-induced locomotion than NH controls but still significantly less than the MS180 group ($P < 0.05$). In females, while significant increases in CIL over NH controls were found only in the MS15 group, they had as great, or greater, CIL than male rats in every group ($P < 0.05$). Female NH controls, however, had elevated CIL that was significantly greater than male NH controls ($F_{(1,65)} = 7.234$, $P < 0.05$). Female rats in the MS15 group also had greater CIL than males ($P < 0.05$).

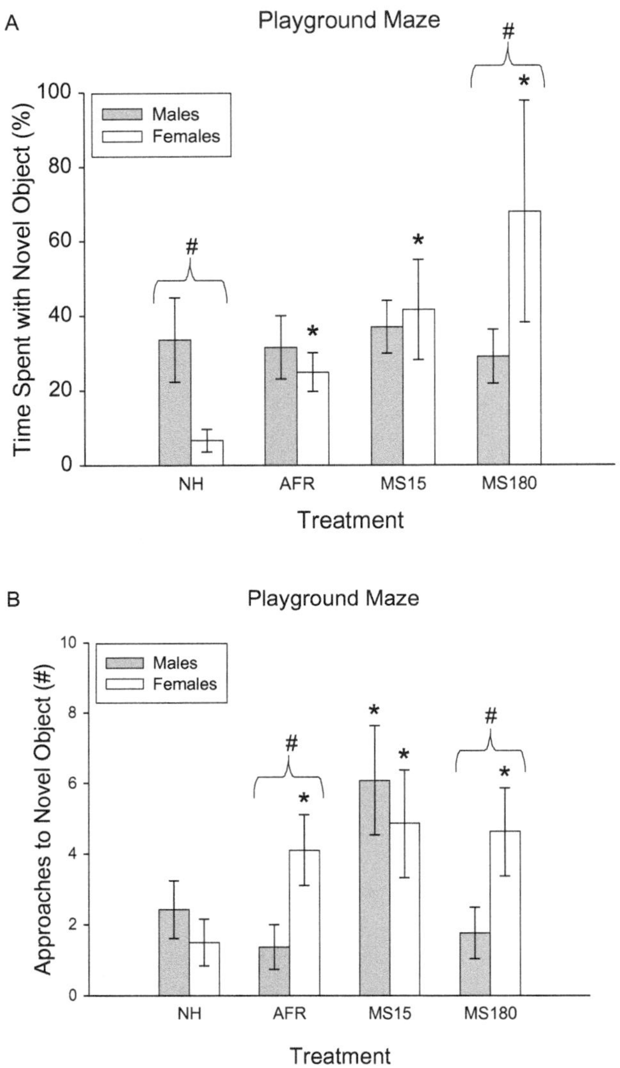

Fig. 2 Effects of early life stress on adolescent novelty-seeking behaviors on the playground maze. (A) Females in the AFR, MS15, and MS180 conditions ($n = 10–13$/condition) all spent significantly more time with the novel object than NH controls. There were no effects for this measure in males. There were significant differences between sexes in the NH and MS180 groups. Data are presented as percentage time spent with novel object (mean ± SEM). (B) AFR, MS15 and MS180 females had significantly more approaches to the novel object than NH controls. Males in the MS15 condition made significantly more approaches to the novel object than NH controls. Sex differences were seen in the AFR and MS180 treatment groups. *Significantly different than NH controls ($P < 0.05$). # Significant sex difference ($P < 0.05$). Data are presented as mean ± SEM.

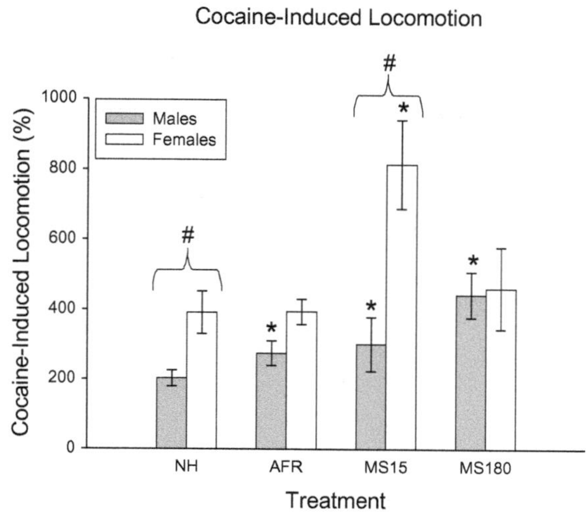

Fig. 3 Early life stress effects on cocaine-induced locomotion in adolescent rats. Males ($n=8$–10/condition) in the AFR, MS15, and MS180 conditions all had significantly increased cocaine-induced locomotion than NH controls. Females ($n=12$–15/condition) in the MS15 condition had significantly increased cocaine-induced locomotion than NH controls. Males in the NH condition had significantly less cocaine-induced locomotion compared to NH females. Males in the AFR condition had significantly less cocaine-induced locomotion compared to AFR females. Males in the MS15 condition had significantly less cocaine-induced locomotion compared to MS15 females. *Significantly different than NH controls ($P<0.05$). #Significant sex difference ($P<0.05$). Data are presented as a percentage baseline locomotion ± SEM.

4.2.1.4 Early life stress alters gene expression of CRF family and 5-HT2A mRNA in reward- and stress-related brain regions

Early life stress (ELS) effects on mRNA expression are summarized in Table 1. The greatest change in gene expression, in any brain region, was that for CRF mRNA in the PVN of female rats in the severely separated group (MS180) with a 6.43-fold increase ($P<0.05$) and a 4-fold increase in the MS15 group ($P<0.055$) (Fig. 4; Table 1). ELS had no effect on CRF expression in the PVN of male rats (Table 1). MS stress caused increased expression of CRF-BP mRNA in the VTA of MS180 female rats (Fig. 5; 2.9-fold; $P<0.05$). CRF-BP mRNA in the MS15 females approached significance ($P\leq 0.07$). Strikingly, the most consistent effect of MS on gene expression in either males or females was a significant upregulation of the 5-HT_{2A} mRNA in males in three of the four brain regions analyzed: VTA (MS15, 2.1-fold, MS180, 3.3-fold; $P<0.05$), NAc (MS15, 2.6-fold; $P<0.05$) and BNST (MS15, 2.7-fold, MS180,

Table 1 Effects of early life stress on gene expression mRNA expression in the NAc, VTA, BNST, PVN and CeA of adolescent rats.

mRNA Expression in Males and Females

	Males	Females
NAc	↑ 5HT-2_A in MS15	NO CHANGES
VTA	↓ CRF in AFR; ↑ 5HT-2_A in MS15; ↑ 5HT-2_A in MS180	↓ CRF in AFR; ↑ CRF-BP in MS180
BNST	↑ 5HT-2_A in MS15; ↑ 5HT-2_A in MS180	NO CHANGES
PVN	NO CHANGES	↑ CRF in MS15, MS180
CeA	NO CHANGES	NO CHANGES

Arrows indicate significance relative to controls ($P < 0.05$).

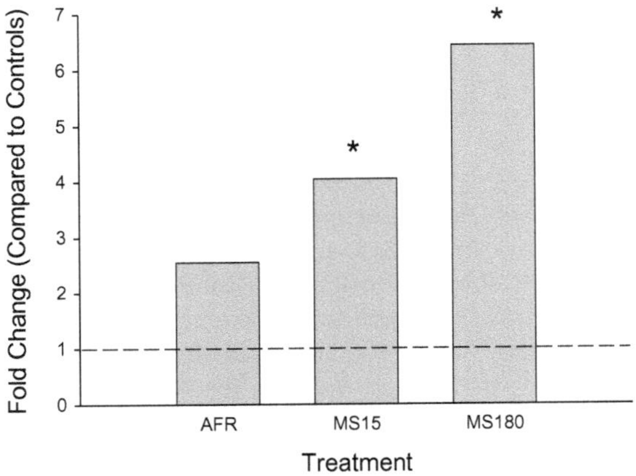

Fig. 4 Early life stress causes increased CRF mRNA expression in the PVN of adolescent female rats. There was a 6.4-fold in the MS180 condition ($n=4$; $*P<0.05$) and a 4.0-fold increase in the MS15 condition over controls that approached significance ($^{\#}P<0.055$). Data are presented as fold-change compared to controls.

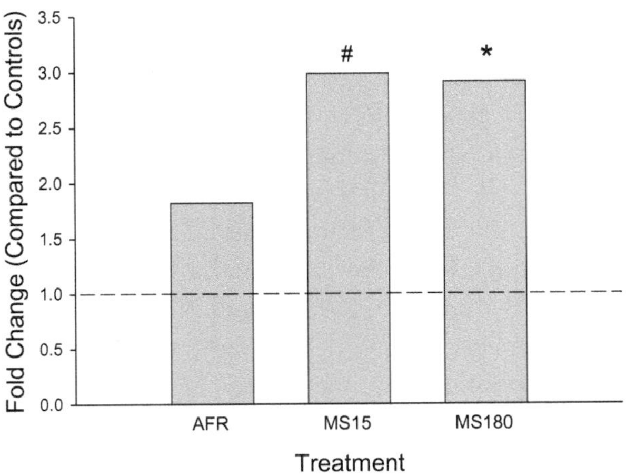

Fig. 5 Increased CRF-BP mRNA expression in the VTA of female adolescent rats that had been subjected to early life stress. CRF-BP in the VTA was upregulated 2.9-fold in the MS180 condition compared to controls ($n=4$; $*P<0.05$). Though there was also increased expression of CRF-BP in the MS15 condition this did not reach significance ($P<0.07$). Data are presented as fold-change compared to controls.

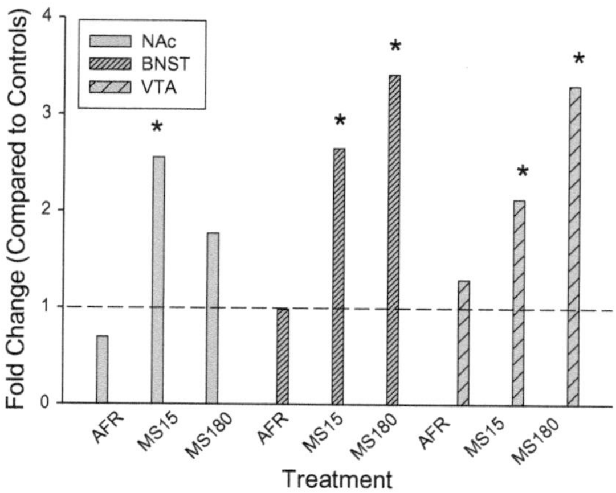

Fig. 6 Early life stress leads to increased 5-HT_{2A} mRNA expression in the NAc, BNST, and VTA of male adolescent rats. There was a 2.6-fold increase in 5-HT2a mRNA in NAc compared to controls. In the BNST maternal separation caused an increase of 2.7-fold (MS15) 3.4-fold (MS180). Maternal separation also led to significant upregulation in both MS15 (2.1-fold) MS180 (3.3-fold) 5-HT_{2A}.*$P < 0.05$, significantly different than controls. Data are presented as fold-change relative to controls.

3.4-fold; $P < 0.05$; Fig. 6). Surprisingly, ELS had no effect on 5-HT_{2A} expression in any brain region in females. The only gene that had decreased expression was CRF in the VTA of both males and females in the AFR group (−0.32-fold females, −0.42-fold males; $P < 0.05$; Table 1). Expression of $CRFR_1$ or $CRFR_2$ was not altered by ELS (data not shown).

4.3 Discussion

Early life stress (ELS) predicts increases vulnerability to psychiatric disorders including addiction in adolescents (Enoch, 2011; Kelley et al., 2004). Here, we demonstrated that ELS leads to the development of a drug-seeking phenotype (increased anxiety-like behavior, risk-taking, novelty-seeking, impulsivity and sensitivity to the locomotor activating properties of a psychostimulant) concomitant with a sex-dependent upregulation of 5-HT_{2A} and CRF system components throughout extended reward- and stress-brain regions in adolescent rats. Together, our results show that increased vulnerability to addiction caused by ELS is potentially mediated by enhanced CRF and 5-HT signaling.

Sex-dependent effects of MS on both behavior and gene expression were found. ELS enhanced measures of novelty-seeking in females more than males. Stress, whether handling or separation, resulted in increased measures of novelty-seeking (% time and approaches). However, novelty-seeking (% time) in the MS180 group was significantly greater than AFR. Novelty-seeking in males (approached significance) was only increased in MS15 group. Interestingly, this same pattern of response was found for CIL, with effects in AFR, MS15 and MS180 males and only MS15 for female rats. Also, MS had greater effect than AFR on CIL in male rats. Though CIL was only increased in the MS15 condition for female rats, it was similar to the increase seen in MS180 males (~2-fold increase over NH). MS had a greater effect on anxiety-like behavior in females (MS15 and MS180) than males (MS180). Another study showed that MS had no effect on anxiety-like behavior in adult female rats (Aisa, Tordera, Lasheras, Del Rio, & Ramirez, 2008). The anxiety-like behavior of female adolescents in our study suggests that adolescence is a period of enhanced vulnerability for females with a history of ELS.

The predominant effect of MS on gene expression in females was increased CRF mRNA in the PVN and CRF-BP in the VTA; in males it increased 5-HT_{2A} levels in the VTA, NAc and BNST. Upregulation of CRF mRNA in the PVN could have a myriad of effects. It could lead to dysfunctional hypothalamic-pituitary-adrenal axis (HPA) activity and enhanced responsiveness to stressors. MS leads to HPA hyperactivity in both female and male adult rats (Aisa et al., 2007, 2008). MS caused an increase in both baseline and stress-induced plasma corticosterone and ACTH levels that were coincident with upregulation of CRF mRNA in the PVN. Similar sex differences in response to early life MS were seen in adult rats of another study that looked at CRF/c-Fos-immunoreactive cells in the PVN after swim stress (Desbonnet et al., 2008). Other studies investigating some form of ELS have found increased CRF mRNA in the PVN of adult male rats (Plotsky & Meaney, 1993; Plotsky et al., 2005). The lack of a significant effect of MS on CRF mRNA in the PVN of adolescent males is surprising and may be due to a difference in temporal expression of this molecular effect or differences in methodology. Enhanced CRF activity in the PVN could also increase susceptibility to addiction. The PVN sends CRF projections to many brain regions including the VTA (Rodaros et al., 2007), where CRF infusion leads to reinstatement of cocaine self-administration (Wang et al., 2005; Wang, You, Rice, & Wise, 2007). Increased CRF mRNA in the PVN is correlated with increased anxiety and neuroendocrine responses to withdrawal from drugs of abuse (c.f., Cleck & Blendy, 2008) and has been directly implicated in stress-induced reinstatement of self-administration for drugs of abuse (Koob & Kreek, 2007). Withdrawal from morphine leads to upregulation of the number of CRF immunoreactive cells of the PVN along with increased co-labeled FosB/ΔFosB (Koob & Kreek, 2007; Nunez, Martin, Foldes, Kovacs, & Victoria Milanes, 2010). Upregulation of ΔFosB has been suggested as a key molecular adaptation initiating and sustaining chronic addictive states (Nestler, Barrot, & Self, 2001). Recently, in other brain regions ΔFosB has been shown to mediate resilience to stressors (Vialou et al., 2010). Thus, there may be other common molecular substrates linking stress and addiction.

The novel finding of increased CRF-BP in the VTA of females exposed to MS is an important finding of this study. Evidence has been building tying CRF-BP mediated signaling to behavioral disorders (Fox & Lowry, 2013; Quadros, Macedo, Domingues, & Favoretto, 2016; Ronan & Summers, 2011; Spierling & Zorrilla, 2017; Vandael & Gounko, 2019; Vasconcelos, Stein, Gallas-Lopes, Landau, & de Almeida, 2020). Originally, CRF-BP was thought to act simply as a buffer that controls CRF signaling by controlling extracellular CRF levels (Ronan & Summers, 2011). More recently, it has been implicated in facilitating active signaling processes in the VTA (Slater, Noches, & Gysling, 2016; Ungless et al., 2003; Wang et al., 2007; Wise & Morales, 2010). Potentiation of NMDA receptor-mediated synaptic transmission in dopamine neurons of the VTA is a process that is critical to addiction (Bonci & Malenka, 1999; Fitzgerald, Ortiz, Hamedani, & Nestler, 1996). Potentiation of this signaling by CRF requires CRF-BP (Ungless et al., 2003). CRF-BP in the VTA has also recently been directly implicated in drug-seeking behaviors and appears to be essential for CRF- and stress (footshock)-induced behavioral and neurotransmitter effects in cocaine-experienced rodents (Wise & Morales, 2010). Also, in a recent study we found that CRF-BP is significantly increased in the VTA of adult male rats both after 2 weeks of cocaine treatment and during subsequent withdrawal (Callahan, Tschetter, & Ronan, 2008). Together these studies suggest further convergence of stress and addiction on a molecular substrate contributing to the processes of addiction. Other substrates also activated by both stress and addiction include phosphorylated transcription factor CREB (pCREB; for review, see Briand & Blendy, 2010). Both CRF and CRF-BP are upregulated by pCREB (Carlezon et al., 1998; Kemp, Woods, & Lowry, 1998) and early life handling stress causes increased pCREB expression in the hippocampi of neonatal rats; an effect that can be prevented by antagonism of 5-HT_{2A} (Garoflos, Stamatakis, Mantelas, Philippidis, & Stylianopoulou, 2005). Thus, increased pCREB may be a mechanism underlying increased gene activity seen in this study.

ELS leads to a widespread upregulation of 5-HT_{2A} receptors in male rats (VTA, NAc, and BNST). The 5-HT_{2A} receptor appears to be a key substrate facilitating addictive behaviors. This receptor is necessary for cocaine-primed reinstatement to drug-seeking, motivational effects of cues paired with cocaine, and mediates anxiety symptoms of cocaine withdrawal (Filip, Alenina, Bader, & Przegalinski, 2010). Additionally, 5-HT_{2A} receptors are involved in hyperlocomotion and sensitization to cocaine (Filip, Bubar, & Cunningham, 2004; Filip, Papla, Nowak, Jungersmith, & Przegalinski, 2002). It is possible that they also mediate initial sensitivity to the locomotor activating properties of cocaine in male rats. This receptor also mediates anxiety behaviors (Blakley, Pohorecky, & Benjamin, 2001) and nefazadone, a potent 5-HT_{2A} antagonist (Olausson et al., 1998), has been shown to have anxiolytic effects along with its antidepressant actions (Nunez et al., 2002). Importantly for our study, CRF-induced anxiety-like behavior was, putatively, potentiated by activation of 5-$HT_{2A/2C}$ receptors. Evidence suggests CRF acts via a $CRFR_1$ mechanism to enhance 5-HT_{2A} signaling by actively recruiting 5-HT_{2A} receptors to the cell surface (Magalhaes et al., 2010). A polymorphism in 5-HT_{2A} has been correlated with the personality trait of novelty-seeking in adult humans (Heck et al., 2009). In another study, neonatal asphyxia in rat pups led to upregulation of 5-HT_{2A} in the striatum of adults and a coincident increase in novelty-seeking behavior (Adriani et al., 2006). Interestingly, efficacy of atypical antipsychotics is thought to be mediated at least in part via antagonism of 5-HT_{2A} receptors (Schmidt, Sorensen, Kehne, Carr, & Palfreyman, 1995). During adolescence, atypical antipsychotics are often used to treat risk-seeking behaviors (Connor et al., 2006; Wheatley, Plant, Reader, Brown, & Cahill, 2004). This work has other potential clinical implications as well.

First, this study helps to elucidate the long-lasting effects of ELS on both behavior and gene expression and highlights the importance of early therapeutic intervention and prevention. Even mild neonatal stress (AFR) caused changes in behavior and gene expression. Findings implicating the CRF and 5-HT systems also provides further evidence that these systems play key roles in stress-induced disorders including addiction. Targeting CRF and 5-HT signaling holds great promise; whether pharmacological or novel approaches such as AAV-mediated optogenetic or RNAi control of their activity and expression.

5. Conclusions

We have discussed the long-lasting effects of early life stress (ELS) on both behavior and molecular changes in reward- and stress-related neurocircuitry. In addition, we have presented methodology that can be used to interrogate these effects in rodents. Furthermore, we have shown, for the first time, that ELS leads to the development of a behavioral profile indicative of increased susceptibility to addiction that is concomitant with a sex-dependent upregulation of CRF and 5-HT system components. The latter were observed in multiple brain regions of reward- and stress-related neurocircuitry in adolescent-treated male and female rats. The novel findings in this study demonstrated that ELS leads to increased CRF-BP expression in the VTA of females, coupled with recent work implicating VTA CRF-BP in addiction (Wang et al., 2005, 2007), suggest an important role for this molecule in translating ELS into later dysfunctional behavior and neurotransmission. Also, increased 5-HT_{2A} receptor expression in males across multiple regions of the extended reward/stress pathway of males suggest that this may be a key mechanism underlying ELS-induced susceptibility to develop substance abuse in males. Overall, this study provides a number of possible mechanisms that warrant further study, especially as they relate to pharmacotherapy.

Acknowledgments

This work was supported with funding from a Department of Veterans Affairs Merit Review Award 1 I01 BX004712 (P.J.R., T.P.B.), NIMH 1 R01 MH122954 (P.J.R.), NIH NCRR P20 RR015567 (P.J.R.), NIH NIGMS U54GM128729 (P.J.R.), Great Plains Veterans Research Foundation (P.J.R.), and the Avera Foundation (P.J.R., S.R.). The authors further acknowledge the Sioux Falls VA Healthcare System for their generous research support. We would like to thank Drs. Jason Petersen, Karen Munger, David Maddox, and Cliff Summers for thoughtful discussions and reviews of this manuscript. The contents of this manuscript do not represent the views of the Department of Veterans Affairs or the United States Government.

Conflict of interest

The authors declare that they have no conflict of interest.

References

Adriani, W., Giannakopoulou, D., Bokulic, Z., Jernej, B., Alleva, E., & Laviola, G. (2006). Response to novelty, social and self-control behaviors, in rats exposed to neonatal anoxia: Modulatory effects of an enriched environment. *Psychopharmacology*, *184*, 155–165. https://doi.org/10.1007/s00213-005-0223-0.

Aisa, B., Tordera, R., Lasheras, B., Del Rio, J., & Ramirez, M. J. (2007). Cognitive impairment associated to HPA axis hyperactivity after maternal separation in rats. *Psychoneuroendocrinology*, *32*, 256–266. https://doi.org/10.1016/j.psyneuen.2006.12.013.
Aisa, B., Tordera, R., Lasheras, B., Del Rio, J., & Ramirez, M. J. (2008). Effects of maternal separation on hypothalamic-pituitary-adrenal responses, cognition and vulnerability to stress in adult female rats. *Neuroscience*, *154*, 1218–1226. https://doi.org/10.1016/j.neuroscience.2008.05.011.
Anastasio, N. C., Stoffel, E. C., Fox, R. G., Bubar, M. J., Rice, K. C., Moeller, F. G., et al. (2011). Serotonin (5-hydroxytryptamine) 5-HT(2A) receptor: Association with inherent and cocaine-evoked behavioral disinhibition in rats. *Behavioural Pharmacology*, *22*, 248–261. https://doi.org/10.1097/FBP.0b013e328345f90d.
Andersen, S. L. (2003). Trajectories of brain development: Point of vulnerability or window of opportunity? *Neuroscience and Biobehavioral Reviews*, *27*, 3–18. https://doi.org/10.1016/s0149-7634(03)00005-8.
Badanich, K. A., Maldonado, A. M., & Kirstein, C. L. (2008). Early adolescents show enhanced acute cocaine-induced locomotor activity in comparison to late adolescent and adult rats. *Developmental Psychobiology*, *50*, 127–133. https://doi.org/10.1002/dev.20252.
Bardo, M. T., Donohew, R. L., & Harrington, N. G. (1996). Psychobiology of novelty seeking and drug seeking behavior. *Behavioural Brain Research*, 77, 23–43. https://doi.org/10.1016/0166-4328(95)00203-0.
Benekareddy, M., Goodfellow, N. M., Lambe, E. K., & Vaidya, V. A. (2010). Enhanced function of prefrontal serotonin 5-HT(2) receptors in a rat model of psychiatric vulnerability. *The Journal of Neuroscience*, *30*, 12138–12150. https://doi.org/10.1523/JNEUROSCI.3245-10.2010.
Blakley, G. G., Pohorecky, L. A., & Benjamin, D. (2001). Bidirectional changes in ethanol consumption in rats with site-specific antisense down-regulation of 5-hydroxytryptamine2A receptors in brain. *The Journal of Pharmacology and Experimental Therapeutics*, *299*, 277–289.
Bonci, A., & Malenka, R. C. (1999). Properties and plasticity of excitatory synapses on dopaminergic and GABAergic cells in the ventral tegmental area. *The Journal of Neuroscience*, *19*, 3723–3730. https://doi.org/10.1523/jneurosci.19-10-03723.1999.
Borgland, S. L., Malenka, R. C., & Bonci, A. (2004). Acute and chronic cocaine-induced potentiation of synaptic strength in the ventral tegmental area: Electrophysiological and behavioral correlates in individual rats. *The Journal of Neuroscience*, *24*, 7482–7490. https://doi.org/10.1523/JNEUROSCI.1312-04.2004.
Briand, L. A., & Blendy, J. A. (2010). Molecular and genetic substrates linking stress and addiction. *Brain Research*, *1314*, 219–234. https://doi.org/10.1016/j.brainres.2009.11.002.
Callaghan, B. L., & Richardson, R. (2011). Maternal separation results in early emergence of adult-like fear and extinction learning in infant rats. *Behavioral Neuroscience*, *125*, 20–28. https://doi.org/10.1037/a0022008.
Callahan, L. B., Tschetter, K. E., & Ronan, P. J. (2008). Binge cocaine use and withdrawal sensitizes brain reward systems to stress/arousal: A possible mechanism driving relapse. In *Society for neuroscencie: Abstract*. online https://doi.org/10.3389/fnins.2013.00195.
Carlezon, W. A., Jr., Thome, J., Olson, V. G., Lane-Ladd, S. B., Brodkin, E. S., Hiroi, N., et al. (1998). Regulation of cocaine reward by CREB. *Science*, *282*, 2272–2275. https://doi.org/10.1126/science.282.5397.2272.
Carobrez, A. P., & Bertoglio, L. J. (2005). Ethological and temporal analyses of anxiety-like behavior: The elevated plus-maze model 20 years on. *Neuroscience and Biobehavioral Reviews*, *29*, 1193–1205. https://doi.org/10.1016/j.neubiorev.2005.04.017.

Carpenter, S. C., Clyman, R. B., Davidson, A. J., & Steiner, J. F. (2001). The association of foster care or kinship care with adolescent sexual behavior and first pregnancy. *Pediatrics, 108*, E46. https://doi.org/10.1542/peds.108.3.e46.

Clark, D. B., Kirisci, L., & Tarter, R. E. (1998). Adolescent versus adult onset and the development of substance use disorders in males. *Drug and Alcohol Dependence, 49*, 115–121. https://doi.org/10.1016/s0376-8716(97)00154-3.

Cleck, J. N., & Blendy, J. A. (2008). Making a bad thing worse: Adverse effects of stress on drug addiction. *The Journal of Clinical Investigation, 118*, 454–461. https://doi.org/10.1172/JCI33946.

Connor, D. F., Carlson, G. A., Chang, K. D., Daniolos, P. T., Ferziger, R., Findling, R. L., et al. (2006). Juvenile maladaptive aggression: A review of prevention, treatment, and service configuration and a proposed research agenda. *The Journal of Clinical Psychiatry, 67*, 808–820. https://doi.org/10.4088/jcp.v67n0516.

Desbonnet, L., Garrett, L., Daly, E., McDermott, K. W., & Dinan, T. G. (2008). Sexually dimorphic effects of maternal separation stress on corticotrophin-releasing factor and vasopressin systems in the adult rat brain. *International Journal of Developmental Neuroscience, 26*, 259–268. https://doi.org/10.1016/j.ijdevneu.2008.02.004.

Dong, M., Anda, R. F., Felitti, V. J., Dube, S. R., Williamson, D. F., Thompson, T. J., et al. (2004). The interrelatedness of multiple forms of childhood abuse, neglect, and household dysfunction. *Child Abuse & Neglect, 28*, 771–784. https://doi.org/10.1016/j.chiabu.2004.01.008.

Edwards, V. J., Holden, G. W., Felitti, V. J., & Anda, R. F. (2003). Relationship between multiple forms of childhood maltreatment and adult mental health in community respondents: Results from the adverse childhood experiences study. *The American Journal of Psychiatry, 160*, 1453–1460. https://doi.org/10.1176/appi.ajp.160.8.1453.

Enoch, M. A. (2011). The role of early life stress as a predictor for alcohol and drug dependence. *Psychopharmacology, 214*, 17–31. https://doi.org/10.1007/s00213-010-1916-6.

Filip, M., Alenina, N., Bader, M., & Przegalinski, E. (2010). Behavioral evidence for the significance of serotoninergic (5-HT) receptors in cocaine addiction. *Addiction Biology, 15*, 227–249. https://doi.org/10.1111/j.1369-1600.2010.00214.x.

Filip, M., Bubar, M. J., & Cunningham, K. A. (2004). Contribution of serotonin (5-hydroxytryptamine; 5-HT) 5-HT2 receptor subtypes to the hyperlocomotor effects of cocaine: Acute and chronic pharmacological analyses. *The Journal of Pharmacology and Experimental Therapeutics, 310*, 1246–1254. https://doi.org/10.1124/jpet.104.068841.

Filip, M., Papla, I., Nowak, E., Jungersmith, K., & Przegalinski, E. (2002). Effects of serotonin (5-HT)(1B) receptor ligands, microinjected into accumbens subregions, on cocaine discrimination in rats. *Naunyn-Schmiedeberg's Archives of Pharmacology, 366*, 226–234. https://doi.org/10.1007/s00210-002-0583-5.

Fitzgerald, L. W., Ortiz, J., Hamedani, A. G., & Nestler, E. J. (1996). Drugs of abuse and stress increase the expression of GluR1 and NMDAR1 glutamate receptor subunits in the rat ventral tegmental area: Common adaptations among cross-sensitizing agents. *The Journal of Neuroscience, 16*, 274–282. https://doi.org/10.1523/jneurosci.16-01-00274.1996.

Forster, G. L., Anderson, E. M., Scholl, J. L., Lukkes, J. L., & Watt, M. J. (2018). Negative consequences of early-life adversity on substance use as mediated by corticotropin-releasing factor modulation of serotonin activity. *Neurobiology of Stress, 9*, 29–39. https://doi.org/10.1016/j.ynstr.2018.08.001.

Fox, J. H., & Lowry, C. A. (2013). Corticotropin-releasing factor-related peptides, serotonergic systems, and emotional behavior. *Frontiers in Neuroscience*, 7, 169. https://doi.org/10.3389/fnins.2013.00169.

Garoflos, E., Stamatakis, A., Mantelas, A., Philippidis, H., & Stylianopoulou, F. (2005). Cellular mechanisms underlying an effect of "early handling" on pCREB and BDNF in the neonatal rat hippocampus. *Brain Research, 1052*, 187–195. https://doi.org/10.1016/j.brainres.2005.06.032.

Graham, Y. P., Heim, C., Goodman, S. H., Miller, A. H., & Nemeroff, C. B. (1999). The effects of neonatal stress on brain development: Implications for psychopathology. *Development and Psychopathology, 11*, 545–565. https://doi.org/10.1017/s0954579499002205.

Green, J. G., McLaughlin, K. A., Berglund, P. A., Gruber, M. J., Sampson, N. A., Zaslavsky, A. M., et al. (2010). Childhood adversities and adult psychiatric disorders in the national comorbidity survey replication I: associations with first onset of DSM-IV disorders. *Archives of General Psychiatry, 67*, 113–123. https://doi.org/10.1001/archgenpsychiatry.2009.186.

Greisen, M. H., Bolwig, T. G., Husum, H., Nedergaard, P., & Wortwein, G. (2005). Maternal separation affects male rat copulatory behaviour and hypothalamic corticotropin releasing factor in concert. *Behavioural Brain Research, 158*, 367–375. https://doi.org/10.1016/j.bbr.2004.09.018.

Gutman, D. A., & Nemeroff, C. B. (2003). Persistent central nervous system effects of an adverse early environment: Clinical and preclinical studies. *Physiology & Behavior, 79*, 471–478. https://doi.org/10.1016/s0031-9384(03)00166-5.

Harro, J., Tonissaar, M., & Eller, M. (2001). The effects of CRA 1000, a non-peptide antagonist of corticotropin-releasing factor receptor type 1, on adaptive behaviour in the rat. *Neuropeptides, 35*, 100–109. https://doi.org/10.1054/npep.2001.0851.

Heck, A., Lieb, R., Ellgas, A., Pfister, H., Lucae, S., Roeske, D., et al. (2009). Investigation of 17 candidate genes for personality traits confirms effects of the HTR2A gene on novelty seeking. *Genes, Brain, and Behavior, 8*, 464–472. https://doi.org/10.1111/j.1601-183X.2009.00494.x.

Huot, R. L., Gonzalez, M. E., Ladd, C. O., Thrivikraman, K. V., & Plotsky, P. M. (2004). Foster litters prevent hypothalamic-pituitary-adrenal axis sensitization mediated by neonatal maternal separation. *Psychoneuroendocrinology, 29*, 279–289. https://doi.org/10.1016/s0306-4530(03)00028-3.

Kelley, A. E., Schochet, T., & Landry, C. F. (2004). Risk taking and novelty seeking in adolescence: Introduction to part I. *Annals of the New York Academy of Sciences, 1021*, 27–32. https://doi.org/10.1196/annals.1308.003.

Kemp, C. F., Woods, R. J., & Lowry, P. J. (1998). The corticotrophin-releasing factor-binding protein: An act of several parts. *Peptides, 19*, 1119–1128. https://doi.org/10.1210/jcem.82.5.3952.

Kirby, L. G., Zeeb, F. D., & Winstanley, C. A. (2011). Contributions of serotonin in addiction vulnerability. *Neuropharmacology, 61*, 421–432. https://doi.org/10.1016/j.neuropharm.2011.03.022.

Klebaur, J. E., & Bardo, M. T. (1999). Individual differences in novelty seeking on the playground maze predict amphetamine conditioned place preference. *Pharmacology, Biochemistry, and Behavior, 63*, 131–136. https://doi.org/10.1016/s0091-3057(98)00258-5.

Klebaur, J. E., Bevins, R. A., Segar, T. M., & Bardo, M. T. (2001). Individual differences in behavioral responses to novelty and amphetamine self-administration in male and female rats. *Behavioural Pharmacology, 12*, 267–275. https://doi.org/10.1097/00008877-200107000-00005.

Koob, G., & Kreek, M. J. (2007). Stress, dysregulation of drug reward pathways, and the transition to drug dependence. *The American Journal of Psychiatry, 164*, 1149–1159. https://doi.org/10.1176/appi.ajp.2007.05030503.

Kuhn, C. (2015). Emergence of sex differences in the development of substance use and abuse during adolescence. *Pharmacology & Therapeutics, 153*, 55–78. https://doi.org/10.1016/j.pharmthera.2015.06.003.

Ladd, C. O., Owens, M. J., & Nemeroff, C. B. (1996). Persistent changes in corticotropin-releasing factor neuronal systems induced by maternal deprivation. *Endocrinology, 137*, 1212–1218. https://doi.org/10.1210/endo.137.4.8625891.

Ladd, C. O., Thrivikraman, K. V., Huot, R. L., & Plotsky, P. M. (2005). Differential neuroendocrine responses to chronic variable stress in adult Long Evans rats exposed to handling-maternal separation as neonates. *Psychoneuroendocrinology, 30*, 520–533. https://doi.org/10.1016/j.psyneuen.2004.12.004.

Litvin, Y., Tovote, P., Pentkowski, N. S., Zeyda, T., King, L. B., Vasconcellos, A. J., et al. (2010). Maternal separation modulates short-term behavioral and physiological indices of the stress response. *Hormones and Behavior, 58*, 241–249. https://doi.org/10.1016/j.yhbeh.2010.03.010.

Lodge, D. J., & Grace, A. A. (2005). Acute and chronic corticotropin-releasing factor 1 receptor blockade inhibits cocaine-induced dopamine release: Correlation with dopamine neuron activity. *The Journal of Pharmacology and Experimental Therapeutics, 314*, 201–206. https://doi.org/10.1124/jpet.105.084913.

Magalhaes, A. C., Holmes, K. D., Dale, L. B., Comps-Agrar, L., Lee, D., Yadav, P. N., et al. (2010). CRF receptor 1 regulates anxiety behavior via sensitization of 5-HT2 receptor signaling. *Nature Neuroscience, 13*, 622–629. https://doi.org/10.1038/nn.2529.

Marcinkiewcz, C. A., Prado, M. M., Isaac, S. K., Marshall, A., Rylkova, D., & Bruijnzeel, A. W. (2009). Corticotropin-releasing factor within the central nucleus of the amygdala and the nucleus accumbens shell mediates the negative affective state of nicotine withdrawal in rats. *Neuropsychopharmacology, 34*, 1743–1752. https://doi.org/10.1038/npp.2008.231.

Marin, M. T., & Planeta, C. S. (2004). Maternal separation affects cocaine-induced locomotion and response to novelty in adolescent, but not in adult rats. *Brain Research, 1013*, 83–90. https://doi.org/10.1016/j.brainres.2004.04.003.

McLaughlin, K. A., Greif Green, J., Gruber, M. J., Sampson, N. A., Zaslavsky, A. M., & Kessler, R. C. (2012). Childhood adversities and first onset of psychiatric disorders in a national sample of US adolescents. *Archives of General Psychiatry, 69*, 1151–1160. https://doi.org/10.1001/archgenpsychiatry.2011.2277.

Mellanby, J., Johansen-Berg, H., Leyland, R., & Milward, A. J. (1999). The effects of tetanus toxin-induced limbic epilepsy on the exploratory response to novelty in the rat. *Epilepsia, 40*, 1058–1061. https://doi.org/10.1111/j.1528-1157.1999.tb00819.x.

Merikangas, K. R., He, J. P., Burstein, M., Swanson, S. A., Avenevoli, S., Cui, L., et al. (2010). Lifetime prevalence of mental disorders in U.S. adolescents: Results from the National Comorbidity Survey Replication—Adolescent Supplement (NCS-A). *Journal of the American Academy of Child and Adolescent Psychiatry, 49*, 980–989. https://doi.org/10.1016/j.jaac.2010.05.017.

Miczek, K. A., Yap, J. J., & Covington, H. E., 3rd. (2008). Social stress, therapeutics and drug abuse: Preclinical models of escalated and depressed intake. *Pharmacology & Therapeutics, 120*, 102–128. https://doi.org/10.1016/j.pharmthera.2008.07.006.

Moffett, M. C., Vicentic, A., Kozel, M., Plotsky, P., Francis, D. D., & Kuhar, M. J. (2007). Maternal separation alters drug intake patterns in adulthood in rats. *Biochemical Pharmacology, 73*, 321–330. https://doi.org/10.1016/j.bcp.2006.08.003.

National Research Council. (2011). *Guide for the care and use of laboratory animals*. Washington, DC: The National Academies Press.

Nestler, E. J. (2005). Is there a common molecular pathway for addiction? *Nature Neuroscience, 8*, 1445–1449. https://doi.org/10.1038/nn1578.

Nestler, E. J., Barrot, M., & Self, D. W. (2001). DeltaFosB: A sustained molecular switch for addiction. *Proceedings of the National Academy of Sciences of the United States of America*, *98*, 11042–11046. https://doi.org/10.1073/pnas.191352698.

Nunez, C., Martin, F., Foldes, A., Luisa Laorden, M., Kovacs, K. J., & Victoria Milanes, M. (2010). Induction of FosB/DeltaFosB in the brain stress system-related structures during morphine dependence and withdrawal. *Journal of Neurochemistry*, *114*, 475–487. https://doi.org/10.1111/j.1471-4159.2010.06765.x.

Nunez, M. J., Rivas, M., Riveiro, P., Suarez, J., Balboa, J., Nunez, L. A., et al. (2002). Effects of nefazodone on voluntary ethanol consumption induced by isolation stress in young and aged rats. *Pharmacology, Biochemistry, and Behavior*, *73*, 689–696. https://doi.org/10.1016/s0091-3057(02)00875-4.

Olausson, P., Ericson, M., Petersson, A., Kosowski, A., Soderpalm, B., & Engel, J. A. (1998). Nefazodone attenuates the behavioral and neurochemical effects of ethanol. *Alcohol*, *15*, 77–86. https://doi.org/10.1016/s0741-8329(97)00101-8.

Palkovits, M., & Brownstein, M. J. (1988). *Maps and guide to microdissection of the rat brain*. New York: Elsevier. https://doi.org/10.1016/0022-510X(88)90044-5.

Pecina, S., Schulkin, J., & Berridge, K. C. (2006). Nucleus accumbens corticotropin-releasing factor increases cue-triggered motivation for sucrose reward: Paradoxical positive incentive effects in stress? *BMC Biology*, *4*, 8. https://doi.org/10.1186/1741-7007-4-8.

Piazza, P. V., & Le Moal, M. (1998). The role of stress in drug self-administration. *Trends in Pharmacological Sciences*, *19*, 67–74. https://doi.org/10.1016/s0165-6147(97)01115-2.

Plotsky, P. M., & Meaney, M. J. (1993). Early, postnatal experience alters hypothalamic corticotropin-releasing factor (CRF) mRNA, median eminence CRF content and stress-induced release in adult rats. *Brain Research. Molecular Brain Research*, *18*, 195–200. https://doi.org/10.1016/0169-328x(93)90189-v.

Plotsky, P. M., Thrivikraman, K. V., Nemeroff, C. B., Caldji, C., Sharma, S., & Meaney, M. J. (2005). Long-term consequences of neonatal rearing on central corticotropin-releasing factor systems in adult male rat offspring. *Neuropsychopharmacology*, *30*, 2192–2204. https://doi.org/10.1038/sj.npp.1300769.

Quadros, I. M., Macedo, G. C., Domingues, L. P., & Favoretto, C. A. (2016). An update on CRF mechanisms underlying alcohol use disorders and dependence. *Frontiers in Endocrinology (Lausanne)*, *7*, 134. https://doi.org/10.3389/fendo.2016.00134.

Rodaros, D., Caruana, D. A., Amir, S., & Stewart, J. (2007). Corticotropin-releasing factor projections from limbic forebrain and paraventricular nucleus of the hypothalamus to the region of the ventral tegmental area. *Neuroscience*, *150*, 8–13. https://doi.org/10.1016/j.neuroscience.2007.09.043.

Ronan, P. J., & Summers, C. H. (2011). Molecular signaling and translational significance of the corticotropin releasing factor system. *Progress in Molecular Biology and Translational Science*, *98*, 235–292. https://doi.org/10.1016/B978-0-12-385506-0.00006-5.

Roos, L. E., Horn, S., Berkman, E. T., Pears, K., & Fisher, P. A. (2018). Leveraging translational neuroscience to inform early intervention and addiction prevention for children exposed to early life stress. *Neurobiology of Stress*, *9*, 231–240. https://doi.org/10.1016/j.ynstr.2018.10.004.

Sarnyai, Z., Shaham, Y., & Heinrichs, S. C. (2001). The role of corticotropin-releasing factor in drug addiction. *Pharmacological Reviews*, *53*, 209–243.

Schmidt, C. J., Sorensen, S. M., Kehne, J. H., Carr, A. A., & Palfreyman, M. G. (1995). The role of 5-HT2A receptors in antipsychotic activity. *Life Sciences*, *56*, 2209–2222. https://doi.org/10.1016/0024-3205(95)00210-w.

Slater, P. G., Noches, V., & Gysling, K. (2016). Corticotropin-releasing factor type-2 receptor and corticotropin-releasing factor-binding protein coexist in rat ventral tegmental area nerve terminals originated in the lateral hypothalamic area. *The European Journal of Neuroscience*, *43*, 220–229. https://doi.org/10.1111/ejn.13113.

Spear, L. P. (2000). The adolescent brain and age-related behavioral manifestations. *Neuroscience & Biobehavioral Reviews*, *24*, 417–463. https://doi.org/10.1016/s0149-7634(00)00014-2.

Spierling, S. R., & Zorrilla, E. P. (2017). Don't stress about CRF: Assessing the translational failures of CRF1antagonists. *Psychopharmacology*, *234*, 1467–1481. https://doi.org/10.1007/s00213-017-4556-2.

Stanis, J. J., Burns, R. M., Sherrill, L. K., & Gulley, J. M. (2008). Disparate cocaine-induced locomotion as a predictor of choice behavior in rats trained in a delay-discounting task. *Drug and Alcohol Dependence*, *98*, 54–62. https://doi.org/10.1016/j.drugalcdep.2008.04.009.

Stansfield, K. H., Philpot, R. M., & Kirstein, C. L. (2004). An animal model of sensation seeking: The adolescent rat. *Annals of the New York Academy of Sciences*, *1021*, 453–458. https://doi.org/10.1196/annals.1308.063.

Tiwari, A. K., Souza, R. P., & Muller, D. J. (2009). Pharmacogenetics of anxiolytic drugs. *Journal of Neural Transmission (Vienna)*, *116*, 667–677. https://doi.org/10.1007/s00702-009-0229-6.

Ueno, K. I., Togashi, H., Mori, K., Matsumoto, M., Ohashi, S., Hoshino, A., et al. (2002). Behavioural and pharmacological relevance of stroke-prone spontaneously hypertensive rats as an animal model of a developmental disorder. *Behavioural Pharmacology*, *13*, 1–13. https://doi.org/10.1097/00008877-200202000-00001.

Ungless, M. A., Singh, V., Crowder, T. L., Yaka, R., Ron, D., & Bonci, A. (2003). Corticotropin-releasing factor requires CRF binding protein to potentiate NMDA receptors via CRF receptor 2 in dopamine neurons. *Neuron*, *39*, 401–407. https://doi.org/10.1016/s0896-6273(03)00461-6.

Valentino, R. J., Lucki, I., & Van Bockstaele, E. (2010). Corticotropin-releasing factor in the dorsal raphe nucleus: Linking stress coping and addiction. *Brain Research*, *1314*, 29–37. https://doi.org/10.1016/j.brainres.2009.09.100.

Vandael, D., & Gounko, N. V. (2019). Corticotropin releasing factor-binding protein (CRF-BP) as a potential new therapeutic target in Alzheimer's disease and stress disorders. *Translational Psychiatry*, *9*, 272. https://doi.org/10.1038/s41398-019-0581-8.

Vasconcelos, M., Stein, D. J., Gallas-Lopes, M., Landau, L., & de Almeida, R. M. M. (2020). Corticotropin-releasing factor receptor signaling and modulation: Implications for stress response and resilience. *Trends in Psychiatry and Psychotherapy*, *42*, 195–206. https://doi.org/10.1590/2237-6089-2018-0027.

Vazquez, D. M., Eskandari, R., Phelka, A., & Lopez, J. F. (2003). Impact of maternal deprivation on brain corticotropin-releasing hormone circuits: prevention of CRH receptor-2 mRNA changes by desipramine treatment. *Neuropsychopharmacology*, *28*, 898–909. https://doi.org/10.1038/sj.npp.1300126.

Vialou, V., Robison, A. J., Laplant, Q. C., Covington, H. E., 3rd, Dietz, D. M., Ohnishi, Y. N., et al. (2010). DeltaFosB in brain reward circuits mediates resilience to stress and antidepressant responses. *Nature Neuroscience*, *13*, 745–752. https://doi.org/10.1038/nn.2551.

Wanat, M. J., & Bonci, A. (2008). Dose-dependent changes in the synaptic strength on dopamine neurons and locomotor activity after cocaine exposure. *Synapse*, *62*, 790–795. https://doi.org/10.1002/syn.20546.

Wang, B., Shaham, Y., Zitzman, D., Azari, S., Wise, R. A., & You, Z. B. (2005). Cocaine experience establishes control of midbrain glutamate and dopamine by corticotropin-releasing factor: A role in stress-induced relapse to drug seeking. *The Journal of Neuroscience*, *25*, 5389–5396. https://doi.org/10.1523/JNEUROSCI.0955-05.2005.

Wang, B., You, Z. B., Rice, K. C., & Wise, R. A. (2007). Stress-induced relapse to cocaine seeking: Roles for the CRF(2) receptor and CRF-binding protein in the ventral tegmental area of the rat. *Psychopharmacology*, *193*, 283–294. https://doi.org/10.1007/s00213-007-0782-3.

Ward, H. E., Johnson, E. A., Salm, A. K., & Birkle, D. L. (2000). Effects of prenatal stress on defensive withdrawal behavior and corticotropin releasing factor systems in rat brain. *Physiology & Behavior, 70*, 359–366. https://doi.org/10.1016/s0031-9384(00)00270-5.

Wheatley, M., Plant, J., Reader, H., Brown, G., & Cahill, C. (2004). Clozapine treatment of adolescents with posttraumatic stress disorder and psychotic symptoms. *Journal of Clinical Psychopharmacology, 24*, 167–173. https://doi.org/10.1097/01.jcp.0000116650.91923.1d.

Wise, R. A., & Morales, M. (2010). A ventral tegmental CRF-glutamate-dopamine interaction in addiction. *Brain Research, 1314*, 38–43. https://doi.org/10.1016/j.brainres.2009.09.101.

Workel, J. O., Oitzl, M. S., Fluttert, M., Lesscher, H., Karssen, A., & de Kloet, E. R. (2001). Differential and age-dependent effects of maternal deprivation on the hypothalamic-pituitary-adrenal axis of brown norway rats from youth to senescence. *Journal of Neuroendocrinology, 13*, 569–580. https://doi.org/10.1046/j.1365-2826.2001.00668.x.

CHAPTER TEN

Future directions for part II: Substance use disorders in adolescence—A vision for a better future

S. Rahman[a,*] and R.L. Bell[b]
[a]Department of Pharmaceutical Sciences, South Dakota State University, Brookings, SD, United States
[b]Department of Psychiatry, Indiana University School of Medicine, Indianapolis, IN, United States
*Corresponding author: e-mail address: shafiqur.rahman@sdstate.edu

Contents

1. Future directions and strategies 303
2. Additional stakeholders for a lasting change 305
References 306

1. Future directions and strategies

Chapter 1: Peri-adolescent Exposure to (Meth)amphetamine in Animal Models *(Phillips & Aldrich)*. This chapter discussed the existing literature on the severity and duration of neurotoxicity induced by methamphetamine. Moreover, the effects of this exposure during adolescence and emerging adulthood are greater than similar exposure during adulthood. Adolescence appears to be a time of reduced sensitivity to acute (meth) amphetamine effects on locomotor and stereotypic behaviors, but increased sensitivity to motivational behavioral and neurotoxic effects. Nevertheless, the neurobiology of these effects needs further investigation.

Chapter 2: The Impact of Adolescent Nicotine Exposure on Alcohol Use during Adulthood: The Role of Neuropeptides *(Chen, Lutfy, & colleagues)*. Future research is required to examine the molecular mechanisms that modulate cross-talk between nicotine exposure and alcohol abuse from adolescent to adulthood. This will pave the way for new therapeutics for this co-morbid disorder.

International Review of Neurobiology, Volume 161
ISSN 0074-7742
https://doi.org/10.1016/bs.irn.2021.09.001

Chapter 3: Cannabis Exposure during Adolescence: A uniquely Sensitive Period for Neurobiological Effects *(Peters, Zlebnik & Cheer)*. Future research is needed to elucidate the specific cellular and molecular mechanisms associated with cannabis abuse and neuropsychiatric disorders

Chapter 4: The Stoned Age: Sex Differences in the Effects of Adolescent Cannabinoid Exposure on Prefrontal Cortex Structure and Function in Animal Models *(Ginder, Wright & McLaughlin)*. Additional translational studies are warranted using preclinical models that mimic human and adolescent brain with long-term effects of cannabis use.

Chapter 5: Adolescent Opioid Abuse: Role of Glial and Neuroimmune Mechanisms *(Rahman & colleagues)*. This chapter has explored evidence that opioid abuse during adolescence can modify brain functions and produce drug–specific neurobehavioral consequences involving glial and neuroimmune signaling pathways. By targeting glial mechanisms and neuroimmune signaling, future research can explore novel therapeutic strategy to prevent and treat opioid use disorder in adolescents.

Chapter 6: Adolescent Neuroimmune function and its Interaction with Alcohol *(Doremus-Fitzwater & Deak)*. A better understanding of adolescent ethanol-neuroimmune interactions will hopefully provide a novel target by which the development of problematic drinking behavior can be identified and circumvented before excessive patterns of use begin and alcohol use disorders emerge. Clearly, this area calls for more research, given the seemingly blunted neuroimmune response to ethanol among adolescents, the involvement of stress in neuroimmune processes, and increased levels of stress that some researchers have argued occurs during adolescence.

Chapter 7: Alcohol, Inflammation, and Blood-Brain-Barrier (BBB) Function in Health and Disease across Development *(Vore & Deak)*. This chapter highlighted many shared CNS mechanisms mediating drugs of abuse with completely different pharmacodynamic properties and associated changes in BBB integrity. Despite progress on BBB function in drug abuse, exact mechanisms of inflammaging need further investigation.

Chapter 8: Age Modifies the Effect of Ethanol on Behavior: Investigation in Adolescent, Adult and Aged rats *(Matthews & Imhoff)*. This chapter describes that aged subjects are often, but not always, more sensitive to the effects of ethanol. By continuing and expanding these lines of research, insights into how normal aging alters the brain and body will help reveal why adolescents are less sensitive than adults to ethanol and why the aged are more sensitive than adults to ethanol.

Chapter 9: Early Life Stress and Susceptibility to Addiction in Adolescence (*Tschetter, Ronan, & colleagues*). This chapter has shown that early life stress leads to the development of a behavioral profile indicative of increased susceptibility to addiction that is concomitant with a sex-dependent upregulation of CRF and 5-HT system. Targeting CRF and 5-HT signaling holds great promise; whether pharmacological or novel approaches such as Adeno associated virus-mediated optogenetic or RNA interference control of their activity and expression.

Taken together, these important research issues and concepts will further our understanding from basic science research to therapeutic drug development in the management of licit and illicit drug abuse and addiction in adolescents. This will foster new ideas and stimulate translational investigations for greater understanding and provide effective strategies for the new or improved treatment of devastating substance use disorder in adolescent population. It is noteworthy to emphasize that adolescence is a critical period during brain development process (Spear, 2016). Evidence suggests that numerous neurobiological and behavioral changes occur due to adolescent nicotine, alcohol, cannabis, and opioid use or their combination (Gray & Squeglia, 2018; Johnston et al., 2020; Kann et al., 2018). Specifically, adolescent exposure to these substances may likely contribute to the development of cognitive deficits, psychopathology, or subsequent substance use disorders during adulthood (cf. Hamidullah, Thorpe, Frie, Mccurdy, & Khokhar, 2020; Kroll et al., 2020; Nock, Minnes, & Alberts, 2017). Interestingly, pediatric care physicians can play an important role in preventing initiation or curb the progression of substance misuse among adolescents (Adams & Denne, 2021; Subramaniam & Volkow, 2014). Thus, we believe that the understanding all these evidences and concepts may help develop new strategies to reduce the harm associated with substance use disorder in adolescence.

2. Additional stakeholders for a lasting change

It is time to change how we as a society address substance use disorders in adolescence. Several novel but practical strategies, including public health approach are essential to curb harm reduction in adolescent substance use disorder (Adams & Denne, 2021; Volkow et al., 2020). These include:

1. *Parents and families*: These groups could set a good example regarding substance use in their respective domain. They can restrict access to

addictive substances. Furthermore, they can communicate clearly with their children and can be consistent with no-use messages enforcing rules based on their value systems.

2. *Health care professionals*: These groups can play pivotal role in preventing substance use in adolescents by utilizing multiple approaches such as having more education and training in addiction services, expanding addiction treatment capacity in the medical system, increasing substance use screenings, brief interventions and treatment, and pressing government and private health care insurers to reimburse for adolescent.
3. *Policymakers*: To get the long-lasting benefit for prevention strategy in adolescent substance use disorder, these stakeholders can work through several models, namely by implementing public awareness campaigns, curbing teen access to addictive substances by raising taxes on tobacco and alcohol products, and raising the minimum age for purchase of tobacco products to 21.

Overall, through parent and family education, engagement with key community stakeholders, numerous prevention, treatment strategies, and policy enforcement outlined above will produce improved outcomes in substance use disorder and adolescents health in our society across the globe.

References

Adams, Z. W., & Denne, S. C. (2021). Pediatric policy council. Improving care for adolescents with substance use disorder: more than screening. *Pediatric Research, 89*, 720–721. https://doi.org/10.1038/s41390-020-01314-4.

Gray, K. M., & Squeglia, L. M. (2018). Research review: What have we learned about adolescent substance use? *Journal of Child Psychology and Psychiatry, 59*, 618–627. https://doi.org/10.1111/jcpp.12783.

Hamidullah, S., Thorpe, H. H. A., Frie, J. A., Mccurdy, R. D., & Khokhar, J. Y. (2020). Adolescent substance use and the brain: Behavioral, cognitive and neuroimaging correlates. *Frontiers in Human Neuroscience, 14*, 298. https://doi.org/10.3389/fnhum.2020.00298.

Johnston, L. D., et al. (2020). *Monitoring the Future national survey results on drug use 1975–2019: Overview, key findings on adolescent drug use*. Ann Arbor: Institute for Social Research, University of Michigan. http://www.monitoringthefuture.org//pubs/monographs/mtf-overview2020.pdf.

Kann, L., et al. (2018). Youth risk behavior surveillance — United States, 2017. *MMWR Surveillance Summaries, 67*, 1–114. https://doi.org/10.15585/mmwr.ss6708a1.

Kroll, D. S., Feldman, D. E., Wang, S. A., Zhang, R., Manza, P., Wiers, C. E., et al. (2020). The associations of comorbid substance use disorders and psychiatric conditions with adolescent brain structure and function: A review. *Journal of the Neurological Sciences, 418*, 117099. https://doi.org/10.1016/j.jns.2020.117099.

Nock, N. L., Minnes, S., & Alberts, J. L. (2017). Neurobiology of substance use in adolescents and potential therapeutic effects of exercise for prevention and treatment of substance use disorders. *Birth Defects Research, 109*, 1711–1729. https://doi.org/10.1002/bdr2.1182.

Spear, L. P. (2016). Consequences of adolescent use of alcohol and other drugs: Studies using rodent models. *Neuroscience and Biobehavioral Reviews*, *70*, 228–243. https://doi.org/10.1016/j.neubiorev.2016.07.026.

Subramaniam, G. A., & Volkow, N. D. (2014). Substance misuse among adolescents: To screen or not to screen? *JAMA Pediatrics*, *168*, 798–799. https://doi.org/10.1001/jamapediatrics.2014.958.

Volkow, N. D., Gordon, J. A., Koob, G. F., Birnbaum, L. S., Clayton, J. A., Koroshetz, W. J., et al. (2020). An examination of child and adolescent neurodevelopment through National Institutes of Health Studies. *Public Health Reports*, *135*, 169–172. https://doi.org/10.1177/0033354919900889.